Power and Energy Engineering

Power and Energy Engineering

Edited by **Helena Walker**

Callisto Reference
New York

Published by Callisto Reference,
106 Park Avenue, Suite 200,
New York, NY 10016, USA
www.callistoreference.com

Power and Energy Engineering
Edited by Helena Walker

International Standard Book Number: 978-1-63239-649-5 (Hardback)

Printed in the United States of America.

Contents

Preface

Power engineering is the subfield of energy engineering that deals with production, transmission and utilization of electrical energy. Energy engineering on the other hand is the amalgamation of mathematics, chemistry, classical and modern physics. Both of these deal with efficient utilization of energy, facility management and energy services. Some of the topics discussed in this book are power generation, power system management, power transmission and distribution, smart grid technologies, etc. This book is compiled in such a manner, that it will provide in-depth knowledge about the theory and practice of power and energy engineering. It will serve as a valuable source of reference for graduate and post graduate students.

All of the data presented henceforth, was collaborated in the wake of recent advancements in the field. The aim of this book is to present the diversified developments from across the globe in a comprehensible manner. The opinions expressed in each chapter belong solely to the contributing authors. Their interpretations of the topics are the integral part of this book, which I have carefully compiled for a better understanding of the readers.

At the end, I would like to thank all those who dedicated their time and efforts for the successful completion of this book. I also wish to convey my gratitude towards my friends and family who supported me at every step.

Editor

1

PI and RST Control Design and Comparison for Matrix Converters Using Venturini Modulation Strategy

Bekhada Hamane[1], Mamadou Lamine Doumbia[1], Hicham Chaoui[2], Mohamed Bouhamida[3], Ahmed Chériti[1], Mustapha Benghanem[3]

[1]Department of Electrical and Computer Engineering, UQTR, Trois-Rivières, Canada
[2]Center for Energy Systems Research, Department of Electrical and Computer Engineering, Tennessee Technological University, Cookeville, USA
[3]Department of Electrical Engineering, University Mohamed Boudiaf, Oran, Algeria
Email: bekhada.hamane@uqtr.ca, mamadou.doumbia@uqtr.ca, hchaoui@tntech.edu, m_bouhamida@yahoo.com, ahmed.cheriti@uqtr.ca, mbenghanem69@yahoo.fr

Abstract

This paper presents a thorough design and comparative study of two popular control techniques, *i.e.*, classical Proportional Integral (PI) and RST, for Matrix Converters (MCs) in terms of tracking the reference and robustness. The output signal of MCs is directly affected by unbalanced grid voltage. Some research works have attempted to overcome this problem with PI control. However, this technique is known to offer lower performance when it is used in complex and nonlinear systems. On the other hand, RST control offers better performance, even in case of highly nonlinear systems. Therefore, the RST can achieve better performance to overcome the limitation of PI control of nonlinear systems. In this paper, a RST control method is proposed as output current controller to improve the performance of the MC powered by unbalanced grid voltage. The overall operating principle, Venturini modulation strategy of MC, PI control and characteristics of RST are presented.

Keywords

Matrix Converter, Unbalanced Grid, Venturini Modulation Strategy, PI Control, RST Control

1. Introduction

Recent advances in power electronics have enabled the emergence of Matrix Converter (MC) for direct AC/AC conversion [1]. Interest in this converter topology was rather academic with efforts provided in many research laboratories [1]. MC uses bidirectional current and voltage power switches that connect converter input and output phases [2]. The direct conversion is performed without intermediate DC link circuit for energy storage [2] [3]. MC was introduced firstly in 1976. To prevent the spread of current harmonics caused by the MC to the supply network, an input LC filter is used. It provides a very low impedance path and absorbs current harmonics [1] [2]. Venturini and Alesina proposed a generalized high-frequency switching strategy in 1980 [3]. The objective of this control strategy is to achieve an ideal electronic transformer capable of varying the voltage, current, frequency and power factor [4]. Another method, known as the direct transfer function approach, proposes the multiplication of the input voltages vectors by the modulation matrix M to obtain a vector of output voltages which correspond to a point of synthesis [4]. However, the simultaneous commutation of controlled bidirectional switches used in MC is very difficult to achieve without generating over current or overvoltage spikes which can destroy the power semiconductors [3]. Also, the load side of the MC is directly affected by the distorted and/or unbalanced input voltages due to the lack of DC intermediate circuit in the MC. The performance of the MC deteriorates, when it is exposed to the harmonic and non-sinusoidal currents and some papers have presented mitigation methods [3] [5]. Conventional PI controller works well only if the mathematical model of the system could be computed. However, it is difficult to implement the conventional PI controller for variable as well as complex systems [5] [6]. So, RST Controller is investigated. This regulator, whose synthesis is purely algebraic, is a sophisticated algorithm based on pole placement method which exploits many numerical resources [7] [8]. The method used to determine the gains of the PI controller is the compensation method of poles, we note here that the interest of the compensation of the poles occurs only if the system parameters are accurately identified as gains K_p and K_i are based on these same parameters. If the actual parameters are different from those used in the synthesis, the compensation is ineffective. In the literature, control law design approaches can be divided into two categories. The first category consists of a nonlinear systems linearization around an operating point of the states. In this case, classical linear control laws are applied for the approximated system. These methods are popular in the industry and are mainly used for their simplicity. However, the control system's performance and stability are not guaranteed for the overall system. The second category deals with nonlinear controllers design based on nonlinear systems dynamics. In this category, the characteristics of nonlinear systems are preserved. However, the design approach difficulties arise with the complexity of the nonlinear systems dynamics. Furthermore, these approaches assume a precise mathematical system model and are able to cope with nonlinearities to a certain degree. But, their performance also degrades in the presence of varying operating conditions, and higher uncertainties and disturbances. Therefore, this paper aims to compare the most popular techniques in the industry with similar design complexity. This work presents a modeling, theoretical analysis and an in-depth comparison of both the classical PI and RST Controller for MCs. Results show the superiority of the RST strategy with faster dynamic response and better robustness. To show the effectiveness of the control methods, the performance of the system is analyzed and compared in various operating conditions.

2. Mathematical Model of Matrix Converter

This part consists of a brief description and modeling of each element of the matrix converter. We start with modeling the MC, then the input filter and it ends with the load RL. Ideal bidirectional switches are represented by S_{ij}, where $i=\{A,B,C\}$ and $j=\{a,b,c\}$ represent respectively the index of input and output voltage [1] [9] [11]:

$$S_{ij}=\begin{cases}1 & \text{If the switch } S_{ij} \text{ is closed}\\ 0 & \text{If the switch } S_{ij} \text{ is opened}\end{cases} \tag{1}$$

$$S_{Aj}+S_{Bj}+S_{Cj}=1 \tag{2}$$

The basic diagram of a MC is represented in **Figure 1**, which the clipping circuit is used to protect the converter against surges that could come from a sudden disconnection of the load [1].

With these restrictions, a 3×3 matrix converter has 27 possible switching states [1]. Let m_{ij} be the duty cycle of switch S_{ij}, defined as [1] [10] [11]:

$$m_{ij}(t)=\frac{t_{ij}}{T_{seq}} \tag{3}$$

where, $0 < m_{ij} < 1$, $T_{seq} = \frac{1}{f_s}$ and f_s is the switching frequency.

The transfer matrix of the converter is defined by [1] [3] [10] [11]:

$$M = \begin{bmatrix} m_{Aa} & m_{Ba} & m_{Ca} \\ m_{Ab} & m_{Bb} & m_{Cb} \\ m_{Ac} & m_{Bc} & m_{Cc} \end{bmatrix} \quad (4)$$

Figure 2 shows an example of the duration of conduction of the switches during a switching sequence T_{seq} of the MC [1] [12].

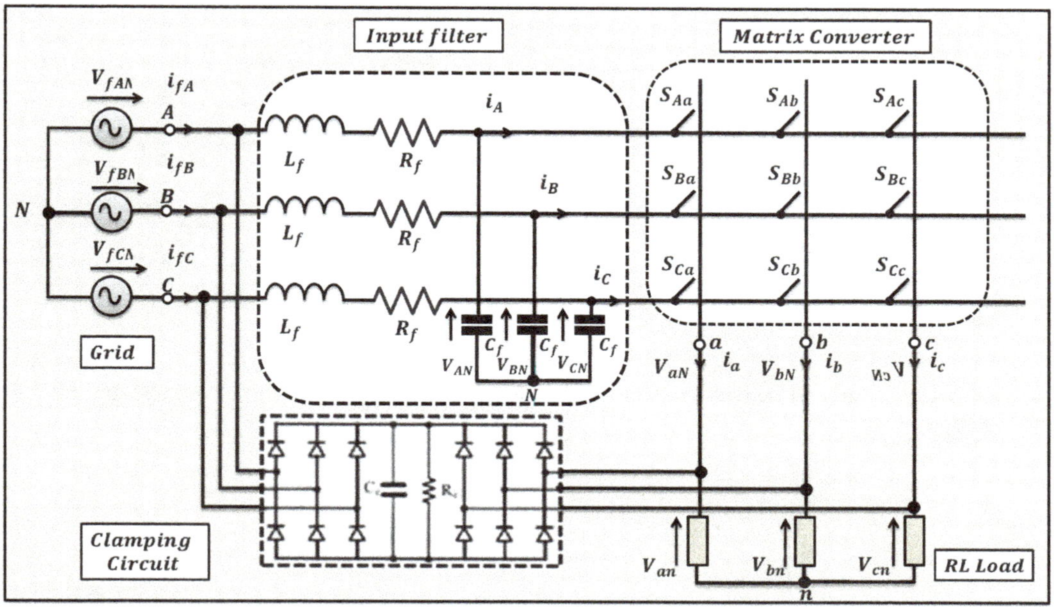

Figure 1. Basic circuit of a Matrix Converter.

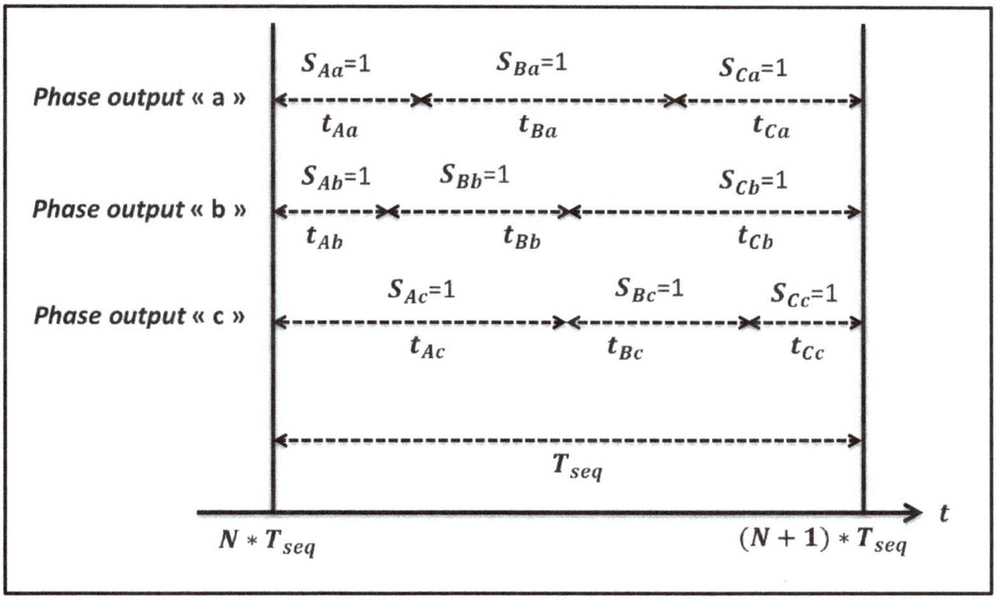

Figure 2. Example of the operation timing of switches during a switching period.

2.1. Modeling of the Matrix Converter

The input voltage and current of the matrix converter are given by [1] [10] [11] [13]:

$$V_i = V_{im}\begin{bmatrix}\cos(\omega_i t)\\ \cos\left(\omega_i t+\frac{2\pi}{3}\right)\\ \cos\left(\omega_i t+\frac{4\pi}{3}\right)\end{bmatrix} \tag{5}$$

$$I_i = I_{im}\begin{bmatrix}\cos(\omega_i t+\varphi_i)\\ \cos\left(\omega_i t+\varphi_i+\frac{2\pi}{3}\right)\\ \cos\left(\omega_i t+\varphi_i+\frac{4\pi}{3}\right)\end{bmatrix} \tag{6}$$

Assuming the relationship between the output and the input signal of the matrix converter [1] [10] [11] [14]:

$$q=\sqrt{\frac{V_o^2}{V_i^2}}=\sqrt{\frac{I_i^2}{I_o^2}} \quad \text{with}: 0<q\le 0.866 \tag{7}$$

The matrix converter will be designed and controlled to provide desired output voltage and output current [1] [10] [11] [13]:

$$V_o = V_{om}\begin{bmatrix}\cos(\omega_o t)\\ \cos\left(\omega_o t+\frac{2\pi}{3}\right)\\ \cos\left(\omega_o t+\frac{4\pi}{3}\right)\end{bmatrix} \tag{8}$$

$$I_o = I_{om}\begin{bmatrix}\cos(\omega_o t+\varphi_o)\\ \cos\left(\omega_o t+\varphi_o+\frac{2\pi}{3}\right)\\ \cos\left(\omega_o t+\varphi_o+\frac{4\pi}{3}\right)\end{bmatrix} \tag{9}$$

The neutral to phase output voltages V_{aN}, V_{bN} and V_{cN} are given by [1] [3] [9] [11]:

$$\begin{bmatrix}V_{an}\\ V_{bn}\\ V_{cn}\end{bmatrix}=\begin{bmatrix}m_{Aa} & m_{Ba} & m_{Ca}\\ m_{Ab} & m_{Bb} & m_{Cb}\\ m_{Ac} & m_{Bc} & m_{Cc}\end{bmatrix}\begin{bmatrix}V_{AN}\\ V_{BN}\\ V_{CN}\end{bmatrix} \tag{10}$$

The input current I_A, I_B and I_C are [3] [9] [11]:

$$\begin{bmatrix}I_A\\ I_B\\ I_C\end{bmatrix}=\begin{bmatrix}m_{Aa} & m_{Ab} & m_{Ac}\\ m_{Ba} & m_{Bb} & m_{Bc}\\ m_{Ca} & m_{Cb} & m_{Cc}\end{bmatrix}\begin{bmatrix}I_a\\ I_b\\ I_c\end{bmatrix} \tag{11}$$

ω_i, V_{im} are respectively the input voltage frequency and amplitude;

I_{im}, φ_i are respectively the input current amplitude and input phase;
ω_o, V_{om} are respectively the output voltage frequency and amplitude.

2.2. Modeling of the Input Filter

The LC input filter [15] (represented as shown in **Figure 3**) is a series resonant circuit tuned to the frequency of harmonics and connected in shunt. It provides a very low impedance path and absorbs harmonic currents [1] [3] [14]. At the fundamental frequency, the filter acts as a reactive power compensator [1] [3]. The LC input filter may be modeled with the equivalent circuit [15]. From the Kirchhoff's laws, node equations and Laplace transformation.

The filter output voltage and input current are obtained as Equation (12) and Equation (13) [1] [10] [12].

$$V_{AN}(p) = \frac{1}{L_f C_f p^2 + R_f C_f p + 1} V_{fAN} - \frac{L_f p + R_f}{L_f C_f p^2 + R_f C_f p + 1} I_A \tag{12}$$

$$I_{fA}(p) = \frac{1}{L_f C_f p^2 + R_f C_f p + 1} I_A + \frac{C_f p}{L_f C_f p^2 + R_f C_f p + 1} V_{fAN} \tag{13}$$

2.3. Modeling of the Load RL

Generally, the neutral at the load (n) is isolated from that of the source (N) as shown in **Figure 1**. Therefore, the objective is calculating the load current, it is necessary to know the potential at the output of the MC corresponding to the neutral of the load. In this case, we have [1] [16]:

$$V_{jn} = V_{jN} - V_{nN} \tag{14}$$

The potential difference between the two neutral is given by [1] [16]:

$$V_{nN} = \frac{V_{aN} + V_{bN} + V_{cN}}{3} \tag{15}$$

As the transfer function of the load current is given by [1] [16]:

$$i_j(p) = \frac{1}{L_l p + R_l} V_{jn}(p) \tag{16}$$

3. Venturini Modulation Strategy of Matrix Converter

This method can produce the sinusoidal input current with unity power factor independently of load [4] [9]. The principle is to synthesize the desired three-phase output voltage from the input during each defined switching period. The initial equations of Venturini method are obtained as the product the ratio q, the voltage amplitude, third harmonic frequency of the input and output voltage as indicated in references [3] [10] [17]:

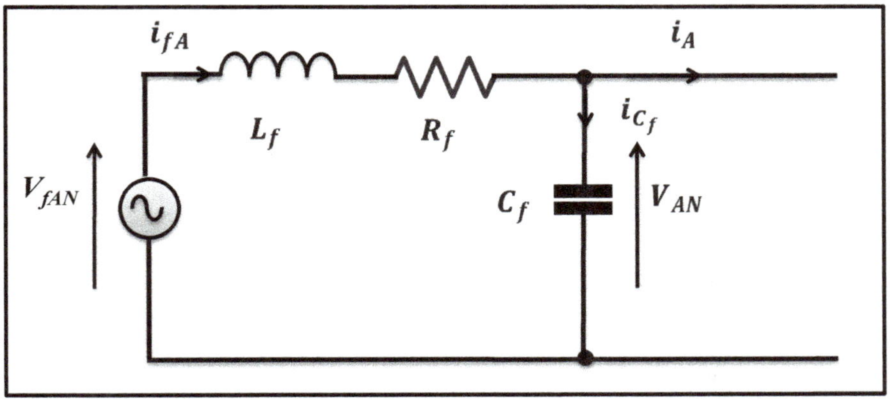

Figure 3. Input filter scheme.

$$V_o(t) = qV_{im} \begin{bmatrix} \cos(\omega_o t) - \frac{1}{6}\cos(3\omega_o t) + \frac{1}{2\sqrt{3}}\cos(3\omega_i t) \\ \cos\left(\omega_o t + \frac{2\pi}{3}\right) - \frac{1}{6}\cos\cos(3\omega_o t) + \frac{1}{2\sqrt{3}}\cos\cos(3\omega_i t) \\ \cos\left(\omega_o t + \frac{4\pi}{3}\right) - \frac{1}{6}\cos\cos(3\omega_o t) + \frac{1}{2\sqrt{3}}\cos\cos(3\omega_i t) \end{bmatrix} \quad (17)$$

According to the optimal amplitude in expression of Venturini, the modulation function is [1] [11] [14] [17]:

$$m_{ij} = \frac{1}{3}\left[1 + \frac{2V_i V_j}{V_{im}^2}\right] \quad (18)$$

The S_{ij} can be obtained according to the logic rules using the activation times t_{ij} [11] [17], as shown in **Figure 4**.

Therefore, only six duty cycles are sufficient to calculate the gate signals of the power switches [10] [11] [13].

$$\left.\begin{array}{l} X = t_{Aj} \\ Y = t_{Aj} + t_{Bj} \end{array}\right\} \Rightarrow \begin{cases} S_{Aj} = (X) \\ S_{Bj} = (\bar{X}) \text{ and } (Y) \\ S_{Cj} = (\bar{X}) \text{ and } (\bar{Y}) \end{cases} \quad (19)$$

The carrier signal is expressed by [10] [11] [13]:

$$U_p = \frac{1}{T_{seq}} t \quad \text{with} : 0 \le t \le T_{seq} \quad (20)$$

4. Control Design

This section deals with the design and synthesis of the PI and RST controllers. Both controllers are designed to achieve current reference tracking with constant and varying current reference signals. This also has to be achieved under both balanced and unbalanced grid voltage conditions.

4.1. PI Controller Design

Current measurements of the load RL using a PI controller is illustrated by **Figure 5**.

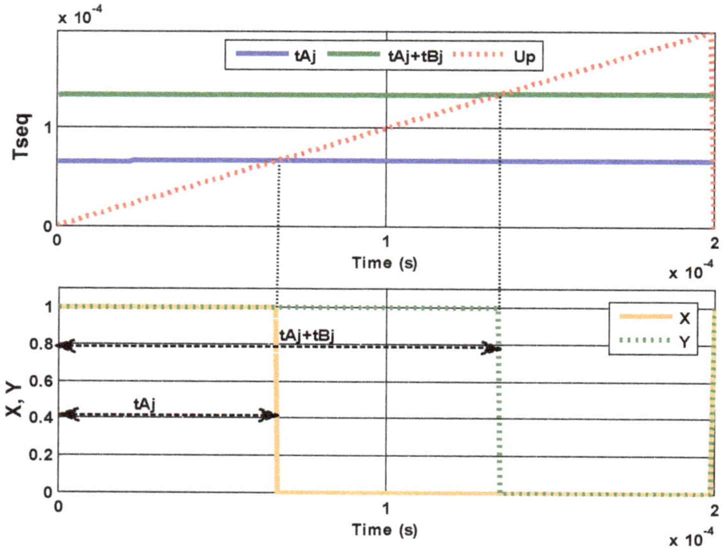

Figure 4. Obtaining logical instructions *X* and *Y*.

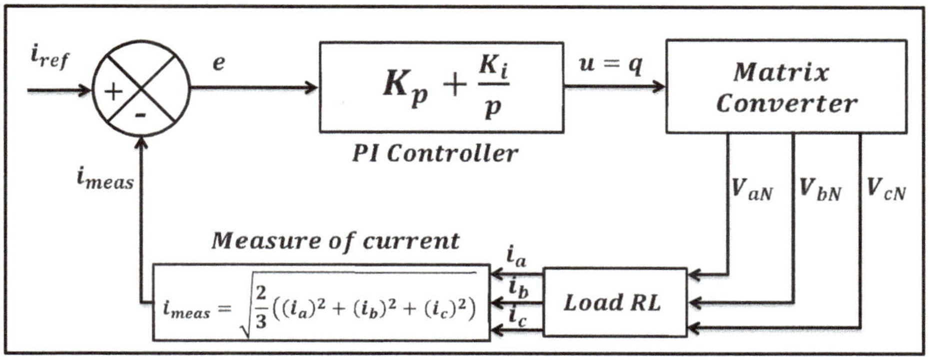

Figure 5. PI Controller for matrix converter.

The transfer function of the system is:

$$T(p) = \frac{B}{A} = \frac{1}{R_l + L_l p} \tag{21}$$

The values of A and B are:

$$\begin{cases} A = R_l + L_l p \\ B = 1 \end{cases} \tag{22}$$

The transfer function of the open-loop including the regulator is:

$$G(p) = \left(\frac{p + \frac{K_i}{K_p}}{\frac{p}{K_p}} \right) \cdot \left(\frac{\frac{1}{L_l}}{p + \frac{R_l}{L_l}} \right) \tag{23}$$

To cancel the pole, a zero was added at the same location as the pole [18]. Equation (24) gives a pole value:

$$\frac{K_i}{K_p} = \frac{R_l}{L_l} \tag{24}$$

The transfer function of the open-loop becomes:

$$G(p) = \frac{K_p \frac{1}{L_l}}{p} \tag{25}$$

The transfer function of the closed loop is expressed by:

$$H(p) = \frac{1}{1 + p\tau_r} \tag{26}$$

Which:

$$\tau_r = \frac{L_l}{K_p} \tag{27}$$

For a response time $\tau_r = 0.66$ s, the K_p and K_i can be expressed by,

$$\begin{cases} K_p = \frac{L_l}{\tau_r} \\ K_i = \frac{R_l}{\tau_r} \end{cases} \tag{28}$$

4.2. RST Controller Design

The closed-loop system of the RST controller for MC is given by the following block diagram in **Figure 6**.

The goal of this section to determinate the RST controller's current. This type of controller is a structure with two freedom degrees and compared to a one degree of freedom structure, it has the main advantage that it allows the designer to specify performances independently with reference trajectory tracking (reference variation) and with regulation [7] [17]. It is based on the pole placement theory [8], which consists in specifying an arbitrary stability polynomial $D(p)$ and calculate $S(p)$ and $R(p)$ according to the Bezout equation [7] [17]:

$$D = AS + BR \tag{29}$$

With:

$$\deg(D) = \deg(A) + \deg(S) \tag{30}$$

For our model, we obtain [17]:

$$\begin{cases} A = a_1 p + a_0 \\ B = b_0 \\ D = d_3 p^3 + d_2 p^2 + d_1 p + d_0 \\ R = r_1 p + r_0 \\ S = s_2 p^2 + s_1 p + d_0 \end{cases} \tag{31}$$

The terms A and B are expressed by Equation (22). According to the robust pole placement strategy [8], the polynomial D is written as [17]:

$$D = \left(p + \frac{1}{T_c}\right)\left(p + \frac{1}{T_f}\right)^2 \tag{32}$$

To accelerate the system, the following conditions were adopted:

$$D = (s - 5P_a)(s - 15P_a)^2 \tag{33}$$

With $P_C = -1/T_C$ pole of polynomial orderCand $P_f = -1/T_f$ double pole of the polynomial filter F [17].

$$\begin{cases} P_c = 5P_a = -5\dfrac{R_l}{L_l} \\ T_c = \dfrac{1}{P_c} \\ T_f = \dfrac{1}{3}T_c \end{cases} \tag{34}$$

By identifying Equation (31) and Equation (34), coefficients of polynomial D were found and are linked to the coefficients of R and S by the Sylvester Matrix [7] [17]. Thus, the parameters of the RST controller can be determined as follows:

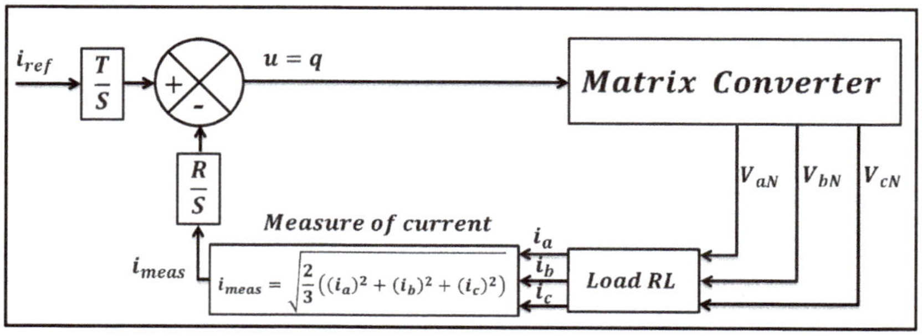

Figure 6. RST Controller for matrix converter.

$$\begin{cases} d_3 = a_1 s_2 \rightarrow s_2 = \dfrac{d_3}{a_1} \\ d_2 = a_1 s_1 \rightarrow s_1 = \dfrac{d_2}{a_1} \\ d_1 = a_0 s_1 + b_0 r_1 \rightarrow r_1 = \dfrac{d_1 - a_0 s_1}{b_0} \\ d_1 = b_0 r_0 \rightarrow r_0 = \dfrac{d_0}{b_0} \\ T = r_0 \end{cases} \quad (35)$$

The reference current is calculated as shown in **Figure 7** [13] [18].

The measured load's current and the reference load's current are given by Equation (36) [13] [18]:

$$\begin{cases} i_{meas} = \sqrt{\dfrac{2}{3}\left((i_a)^2 + (i_b)^2 + (i_c)^2\right)} \\ i_{ref} = \sqrt{\dfrac{2}{3}\left((i_{aref})^2 + (i_{bref})^2 + (i_{cref})^2\right)} \end{cases} \quad (36)$$

5. Simulations Results

The PI and RST are used to control a matrix converter and a set of simulation runs is performed using SimPowerSystems toolbox of Matlab/Simulink software. The input filter parameters are calculated as given in [14]. Bidirectional switches MOSFET are considered ideal and ode23tb simulation solver was used. The MC system's parameters are listed in **Table 1**.

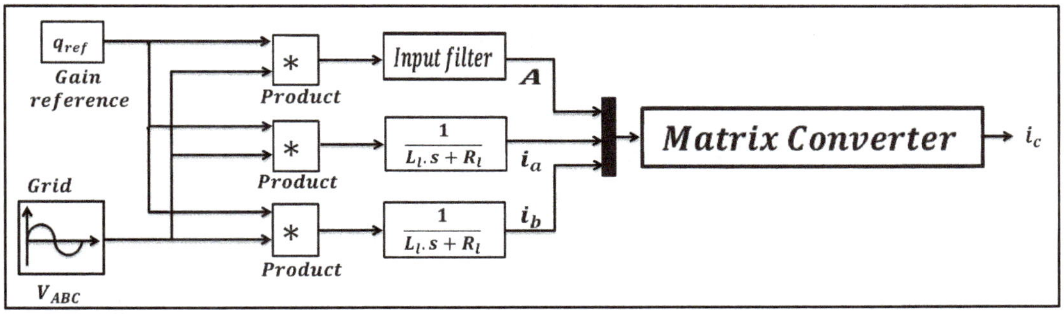

Figure 7. Load reference current.

Table 1. System Parameters.

Parameters	Values
Input voltage phase to neuter RMS	$V_{im} = 220$ V
Input frequency	$f_i = 50$ Hz
Switching frequency	$f_s = 5$ KHz
Input filter resistance	$R_f = 0.08\ \Omega$
Input filter inductance	$L_f = 30$ mH
Input filter capacitor	$C_f = 25\ \mu$F
Load resistance	$R_l = 10\ \Omega$
Load inductance	$L_l = 55$ mH
Input voltage phase to neuter RMS	$V_{im} = 220$ V
Input frequency	$f_i = 50$ Hz

5.1. Balanced Grid Case with PI Controller

Figure 8 shows the balanced grid voltage.

- Constant reference current I_{ref} :

Figure 9 shows the output voltage and linear load current using PI controller for balanced grid voltage with constant current reference. **Figure 10** presents load current and variation of the ratio q . PI controller is used and the grid voltage balanced. **Figure 11** shows the THD of load current with constant current reference.

- Time-varying reference current I_{ref} :

Figure 12 shows the output voltage and linear load current using PI controller for balanced grid voltage with stepped changing reference current. **Figure 13** presents load current and variation of the ratio q . PI controller is used and the grid voltage balanced. **Figure 14** shows the THD of load current with stepped changing reference current.

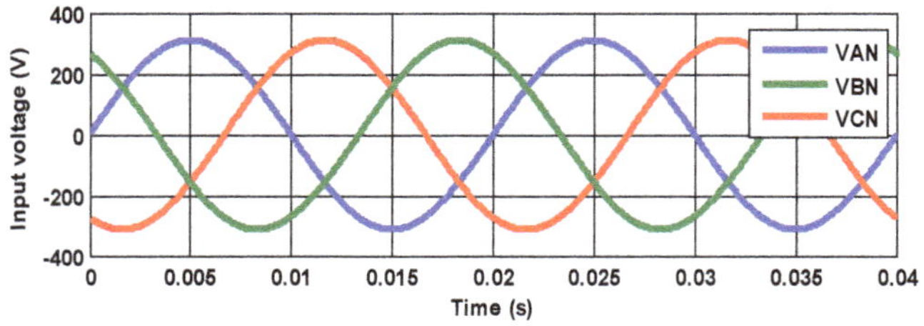

Figure 8. Balanced grid voltage.

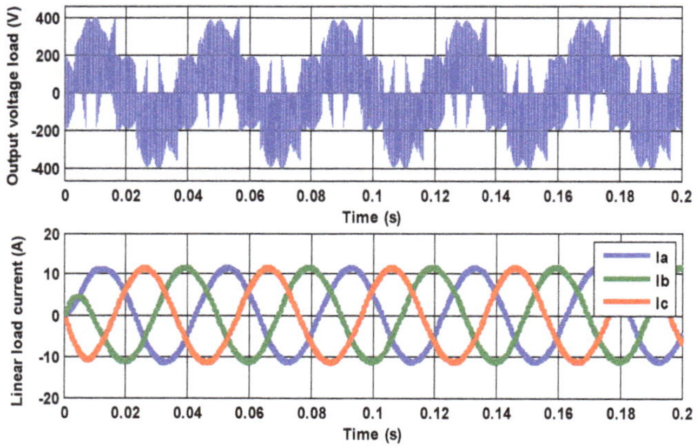

Figure 9. Output voltage and load current (PI, balanced grid and with constant I_{ref}).

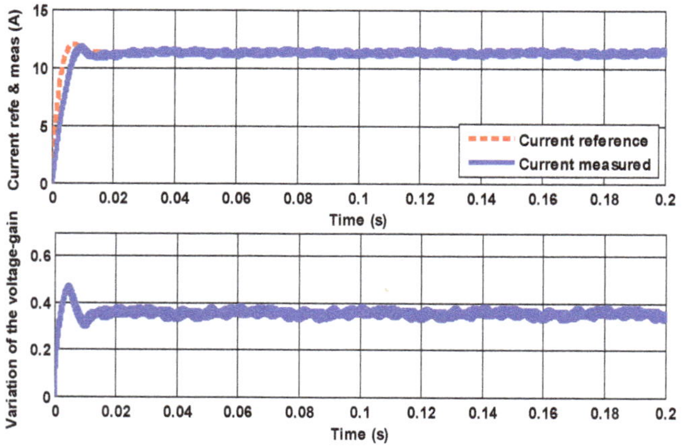

Figure 10. Load current and variation of the q (PI, balanced grid and with constant I_{ref}).

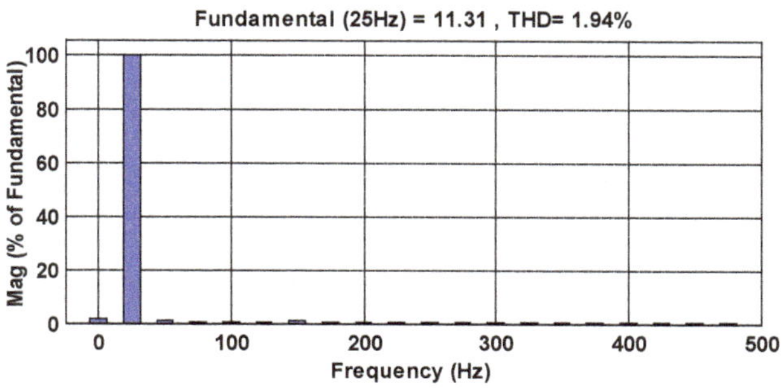

Figure 11. Harmonics spectrum of load current (PI, balanced grid and with constant I_{ref}).

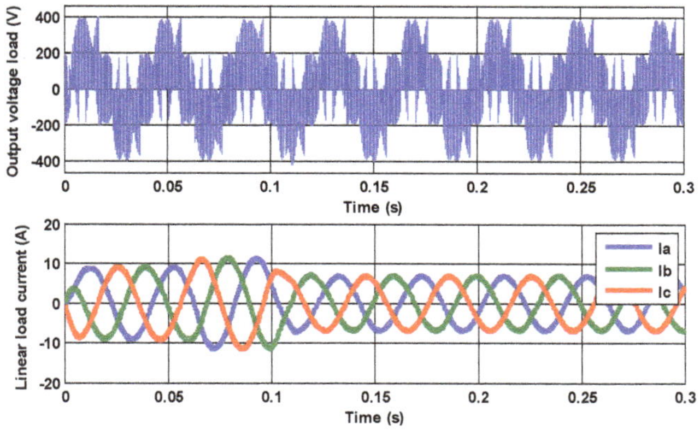

Figure 12. Output voltage and load current (PI, balanced grid and with stepped changing I_{ref}).

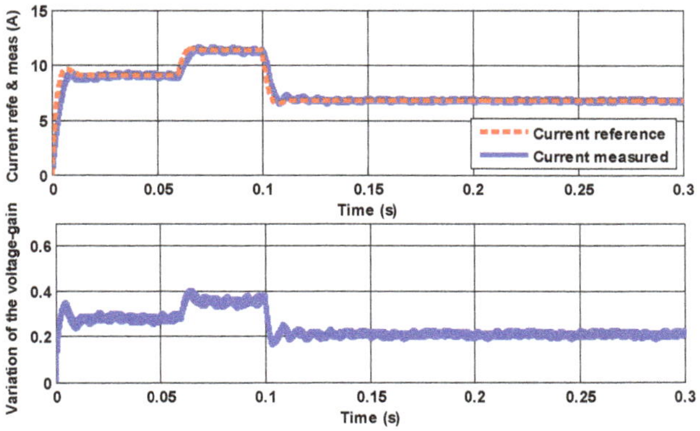

Figure 13. Load current and variation of the q (PI, balanced grid and with stepped changing I_{ref}).

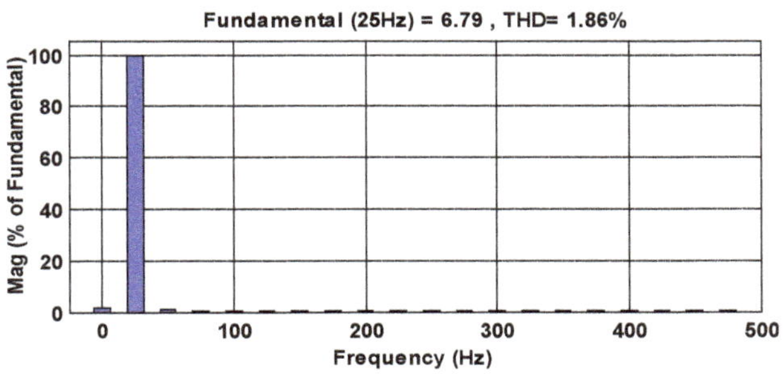

Figure 14. Harmonics spectrum of load current (PI, balanced grid an with stepped changing I_{ref}).

5.2. Balanced Grid Case with RST Controller

- Constant reference current I_{ref}:

Figure 15 shows the output voltage and linear load current using RST controller for balanced grid voltage with constant current reference. **Figure 16** presents load current and variation of the ratio q. RST controller is used and the grid voltage balanced. **Figure 17** shows the THD of load current with constant current reference.

- Time-varying reference current I_{ref}:

Figure 18 shows the output voltage and linear load current using RST controller for balanced grid voltage with stepped changing reference current. **Figure 19** presents load current and variation of the ratio q. RST controller is used and the grid voltage balanced. **Figure 20** shows the THD of load current with stepped changing reference current.

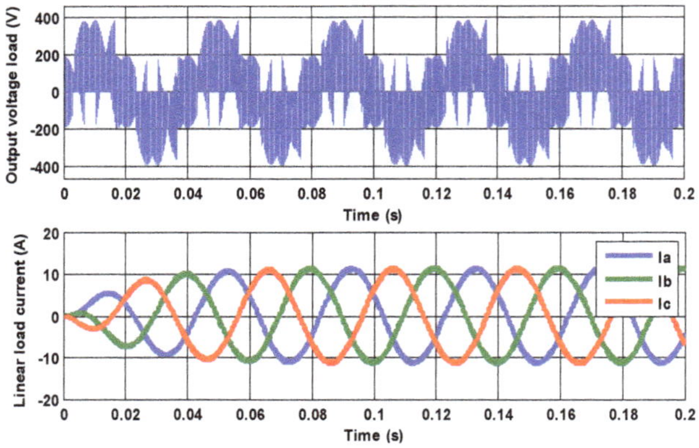

Figure 15. Output voltage and load current (RST, balanced grid and with constant I_{ref}).

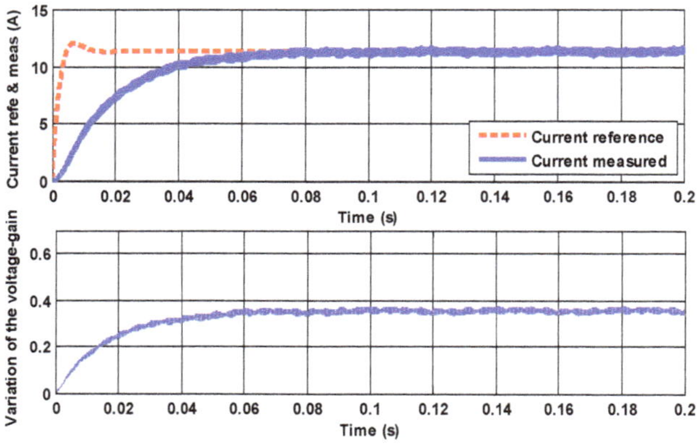

Figure 16. Load current and variation of the q (RST, balanced grid and with constant I_{ref}).

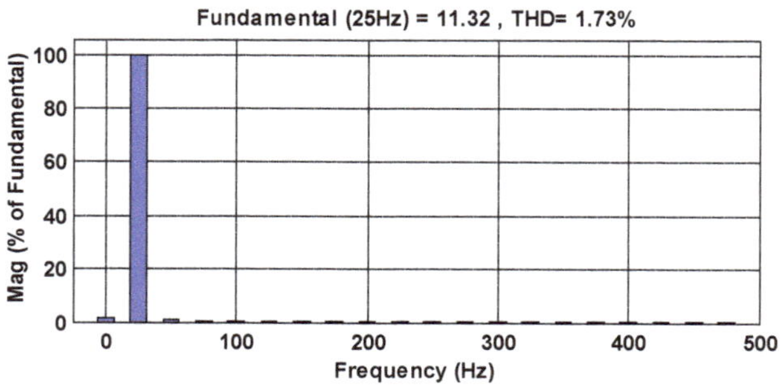

Figure 17. Harmonics spectrum of load current (RST, balanced grid and with constant I_{ref}).

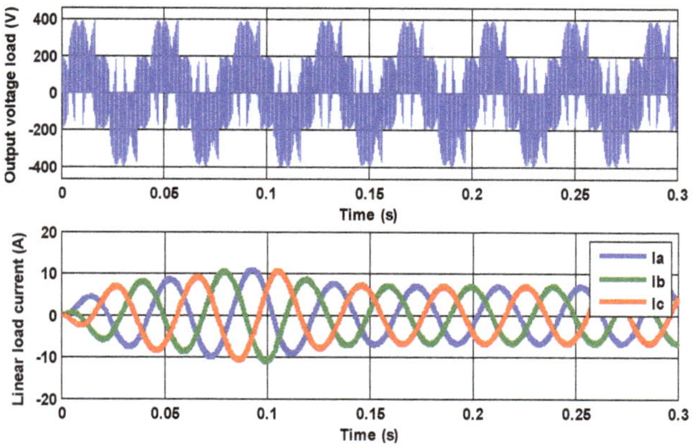

Figure 18. Output voltage and load current (RST, balanced grid and with stepped changing I_{ref}).

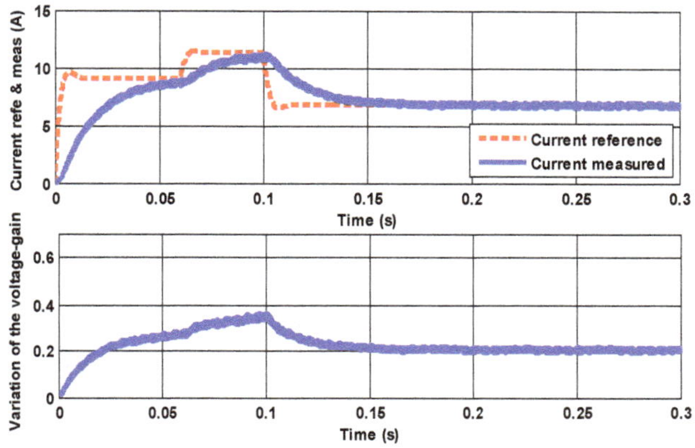

Figure 19. Load current and variation of the q (RST, balanced grid and with stepped changing I_{ref}).

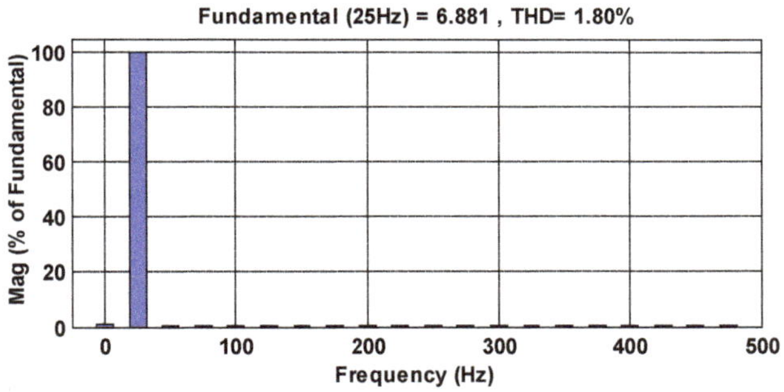

Figure 20. Harmonics spectrum of load current (RST, balanced grid an with stepped changing I_{ref}).

5.3. Unbalanced Grid Case with PI Controller

In this case, the amplitude of the input voltage of phase b is reduced to 20% relative to the phases a and c (**Figure 21**).

- Constant reference current I_{ref} :

Figure 22 shows the output voltage and linear load current using PI controller for unbalanced grid voltage with constant current reference. **Figure 23** presents load current and variation of the ratio q . PI controller is used and the grid voltage unbalanced. **Figure 24** shows the THD of load current with constant current reference.

- Time-varying reference current I_{ref} :

Figure 25 shows the output voltage and linear load current using PI controller for unbalanced grid voltage with stepped changing reference current. **Figure 26** presents load current and variation of the ratio q. PI controller is used and the grid voltage unbalanced. **Figure 27** shows the THD of load current with stepped changing reference current.

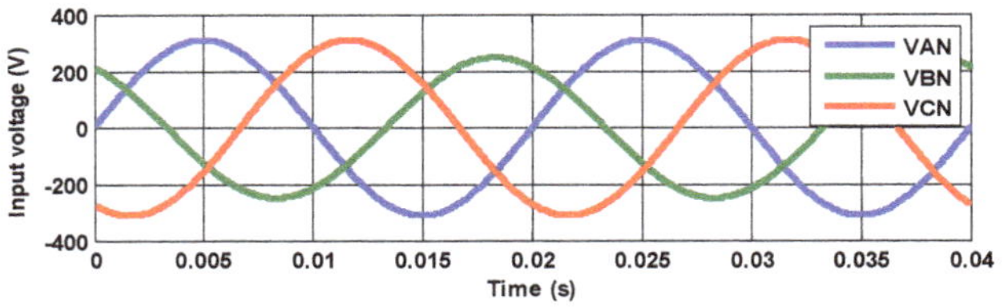

Figure 21. Unbalanced grid voltage.

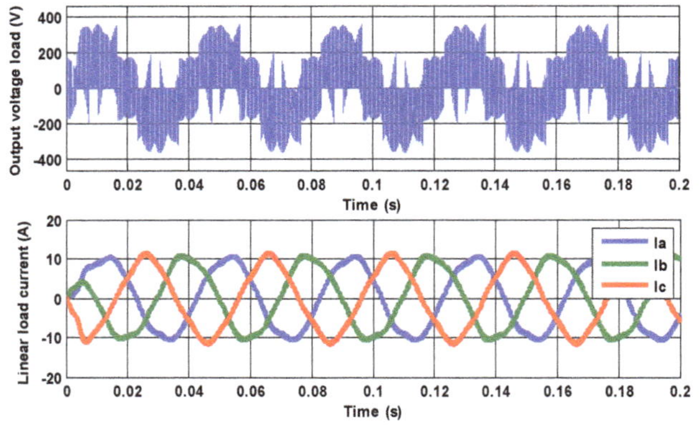

Figure 22. Output voltage and load current (PI, unbalanced grid and with constant I_{ref}).

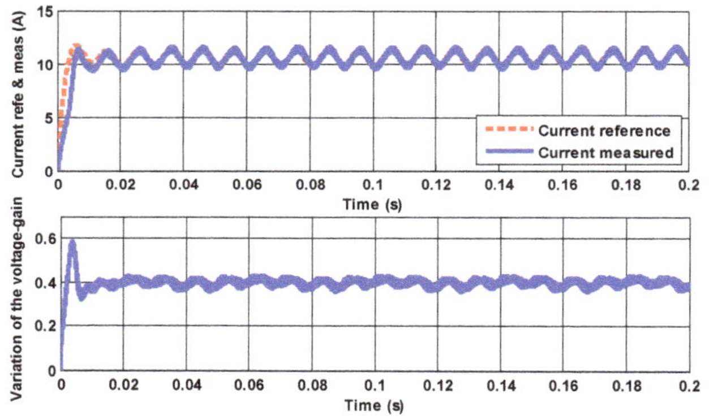

Figure 23. Load current and variation of the q (PI, unbalanced grid and with constant I_{ref}).

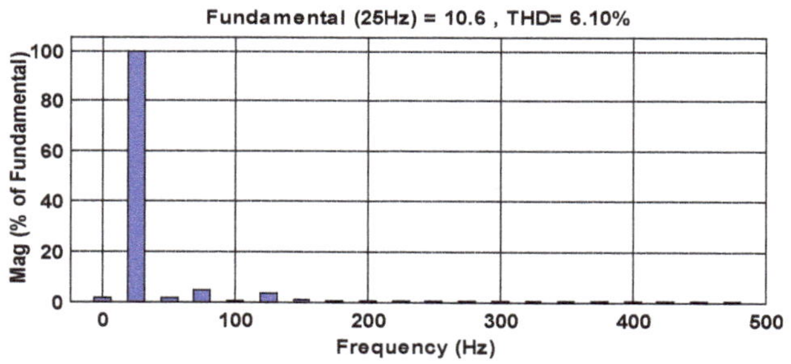

Figure 24. Harmonics spectrum of load current (PI, unbalanced grid and with constant I_{ref}).

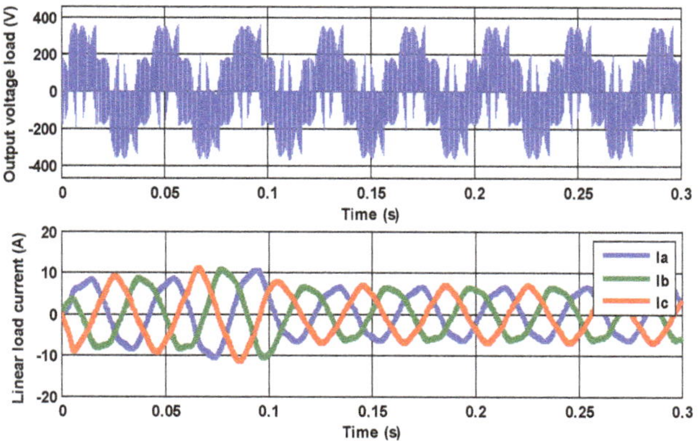

Figure 25. Output voltage and load current (PI, unbalanced grid and with stepped changing I_{ref}).

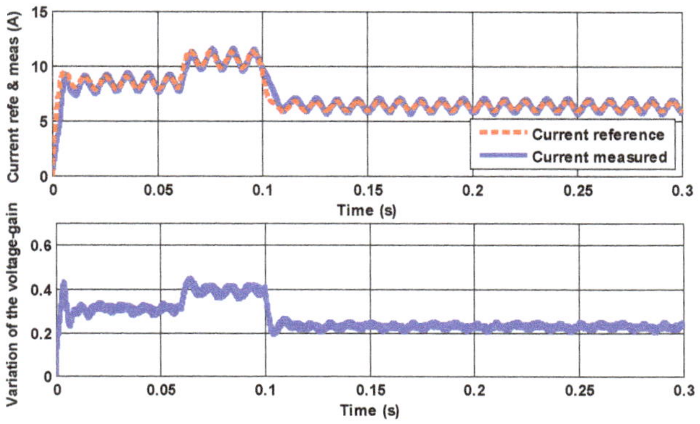

Figure 26. Load current and variation of the q (PI, unbalanced grid and with stepped changing I_{ref}).

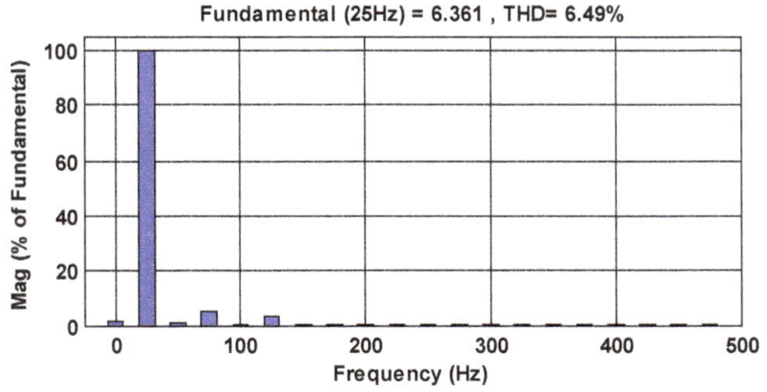

Figure 27. Harmonics spectrum of load current (PI, unbalanced grid an with stepped changing I_{ref}).

5.4. Unbalanced Grid Case with RST Controller

- Constant reference current I_{ref} :

Figure 28 shows the output voltage and linear load current using RST controller for unbalanced grid voltage with constant current reference. **Figure 29** presents load current and variation of the ratio q . RST controller is used and the grid voltage unbalanced. **Figure 30** shows the THD of load current with constant current reference.

- Time-varying reference current I_{ref} :

Figure 31 shows the output voltage and linear load current using RST controller for unbalanced grid voltage with stepped changing reference current. **Figure 32** presents load current and variation of the ratio q . RST controller is used and the grid voltage unbalanced. **Figure 33** shows the THD of load current with stepped changing reference current.

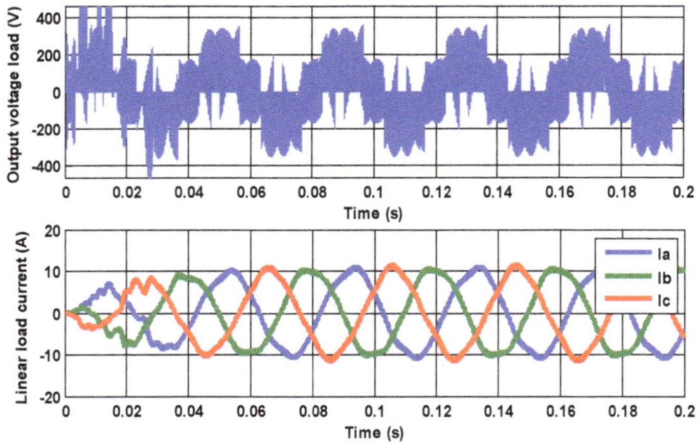

Figure 28. Output voltage and load current (RST, unbalanced grid and with constant I_{ref}).

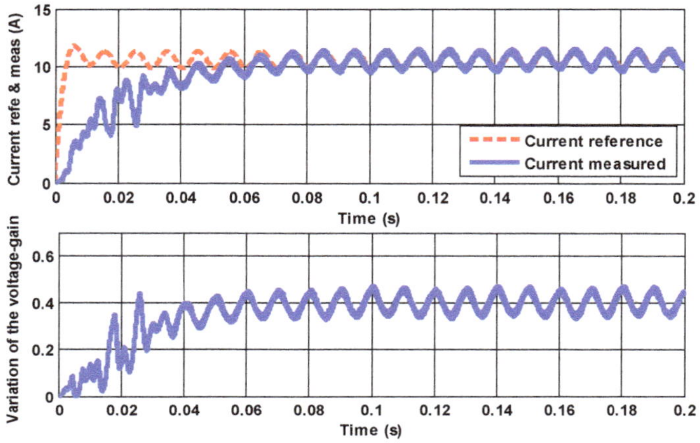

Figure 29. Load current and variation of the q (RST, unbalanced grid and with constant I_{ref}).

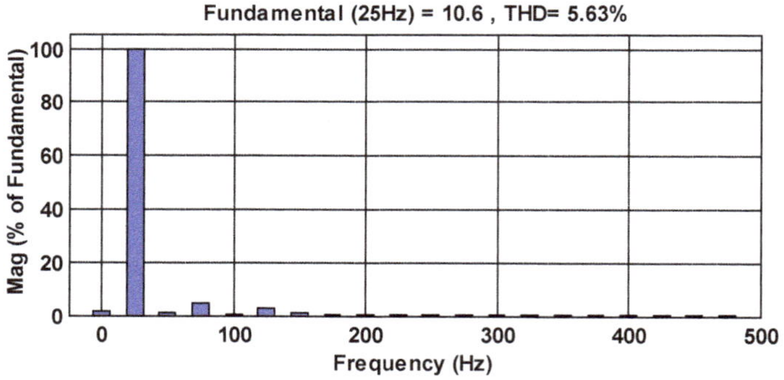

Figure 30. Harmonics spectrum of load current (RST, unbalanced grid and with constant I_{ref}).

5.5. Discussion the Results of Simulations

In **Figure 9** and **Figure 15**, the voltage at the output of the matrix converter is formed by a succession of pulse widths conversely proportional to the frequency of the reference voltage $f_o = 25$ Hz, and the RL load's current is almost sinusoidal with low Total Harmonic Distortion (THD) values. In **Figure 22** and **Figure 28**, the voltage at the output of the matrix converter is formed by a succession of patterns which widths are proportional to the frequency of the reference voltage and the amplitude is $V_{im} = 300$ V.

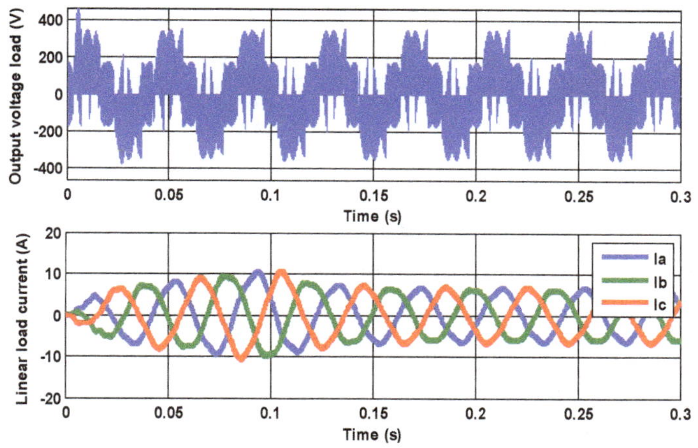

Figure 31. Output voltage and load current (RST, unbalanced grid and with stepped changing I_{ref}).

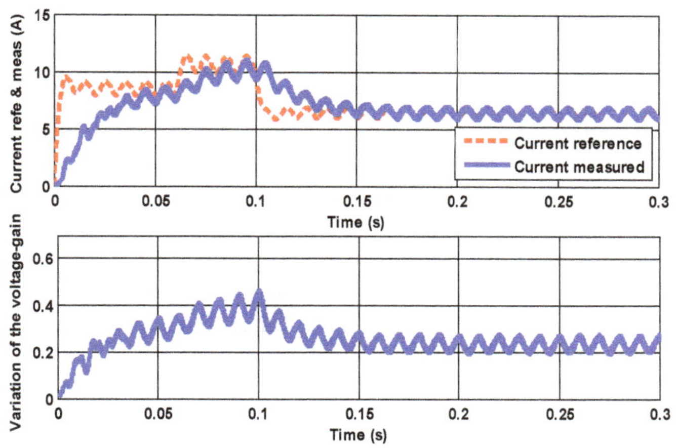

Figure 32. Load current and variation of the q (RST, unbalanced grid and with stepped changing I_{ref}).

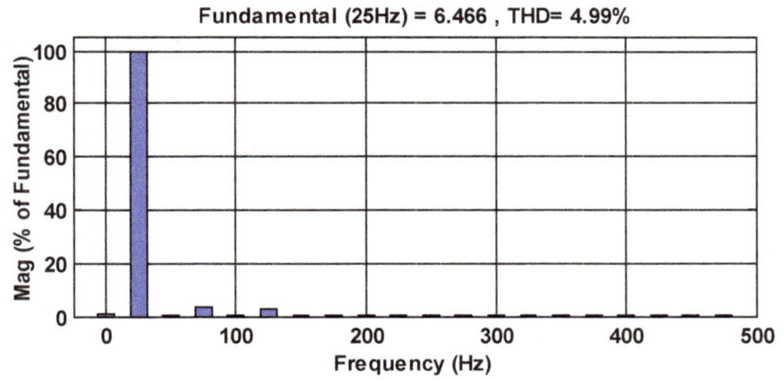

Figure 33. Harmonics spectrum of load current (RST, unbalanced grid an with stepped changing I_{ref}).

The THD increases for the unbalanced grid unlike in the balanced case (**Figure 11** and **Figure 17**). However, the output currents are almost balanced, but are distorted. With the RST strategy, the signal quality of load current is much better than PI. Indeed, the THD is improved by 10.82% in the case of balanced grid, while this improvement is around 7.70% in the case of unbalanced grid Constant reference current I_{ref}. Note that in all the investigated cases, the gain q does not exceed 0.866.

Table 2 and **Table 3** show the values of THD for balanced and unbalanced cases presented above.

Table 2. THD of load current with balanced grid.

Case with balanced grid	Values THD	IMP%
Constant I_{ref} (PI)	1.94%	10.82%
Constant I_{ref} (RST)	1.73%	
Time-varying I_{ref} (PI)	1.86%	3.220%
Time-varying of I_{ref} (RST)	1.80%	

Table 3. THD of load current with unbalanced grid.

Case with unbalanced grid	Values THD	IMP%
Constant I_{ref} (PI)	6.10%	7.700%
Constant I_{ref} (RST)	5.63%	
Time-varying I_{ref} (PI)	6.49%	23.11%
Time-varying of I_{ref} (RST)	4.99%	

Table 4. SSE with balanced grid.

Case with balanced grid	Values SSE
Constant I_{ref} (PI)	6.4827e + 04
Constant I_{ref} (RST)	1.2164e + 06
Time-varying I_{ref} (PI)	5.2185e + 04
Time-varying of I_{ref} (RST)	1.0137e + 06

Table 5. SSE with unbalanced grid.

Case with unbalanced grid	Values SSE
Constant I_{ref} (PI)	8.5052e + 04
Constant I_{ref} (RST)	1.0992e + 06
Time-varying I_{ref} (PI)	6.4179e + 04
Time-varying of I_{ref} (RST)	9.4901e + 05

Table 4 and **Table 5** show the Sum Squared Error (SSE).

In terms of the response of the system and the static error, the PI controller gives little better results than RST controller as it can be seen the **Table 4** and **Table 5**.

6. Conclusion

In this paper, a thorough theoretical modeling, analysis and comparison are presented for PI and RST control of MCs. A comprehensive control compensation method is used to find the PI gains. Moreover, the use of the pole placement technique is also shown to determine the RST's polynomial coefficients. Results for a balanced grid show lower load current THD as opposed to the unbalanced grid case, which is expected. However, RST control shows better performance. Nonlinear controllers tend to outperform these techniques at the expense of added complexity and computation. However, it is noteworthy that compared controllers are known for similar design complexity, which has been driving their use in the industry.

References

[1] Dendouga, A. (2010) Contrôle des puissances active et réactive de la machine asynchrone à double alimentation (DFIM). PhD Thesis, University of Batna, Batna.

[2] Luis, F.P.A. (2011) Maximum Power Point Tracker of Wind Energy Generation Systems using Matrix Converters. Master's Thesis, Technical University of Lisbon, Lisbon.

[3] Hulusi, K., Ramazan, A., Hüseyin, D., *et al.* (2008) A Novel Compensation Method Based on Fuzzy Logic Control for

Matrix Converter under Distorted Input Voltage Conditions. *Proceedings of the* 2008 *International Conference on Electrical Machines*, Vilamoura, 6-9 September 2008, 1-5.

[4] Venturini, M., Alesina, A., *et al.* (1980) The Generalised Transformer: A New Bidirectional Sinusoidal Waveform Frequency Converter with Continuously Adjustable Input Power Factor. *Proceedings of the Power Electronics Specialists Conference* (*PESC*' 80), Atlanta, 16-20 June 1980, 242-252.

[5] Filho, M.E.O., Filho, E.R., Quindere, K.E.B., Gazoli, J.R., *et al.* (2006) A Simple Current Control for Matrix Converter. *Proceedings of the International Symposium on Industrial Electronics*, Montreal, 9-13 July 2006, 2090-2094.

[6] Ram, G., Lincoln, S.A., *et al.* (2012) Fuzzy Adaptive PI Controller for Single Input Single Output Non-Linear System. *ARPN Journal of Engineering and Applied, Sciences*, **7**, 1273-1280.

[7] Hachicha, F., Krichen, L., *et al.* (2011) Performance Analysis of a Wind Energy Conversion System Based on a Doubly-Fed Induction Generator. *Proceedings of the* 8*th International Multi-Conference on Systems, Signals & Devices*, Sousse, 22-25 March 2011, 1-6.

[8] Bouhamida, M., Denai, M.A., *et al.* (2005) Robust Stabilizer of Electric Power Generator Using H∞ with Placement Constraints. *Journal of Electrical Engineering*, **56**, 176-182.

[9] Oubelli, A.L. (2011) Mise En œuvre d'un modèle générique du convertisseur matriciel dans les environnements EMTP-RV et MATLAB-SIMULINK. Master's thesis, Ecole Polytechnique de Montréal, Montréal.

[10] Hamane, B., Doumbia, M.L., Cheriti, A., Belmokhtar, K., *et al.* (2014) Comparative Analysis of PI and Fuzzy Logic Controllers for Matrix Converter. *Proceedings of the 9th International Conference on Ecological Vehicles and Renewable Energies* (*EVER*), Monte-Carlo, 25-27 March 2014, 25-27.

[11] Hamane, B., Doumbia, M.L., Cheriti, A., Belmokhtar, K., *et al.* (2013) Modeling and Control of a Matrix Converter Using Fuzzy Supervisory Controller. *Proceedings of the* 3*rd International Conference on Systems and Control* (*ICSC*), Algiers, 29-31 October 2013, 433-438.

[12] Afonso, L.P. (2011) Maximum Power Point Tracker of Wind Energy Generation Systems Using Matrix Converters. Master's Thesis, Higher Technical Institue of Technical University of Lisbon, Lisbon.

[13] Boukadoum, A., Bahi, T., Oudina, S., Souf, Y., Lekhchine, A.S., *et al.* (2012) Fuzzy Control Adaptive of a Matrix Converter for Harmonic Compensation Caused by Nonlinear Loads. *Energy Procedia*, **18**, 715-723.

[14] Ghedamsi, K. (2008) Contribution à la modélisation et la commande d'un convertisseur direct de fréquence Application à la conduite de la machine asynchrone. PhD Thesis, National Polytechnic School of Process Control Laboratory, El-Harrach.

[15] Dendouga, A., Abdessemed, R., Essounbouli, N., Megherbi, A.C., *et al.* (2013) Robustness Evaluation of Vector Control of Induction Motor fed by SVM Matrix Converter. 3*rd International Conference on Systems and Control* (*ICSC*), Algiers, 165-170.

[16] Rodriguez, S.E., Blaabjerk, F., *et al.* (1985) Modelling, Analysis and Simulation of Matrix Converters. *Applications*, **IA-21**, 1337-1342.

[17] Belabbes, A., Hamane, B., Bouhamida, M., Draou, A., Benghanem, M., *et al.* (2012) Power Control of a Wind Energy Conversion System based on a Doubly Fed Induction Generator using RST and Sliding Mode Controllers. *Proceedings of the International Conference on Renewable Energies and Power Quality* (ICREPQ'12), Santiago de Compostella, 28-30 March 2012. http://www.icrepq.com/icrepq%2712/298-belabbes.pdf

[18] Mai, T.D., Mai, B.L., Pham, D.T., Nguyen, H.P., *et al.* (2007) Control of Doubly-Fed Induction Generators Using Dspace R&D Controller Board—An Application of Rapid Control Coordinated with Matlab/Simulink. *Proceedings of the International Symposium on Electrical & Electronics Engineering*, 3, 302-307.

2

Power Line Monitoring Data Transmission Using Wireless Sensor Network

Lifen Li[1], Huaiyu Zhao[2]

[1]School of Control and Computer Engineering, North China Electric Power University, Baoding, China
[2]School of Electrical & Electronic Engineering, North China Electric Power University, Baoding, China
Email: Lilifen70@163.com

Abstract

The WSN used in power line monitoring is long chain structure, and the bottleneck near the Sink node is more obvious. In view of this, A Sink nodes' cooperation mechanism is presented. The Sink nodes from different WSNs are adjacently deployed. Adopting multimode and spatial multiplexing network technology, the network is constructed into multi-mode-level to achieve different levels of data streaming. The network loads are shunted and the network resources are rationally utilized. Through the multi-sink nodes cooperation, the bottlenecks at the Sink node and its near several jump nodes are solved and process the competition of communication between nodes by channel adjustment. Finally, the paper analyzed the method and provided simulation experiment results. Simulation results show that the method can solve the funnel effect of the sink node, and get a good QoS.

Keywords

Wireless Sensor Network (WSN), Power Line Monitoring, Data Transmission, Multimode Network

1. Introduction

In our country, power transmission line is all over the land. Overhead lines exposed to the atmosphere are prone to failure, or even lead to disaster, and we need to find a simple method to monitor the line in real-time [1]. Up till the present, the entire state of automatic monitoring of overhead transmission lines is only mentioned in reference [2], it described the insulator leakage of electric current online monitoring system, but its drawback is the transmission of sensor data for monitoring use of the GPRS of telecommunications company.

The wireless sensor network (WSN) technology penetrated from the field of military reconnaissance to industrial areas in recent years, which became the most popular areas of wireless networks. At present, the main practical applications of wireless sensor networks are ZigBee [3] wireless sensor networks that base on the IEEE 802.15.4 standard. It is a low complexity, and low power, low-cost wireless network technology. This will provide new ideas for data transmission in some industrial monitoring systems. Using wireless sensor network in power line monitoring and warning system is a promising way [4] [5].

2. The Transmission Line Monitoring System Design Based on WSN

The distance between adjacent towers is from dozens of meters to hundreds of meters, even across the valley, rivers and other special cases, it is no more than one thousand meters [6]. Sensor nodes deployed in high or low voltage transmission line tower, which form a long chain structure [7]. These sensor nodes are responsible for data perception, acquisition, computing and communication in real time. Several long chains form a long chain tree topology with the Sink node on the root. Each long chain presents a high voltage or low voltage transmission lines, and the Sink node generally located in the substation. In this way, a transmission line has three chains, as shown in **Figure 1**.

WSN based power transmission line monitoring system has characteristics as following: 1) The system needs to manage a long chains tree-like topological structures so that transmission is essentially multi-hop. 2) They all periodically collect data and transfer them to a concentrator station connected to a fixed network. 3) From the mobility point of view, all the systems can be considered to be stable. Sensors are not mobile and the network is stationary during the transmission. 4) Nodes are equipped with GPS. And the node localization, time synchronization can be solved by GPS. Each node has a unique ID number, uniformly distributed in the network initialization. 5) The node energy is not restricted. 6) For real-time data transmission, the system have higher demand to the reliability.

3. WSN Based Power Line Monitoring Data Transmission

3.1. Multimode Hierarchical Network Model

To build Multi-mode layer network, the multiple Sink nodes are set in the network. The data collecting sensor network is divided into several levels [8]. Different Sink nodes are responsible for the different levels. Each level uses different band to transmit data. After dividing levels, we can construct a load balancing network to realize the multimode traffic distribution, and solve the funnel effect [9].

Network with long chain type as an example, we assume that the node communication coverage 3 units. If the network is not layered, the number of the nodes within a node communication scope is about 20, and the 20 nodes compete wireless channel in the Communication. Through the network Level division, some nodes within the coverage area will be in the level of other bands, and only communicate with the edge nodes. Thus nodes to competition channel will be reduced to eight.

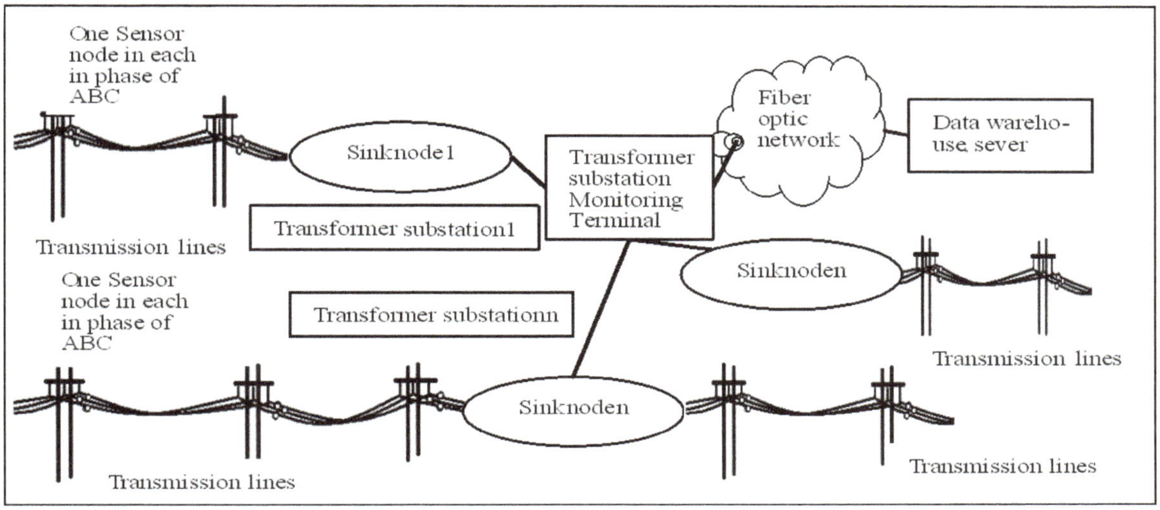

Figure 1. Long chain tree-like WSN topology.

In the distributed coordination of DCF model access control mode [10], the MAC layer using virtual carrier listening to determine what the wireless channel state, based on CSMA/CA protocol, when a node to transmit a packet, it need to pass a interframe spacing DIFS to determine whether the channel free [11]. Twenty nodes under the condition of channel competition, will be 20 nodes of network channel caused by the pressure of competition placed within range of a node, each node as forwarding used by a group of average channel listener time to $(1 + 2 + 3 + \ldots + 20)/(2 \times 20) \approx 10$ (DIFS), if reducing to eight node average channel the listening time is 4 (DIFS) [12].

3.2. The Implementation of Multimode Hierarchical Network

Sensor network as a graph $G = (V, E, SC)$, among them, the Sink node set, SC said V sensor nodes, the nodes if u and V can direct communication, are contained in the collection of E an edge (u, V). For each node in this paper, the modeling algorithm need the support of four list: NL neighbor node list, the parent node list PL, list of CL and brother child node SL.

In the node deployment, each node set up their own hop count to infinity, the node transmission power control unit is in a jump distance, network connection with probability 1. Then, each node broadcast network news, so that a node can know the existence of all its neighbor nodes, and the information stored in the NL list. Jump and then set the Sink node values, such as network exists in three Sink node, it will Sink node hop count is set to 0, 1, 2 in turn. Then choose the hop of 0 to the Sink node in the network broadcast a hello message, the message contained in the Sink node jump numerical and Sink node number value. When a node u v from the node receives a

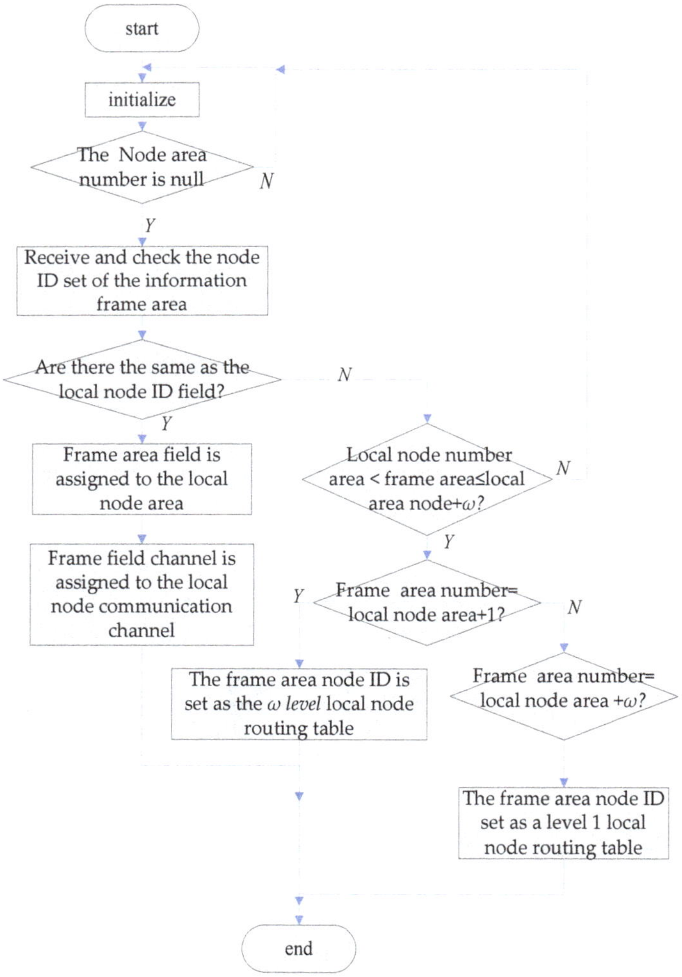

Figure 2. Local routing table formation process.

hello message, after get jump numerical *Nv*, then use your own jump numerical *Nu* compare the following and perform the corresponding operation:

If $Nv < Nu - 1$, set the $Nu = Nv + 1$, broadcast hello message and *Nu* to the neighbor nodes. If $Nu \neq \infty$, the node transmission power will be raised to the largest and the node *NL* list will be empty.

Again after a certain time delay after each node broadcasting notice sending its neighbor nodes report its existence, the purpose is to obtain the node communication coverage expanded new *NL* list, then each Sink node to send a hello message in networking on a network, with reference to the number of Sink node modulus value [13]. Using the hop count, obtained from the first step on node hop count as its take over, take the same remainder of nodes of the same level. Sink node is responsible for the different layers, and work with different frequencies. Routing table establishment process is as follows, when a node *u v* from the node receives a hello message, after get jump numerical *Nv*, then use your own jump numerical *Nu* compare the following and perform the corresponding operation:

if $Nu/Ns = Nv/Ns$,Add the node *u*'s and *v*'s id number to the *NL* list.

if $Nu/Ns > Nv/Ns$ and $Nu\%Ns = Nv\%Ns$, Add the node *u*'s and *v*'s id number to the *PL* list.

if $Nu/Ns < Nv/Ns$ and $Nu\%Ns = Nv\%Ns$, Add the node *u*'s and *v*'s id number to the *CL* list.

if $Nu/Ns = Nv/Ns$ and $Nu\%Ns = Nv\%Ns$, Add the node *u*'s and *v*'s id number to the *SL* list.

The uplink process routing table is shown as **Figure 2** (descending table creation process).

The ω grade in routing table can ensure that there was a fault in the network, the channel adjustment mechanism to fault to connect with the other layer, reach the purpose of rapid, complete data transmission, embodied in the local node transmission data channel adjustment algorithm [14], such as in **Figure 3**.

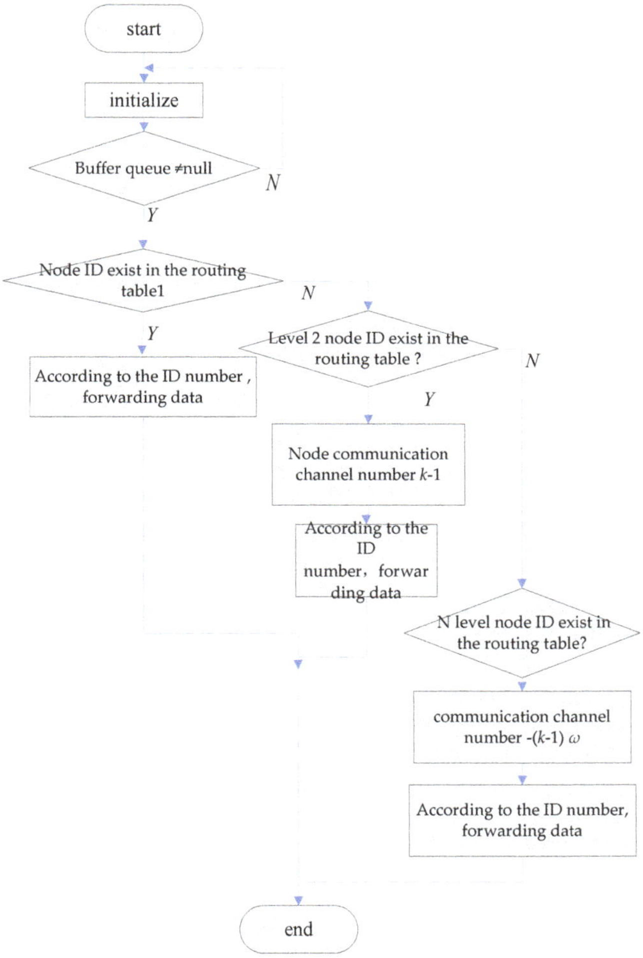

Figure 3. Channel adjustment algorithm.

4. The Simulation Experiment and Analysis

Simulation experiment platform for OMNeT++ of selection will be used in transmission line monitoring wireless sensor network model is defined as shown in **Figure 3** long chain tree structure. One set of sensor nodes coverage for km, by increasing the transmitted power, more increase coverage.

According to the method described in this article, We will use the hierarchical network construction of multimode hierarchical load balance network, which is shown in **Figure 4**. Corresponding to different sink node number, the use of 6 kinds of data acquisition time interval: 10, 20, 30, 40, 50, 60 seconds. By the average lifetime of sensor networks with the heaviest load node packet traffic. **Figure 4** and **Figure 5** show the configuration different sink node corresponding to the number of network lifetime and network traffic. *Ns* in the figure represents the number of sink nodes in the network configuration.

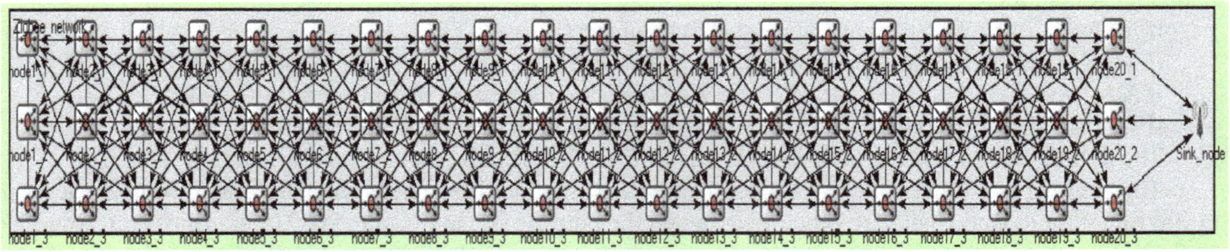

Figure 4. The diagram of the network topology.

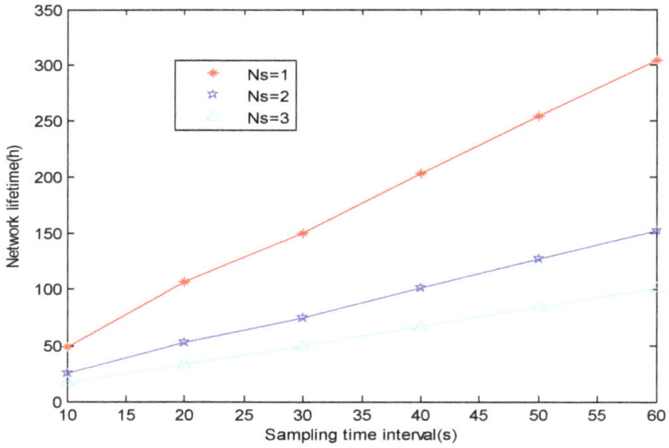

Figure 5. The comparison of network lifetime.

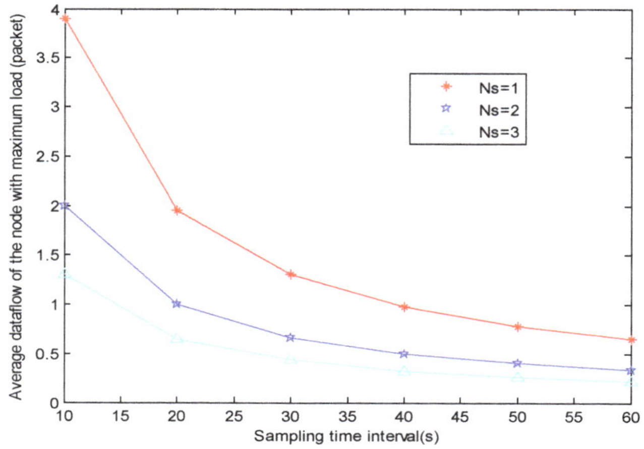

Figure 6. The comparison of network dataflow.

In **Figure 5**, with the increase of sink node number, different layers of the network, the more the shorter the network lifetime. Reflected in order to improve the energy of the price of single hop transmission distance. From **Figure 6**, you can see that with the increase of number of sink nodes in the network configuration, congestion of the heaviest load node is $1/Ns$ times to fall, decreases in the packet queuing delay of high efficiency in the network. Embodies the algorithm of the thought of distribution of average flow rate, different levels and flow distribution is not affected by other factors only concerned with Ns. Such as adopting multimode network grade, can solve the problem of the bottleneck of sink node place, but a few jump for near the sink node within the scope of the sensor nodes there is still a bottleneck problem, the multimode multi-level model can guarantee the sink node data into $1/Ns$ times decreased, guaranteeing each layer nodes forward data into $1/Ns$ times down. Data collected in this experiment with the method of periodic query, query is frequent, rapid energy consumption. For the query frequency is low or emergency is mainly used to collect the network packets sensor networks, network lifetime will be much longer.

5. Conclusion

Combined with the characteristics of the transmission system, the long chain tree network structure is put forward. The recent convergence node of sensor network is easy to form a "funnel", the multimode hierarchical network is built for load balance. Monitoring sensor network was applied to transmission line which has the advantages, and faces many difficulties, however. There are still some technical problems to be studied, mainly including transmission bandwidth, transmission distance, node power supply and electromagnetic compatibility, network security etc.

Acknowledgements

This work is supported by National Natural Science Foundation of China (No. 60974125).

References

[1] Lu, J.-Z., Zhang, H.-X., Fang, Z. and Li, B. (2009) Result and Its Analysis of Ice Disaster Monitoring of Hunan Power System. *Power System Protection and Control*, **37**, 99-105.

[2] Huang, X.B., *et al.* (2008) Transmission Line On-Line Monitoring and Fault Diagnosis. China Electric Power Press, Beijing.

[3] Lv, Z.-A. (2008) ZigBee Network Theory and Applications Development. Beihang University Press, Beijing.

[4] Wang, Y.-G., Yin, X.-G., You, D.-H., *et al.* (2009) A Real-Time Monitoring and Warning System for Electric Power Facilities Icing Disaster Based on Wireless Sensor Network. *Power System Technology*, **33**, 14-19.

[5] Zhao, Z.-H., Shi, G.-T., Han, S.-L., *et al.* (2009) A Heterogeneous Wireless Sensor Network Based Remote District High-Voltage Transmission Line On-Line Monitoring System. *Automation of Power System*, **33**, 80-84.

[6] Hull, B., Jamieson, K. and Balakrishnan, H. (2014) Mitigating Congestion in Wireless Sensor Networks. *Proceedings of the 2nd ACM Conference on Embedded Networked Sensor Systems (SenSys)*, Baltimore, Vol. 3, 556-664.

[7] El Gamal, H. (2013) On the Scaling Laws of Dense Wireless Sensor Networks. *IEEE Transactions on Information Theory*, April 2013, Submitted to Publication.

[8] Guo, R.C., Xu, Z.Z. and Li, X.L. (2007) Typical Design of Transmission Lines with Voltage Grades from 110 kV to 500 kV and Its Application. *Power System Technology*, **31**, 56-64.

[9] Gopala, P.K. and El Gamal, H. (2010) On the Scaling Laws of Multi-Modal Wireless Sensor Networks. *Annual Joint Conferences of the IEEE Computer and Communication Societies (INFOCOM)*, Hong Kong, Vol. 6, 234-243.

[10] Zheng, P., Zheng, G.S., Gong, Z.Y. and He, G.M. (2006) Research for Structure of Wireless Sensor Networks Based on Power Transmission Hallway. *Engineering Journal of Wuhan University*, **39**, 115-118.

[11] Zhang, C.-Q., Li, M.-L. and Wu, M.-Y. (2007) An Approach for Constructing Load-Balancing Networks for Data Gathering Wireless Sensor Networks. *Journal of Software*, **18**, 1110-1121. http://dx.doi.org/10.1360/jos181110

[12] Mohamed, R., Fahmy, S. and Pandurangan, G. (2013) Latency-Sensitive Power Control for Wireless Ad Hoc Networks. In: Boukerche, A., Ed., *Proc. of the MS-WiM* 2005, ACM Press, Montreal, 31-38.

[13] Wang, J.-M. (2009) Energy Level Selection for Wireless Sensor Network with Variable Transmit Power. *Computer Engineering*, **35**, 108-110.

[14] Li, F.M., Xu, W.J. and Liu, X.H. (2008) Power Control for Wireless Sensor Networks. *Journal of Software*, **19**, 716-732. (In Chinese) http://dx.doi.org/10.3724/SP.J.1001.2008.00716

3

Optimal Design of a Multibody Self-Referencing Attenuator

Dongmei Zhou, Jennifer A. Eden

Department of Mechanical Engineering, California State University, Sacramento, CA, USA
Email: zhoud@ecs.csus.edu

Abstract

The purpose of this paper is to determine the optimal size and number of tubes for a generic attenuator that is similar to Pelamis P2, the wave energy converter. Simulations using ANSYS Workbench, Design Modeler, and AQWA are performed to study the energy absorption at the nodes between the tubes. The analysis is limited to linearized hydrodynamic fluid waves loading on floating bodies by employing three-dimensional radiation/diffraction theory in regular waves in the frequency domain. Three sets of tests are conducted by varying total tube number, each tube length and the order of tubes with different lengths. After a systematic study in the frequency domain, the optimal size and number of the genetic attenuator is recommended.

Keywords

Attenuator, Pelamis, Extraction Efficiency & Power, Numerical Simulation, Parametric Study

1. Introduction

Hydrokinetic wave energy is interchanging potential and kinetic energy, carried by the wave away from its origin as wave travels through space in time. Wind, creates the most common waves, surface wave and then sea swells. In order to convert the constant motion of ocean waves into usable energy such as electricity, the wave energy converter (WEC) must survive the hostile environment at extreme sea states, absorb the maximum wave power, and be cost-effective for commercial market. Considerable research and effort continue to focus on the following aspects: 1) design new wave energy converter [1]; 2) create generic mode like wind energy [2]; 3) conduct numerical benchmarking [3] [4]; and 4) improve the design to be more cost-effective [5]. Excellent literature reviews on WEC systems can be found in the book edited by Cruz [6] and in the reviews on WEC technology by Falcão [7], on hydrodynamic modeling methods by Ye and Yu [8], and on numerical modeling of WEC arrays by Folley *et al.* [9]. This paper concentrates on the optimal design to improve cost efficiency of a generic attenuator that is similar to the wave energy converter, Pelamis.

The Pelamis is a particular type of ocean attenuator consisting of either four (P1 model) or five (P2 model) cylindrical sections linked together by universal joints that allow for motion with four degrees of freedom, as seen in **Figure 1**. The Pelamis faces the direction of the waves. As the waves pass down the machine, the sections bend. It is this bending movement that converts the energy from the waves into electricity by hydraulic jack systems housed at each joint. The P1 model was the world's first full-scale offshore WEC to generate electricity and the first wave energy farm to successfully supply electricity to the national grid. The new P2 model has five tube sections with the power conversion modules integrated into them and it has an overall length of 180 m and a diameter of 4 m; each joint has four hydraulic motors. Both P1 and P2 models create the same amount of electricity, 750 kW.

Pelamis Wave Power Corporation has published several papers [10]-[14], reporting the development progress from numerical modeling, experimental test to grid connection. The Pelamis Wave Power Corporation conducted their own in-house simulations using their own proprietary software to analyze various aspects of the structure in both linear and non-linear hydrodynamics. Yet, limited data on optimization from numerical modeling has been published. Most studies in other papers have comprised of performance comparisons between different types of WECs. It was found by O'Connor *et al.* [15] and Rusu *et al.* [16] that the Pelamis devices had the highest energy and economic returns at high resource locations but produced poor results at poor resource locations; therefore the location determines which WEC is the best option. The impact of electricity from the Portugal Pelamis Farm on current market prices was found to be negligible for wholesale electricity prices [17]. There was also research [18] done to create an optimum model for the hydraulic power take-off system of the Pelamis, yet no research has been done into optimizing the attenuator Pelamis device. This lack of knowledge and research, along with the excitement of a new technology lead to this paper.

The objective of this study is to determine the optimal design of a multibody self-referencing attenuator that is similar to the wave energy converter Pelamis P2. The software programs, ANSYS Workbench, Design Modeler, and AQWA [19] are used to determine the optimal length, diameter, and number of individual tubes for the attenuator based on parametric study. The analysis is limited to linearized hydrodynamic fluid waves loading on floating bodies by employing three-dimensional radiation/diffraction theory in regular waves in the frequency domain. Not only will this analysis provide the model with the overall highest energy extraction, it will also provide the model with the greatest efficiency over the range of frequencies experienced off the coast of Scotland.

2. Numerical Method

The development of a WEC starts from concept and design, then numerical modeling, model testing in wave basin, testing under real sea conditions, and finally to commercial stage. It is a long, difficult, and expensive process. Numerical modeling has the advantage of providing quick and inexpensive evaluation and optimization

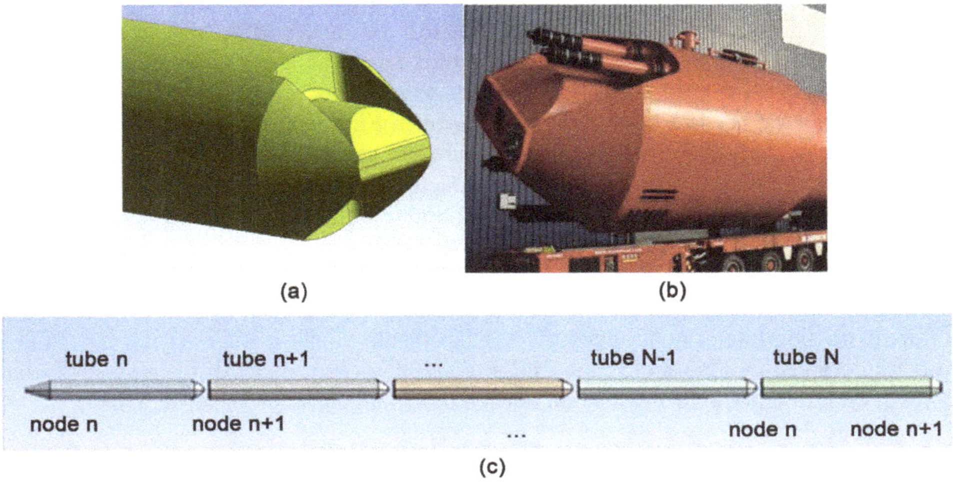

Figure 1. Attenuator view: (a) created simulation model; (b) actual device; (c) control model.

of designs. In this study, ANSYS AQWA, an industry standard hydrodynamic software package, is employed to capture the behavior of the attenuator by simulating the interaction between waves, the WEC device and the power take-off mechanism. The software ANSYS AQWQ can simulate linearized hydrodynamic fluid wave loading on floating bodies by employing three-dimensional radiation/diffraction theory in regular waves in the frequency domain. Frequency domain analysis is the first step in the hydrodynamic modeling process by assuming everything is linear. This analysis is especially useful in geometry optimization routines and thus it is employed in this paper. However, it should be fully aware the limitations of this analysis method compared to realistic conditions.

2.1. Linear (Airy) Theory and Wave Data

The resource, surface gravity waves can be linearized if $a/\lambda << 1$ and $a/H << 1$, where the variable, a, is the wave amplitude (m), λ is the wavelength (m) and H is the uniform water depth (m). Typically the surface gravity waves of oceans have wavelengths between 30 - 40 m, enabling the water surface tension to be neglected as it pertains to wavelengths of less than 5 - 10 cm. As the attenuator is deployed at water depths greater than 50 m it is classified as a deep-water device due to $H > \lambda/3$. It is also noted that as the frequency is much larger than the Coriolis frequency, the wave motion is unaffected by Earth's rotation. Assumptions made here are that 1) the water is incompressible (constant density) and irrotational; 2) gravity is the only external force; and 3) viscosity can be neglected. For the regular wave based on linear theory, the power obtained from ocean per meter of wave front is the total energy of the system (kinetic plus potential) per unit horizontal area multiplied by the phase speed to yield the following relation:

$$P_t = \frac{\rho g^2 a^2}{2\omega} \quad (1)$$

where, P_t is power per unit length of wave front (W/m), ρ is density of seawater (kg/m^3), g is gravity (m/s^2), a is the wave amplitude (m), and ω is angular frequency (rad/s).

The European Marine Energy Center (EMEC) [20] compiled the wave data from three data-well directional waverider buoys. The EMEC determines the instantaneous data for the maximum wave height, the significant wave height, the maximum wave period, and the significant wave period. It also provides a graph for a 24 hour time period and this data can be used to determine the current energy production of the device. As the actual attenuator Pelamis P2 is located off the west coast of Orkney, Scotland, according to the Scottish Government website [21] the annual mean wave height for the region is 2.0 - 2.4 m; therefore from here on, a significant wave height of 2.2 m, resulting in an amplitude a of 1.1 m, is used in the simulations.

2.2. Evaluation of Pelamis Performance

The multibody self-referencing attenuator generates electricity based on the relative motion between its tubes. The attenuator moves with two degrees of freedom to capture energy: pitch and heave, given that this study is restricted to the case in the x-z plane, with z as the vertical axis and x as the axis passing through the head and tail of a tube. Thus for each tube there are two dynamic equations, one for moment due to pitch and one for force due to heave. In order to find the potential power output from the attenuator, the forces and moments acting on the attenuator tube need to be defined. As shown in **Figure 2**, the vertical hydrodynamic force, F_n, on tube n is a resultant force of wave pressure. The pressure can be resolved from the Bernoulli's equation after the velocity potential is determined. The reaction force, R_n, exerted on node n represents the power take-off mechanism. The hydrodynamic moment, M_n, about the midpoint of tube n can be expressed in terms of the angular displacement, Θ_n.

In rotational systems, the power is derived as the product of angular velocity and torque, or in our case the product of angular frequency ω and the moment M_n. Thus the power absorbed by attenuator at hinge n is the product of the extraction rate at node n and the angular displacement of tubes n and $n-1$. The time averaged power extraction derived by Flanes [22] and Mei *et al.* [23], is the product of the angular displacement squared divided by two, the extraction rate and the real portion of the angular rotation squared as:

$$P_n = \frac{\omega^2}{2}\sum_{n=2}^{N}\left(\bar{\alpha}_n\left|\frac{2B_{n-1}}{L_{n-1}} - \frac{2B_n}{L_n}\right|^2\right) \quad (2)$$

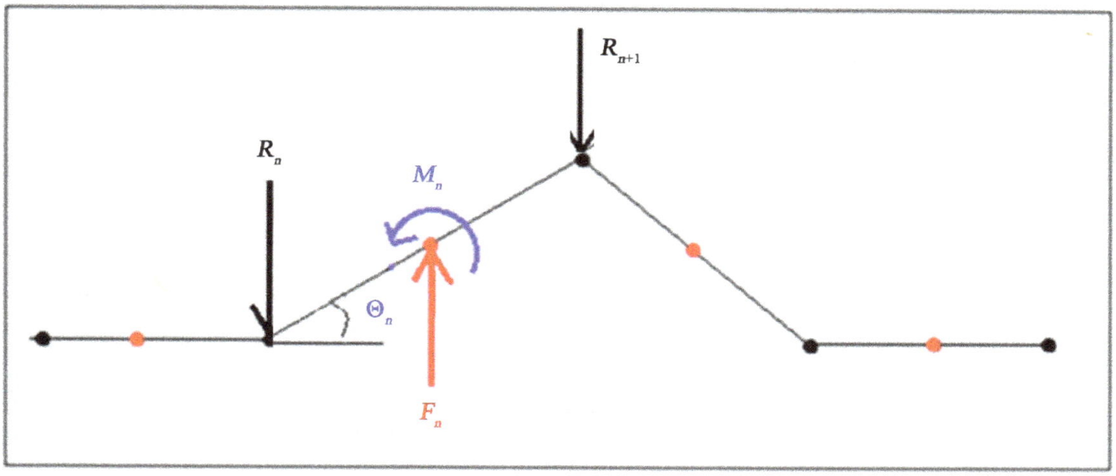

Figure 2. Illustration for force F_n; moment M_n; reaction force R_n; angular displacement Θ_n.

where $\bar{\alpha}_n$ in the unit of Nm/(m/s) is the extraction rate between tubes and B_n/L_n(m/m) is the tube amplitude. The extraction rate $\bar{\alpha}_n$ is equivalent to the amount of radiation damping in the global x-direction. The overall efficiency of the system at a particular frequency is defined as the mean power P_n that is extracted at a node from Equation (2) divided by the total power of the system P_t that is defined in Equation (1), yielding

$$eff = \frac{P_n}{P_t} \tag{3}$$

2.3. Modeling Using ANSYS AQWA and Validation

In the simulation process using computer software ANSYS AWQA, a computer model of attenuator is first created by use of the ANSYS Design Modeler software to generate three-dimensional tubes as shown in **Figure 1**. This model is then fed to the hydrodynamics package. Next, the hydrodynamics package is applied and numerical simulations are performed. Finally, Excel is used to post process the data provided by ANSYS AQWA software. A summary of input parameters to calculate the power, P_n, extracted from the attenuator using Equation (2) and the efficiency, eff, using Equation (3) is displayed in **Table 1**. The extractions rates $\bar{\alpha}_n$ are determined from the ANSYS AQWA software via the Radiation Damping. The ratios of amplitude of the center of mass for tube n (midpoint of the tube) to the length of tube n, B_n/L_n, are also determined from the ANSYS AQWA software via the RAOs (Response Amplitude Operators); the results are generated in terms of m/m. A large enough range for the frequency f is selected to encompass the range of efficient energy capture for all design models of attenuator.

To validate the numerical model, ANSYS AQWA is applied to the control model T5L36D4, an attenuator with 5 tubes, a 36 m tube length, and a 4 m tube diameter as defined in **Table 3**. The actual Pelamis P2 currently in production off the west coast of Scotland at a frequency of 0.15 Hz. Pelamis Wave Power [24] states that the actual Pelamis P2 model experiences efficiencies around 70% in all sea states. The validation here is to examine how varying the mesh selection affects the efficiency value of the WEC system. It should be noted that the ANSYS program demands the default tolerance be no greater than 0.6 times the maximum element size. The validation results are presented in **Table 2**, and it can be observed that the efficiencies are all within 2.5% of one another and in agreement with the measured efficiency data 70%. The amount of difference appears to relate with the difference between the maximum element size and the default tolerance; the closer the maximum element size and the default tolerance values are, the larger the change in efficiency is. **Figure 3** shows the effect of altering mesh size on the efficiency eff (%) of the attenuator at different frequencies. From **Figure 3**, it can be seen that the shape of the curves is consistent because they all appear to overlap one another. As the efficiency curves do not change and 2.5% difference in efficiency is such a small value, it is concluded that the numerical simulation results are independent of the mesh size.

Table 1. Input parameters to calculate overall efficiency.

Parameter	Value(s)	Units
Density, ρ	1023.485	kg/m^3
Gravity, g	9.80665	m/s^2
Amplitude, a	1.1	m
Frequency, f	0.05 - 0.30	Hz
Extraction rate, $\bar{\alpha}_n$	Program generated	Nm/(m/s)
Tube amplitude, B_n/L_n	Program generated	m/m

Table 2. Validation results with altering mesh size at frequency $f = 0.15$ Hz.

	Control model Pelamis P2	Element size 1.5 tolerance of 0.6	Element size 1.5 tolerance of 0.8	Element size 1.0 tolerance of 0.5	Element size 0.8 tolerance of 0.3
Model number	T5L36D4	E1.5 T0.6	E1.5 T0.8	E1.0 T0.5	E0.8 T0.3
Efficiency eff (%)	73.27	71.15	70.84	71.84	72.05

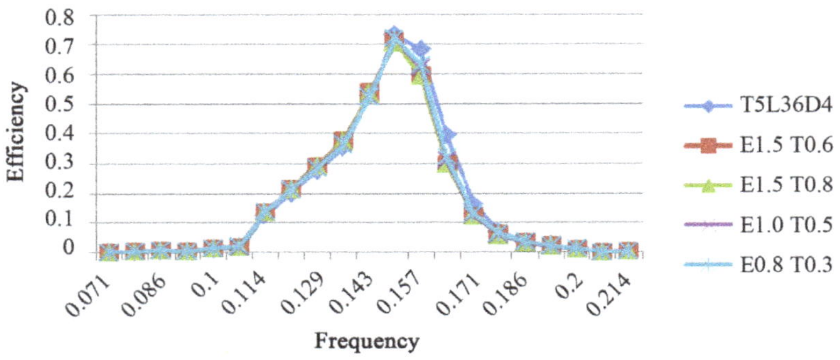

Figure 3. Effect of altering mesh size on efficiency of attenuator at different frequencies.

3. Results and Discussion

3.1. Test Models of Attenuator Design

The attenuator uses resonance to increase power capture of small waves. The default setting of the attenuator is non-resonant, allowing it to withstand large swells. However, the joints can be actively controlled by its power take off system to create a cross-coupled resonant response. Optimal design of attenuator is based on such a principle that the device must be designed to operate efficiently within the frequency range and power level. In an attempt to standardize the amount of raw materials required for each model, the overall length of 180 m and weight of 1350 tons are chosen to maintain the same overall length and weight as the current Pelamis P2 model.

A summary of all test models is provided in **Table 3**. The first set of test models varies the number of tubes for each structure while maintaining a diameter of 4 m. The second set of test models uses the same number of tubes and lengths as the first set, but it varies the tube diameter. The third set of models takes the most efficient model from sets one and two, which is the model T6L30D5 as shown later, and then determines how altering the length of each tube affects the efficiency. The third set contains the model T6D5 Increasing, where the tubes increase in length ($L1 < L2 < L3 < L4 < L5 < L6$) and the model T6D5 Decreasing, where the tubes decrease in length ($L6 < L5 < L4 < L3 < L2 < L1$). For these two models (T6D5 Increasing and Decreasing), the ratio of 1:2:3:4:5:6 with a total of 21 units, does not divide evenly into the total length of the structure (180 m), resulting in 8 - 12/21. To overcome this, this remainder of 12 in 8 - 12/21 is split evenly amongst the six tubes. Thus, the length of each tube in meter is 10:18:26:34:42:50 for model T6D5 Increasing and 50:42:34:26:18:10 for model T6D5 Decreasing. The model T6D5 SBS has the end tubes that are longer than the center tubes ($L1 > L2 > L3 = L4 < L5 < L6$) and the model T6D5 BSB has the middle tubes that are longer than the end tubes ($L1 < L2 < L3 =$

Table 3. Summary of three sets of test models including control model T5L36D4.

	Number of tubes	Tube length (m)	Tube diameter (m)	Model number
Test set one	3	60	4	T3L60D4
	4	45	4	T4L45D4
	5	36	4	T5L36D4
	6	30	4	T6L30D4
	8	22.5	4	T8L22.5D4
Test set two	3	60	3	T3L60D3
	4	45	3	T4L45D3
	5	36	3	T5L36D3
	6	30	3	T6L30D3
	8	22.5	3	T8L22.5D3
	3	60	5	T3L60D5
	4	45	5	T4L45D5
	5	36	5	T5L36D5
	6	30	5	T6L30D5
	8	22.5	5	T8L22.5D5
Test set three	6	10:18:26:34:42:50	5	T6D5 increasing
	6	50:42:34:26:18:10	5	T6D5 decreasing
	6	15:30:45:45:30:15	5	T6D5 SBS
	6	45:30:15:15:30:45	5	T6D5 BSB

L4 > *L5* > *L6*). For these two models (T6D5 SBS and BSB), the ratio of 1:2:3:3:2:1 with a total of 12 units does divide evenly into the total length, resulting in the lengths of each tube to be, in meters, 15:30:45:45:30:15 for the model T6D5 SBS and 45:30:15:15:30:45 for the model T6D5 BSB.

3.2. Test Set One Results

Test set one models vary the number of tubes for each structure while maintaining a diameter of 4 m. The peak efficiencies for each model, along with the corresponding frequencies are provided in **Table 4**. The efficiency of the actual Pelamis P2 model has been stated to be around 70%. The maximum efficiency for the control model of our attenuator is calculated to be 73.27%. This is the theoretical maximum of the device and does not account for additional efficiency losses generated by the conversion of mechanical work to electrical work. With the exception of model T8L22.5D4, the efficiency for every other model is lower than that of the control model. As shown in **Table 4** and **Figure 4**, as the tube length increases (*i.e.*, decrease in tube number), the efficiency of the system drastically decreased. The results do not show any correlation for the decrease in tube length (or increase in tube number) and efficiency; the efficiency of T6L30D4 is slightly lower than that of control model T5L3D4, while the efficiency of T8L22.5D4 is slightly higher than that of control model by 1.88%. Also of note is that the frequency corresponding to the peak efficiency gradually increases for the first four models, and decreases slightly for the last model. As is evident in **Figure 4** that the range of frequency for available energy capture is widest for models T4L45D4, T5L36D4, and T6L30D4.

Given the restrictions of maintaining the overall length and weight of the attenuator structure, the efficiencies are highest with a smaller tube length (*i.e.*, larger number of tubes). The T8L22.5D4 model would be the best choice from this test set at 1.88% more efficient than the control model. However, devices with more moving parts require more maintenance and can experience a greater loss in efficiency due to friction when converting the mechanical energy to electrical energy, therefore it may not be cost effective from this perspective to increase the amount of moving parts associated with additional nodes while only achieving a 1.88% increase in

Table 4. Test set-one: maximum efficiencies and corresponding frequency for attenuators with 4 m diameter.

Model number	T3L60D4	T4L45D4	T5L36D4	T6L30D4	T8L22.5D4
Efficiency (%)	24.93	50.93	73.27	72.51	75.14
Frequency (Hz)	0.121	0.143	0.150	0.157	0.150

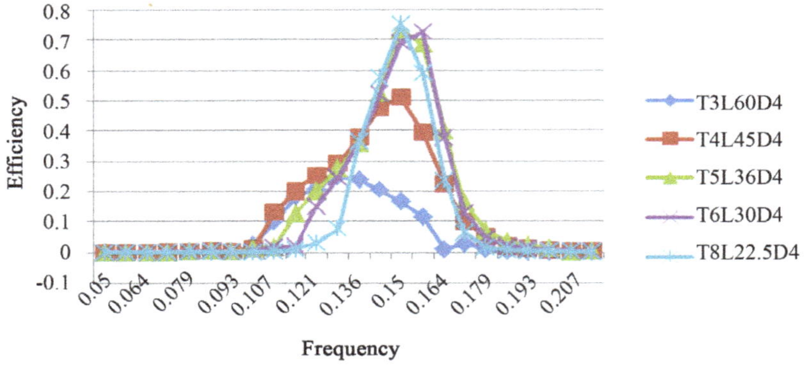

Figure 4. Test set one: efficiency curves for attenuators with 4 m diameter (T = tube number, L = tube length).

efficiency from the control model.

3.3. Test Set Two Results

Test set two demonstrates the effects of tube length at different diameter of the structures. The peak efficiencies for each model, along with the corresponding frequencies are provided in **Table 5**. For the set of models with a 3 m diameter, there is a direct correlation between the tube length and the maximum efficiency; as the tube length decreases (*i.e.*, increase in tube number) the efficiency increases. Yet all these values are significantly lower than those seen in Test Set One. By observing **Table 5** and **Figure 5**, no direct correlation is found between tube length and the frequency where maximum efficiency occurs; T4L45D3, T5L36D3, and T6L30D3 occur at the same frequency of 0.143 Hz, while T3L60D3 and T8L22.5D3 occur the same (but higher) frequency of 0.163 Hz.

For the set of models with a diameter of 5 m, the model with tube length of 30 m (six tubes, model T6L30D5) has the highest efficiency. It is interesting to note that the efficiencies for T3L60D5, T4L45D5, and T5L36D5 are lower than their counterparts in Test Set One, yet the efficiencies for T6L30D5 and T8L22.5D5 are higher than those in Test Set One, by 6.39% and 2.80% respectively. Therefore, additional models are created for tube lengths of 30 m and 22.5 m to have a diameter of 6 m; these efficiencies are lower than those of the 5 m diameter models. From **Table 5** and **Figure 6**, it is observed that as the tube length decreases (*i.e.*, the tube number increases) at the same diameter 5 m, the frequency where the maximum efficiency occurs, increases for the first two models and then holds at the same value of 0.157 Hz.

For the effect of tube diameter, it is observed that for a given tube length, there is an optimum tube diameter. By making the structure narrower, the tube slices through the oncoming waves rather than the wave lifting the tubes to create sufficient angular rotation about the hinges. The narrower models experience less hull pressure from the water, and thus it creates less of a hydrodynamic moment and therefore it absorbs less energy. This is the same principle used when oceangoing vessels are designed for speed. The reason that an increase in diameter reduces the efficiency of the system is theorized here that the first tube(s) does experience a higher vertical pressure but then it experiences an effect called "slamming" [25], wherein the tubes experience a large force upon impact with the water surface. This slamming would then result in the generation of a large amount of radiation waves, thereby taking the structure out of resonance and decreasing its overall efficiency.

3.4. Test Three Results

Based solely on maximum efficiency, from Test Set One and Test Set Two, the model T6L30D5 with 6 tubes, a

Table 5. Test set two: maximum efficiencies and corresponding frequency for attenuators at different tube diameters.

	T3L60D3	T4L45D3	T5L36D3	T6L30D3	T8L22.5D3
Efficiency (%)	24.39	49.76	52.19	56.93	58.66
Frequency (Hz)	0.163	0.143	0.143	0.143	0.163
	T3L60D4	**T4L45D4**	**T5L36D4**	**T6L30D4**	**T8L22.5D4**
Efficiency (%)	24.93	50.93	73.27	72.51	75.14
Frequency (Hz)	0.121	0.143	0.150	0.157	0.150
	T3L60D5	**T4L45D5**	**T5L36D5**	**T6L30D5**	**T8L22.5D5**
Efficiency (%)	25.12	46.45	67.06	78.90	77.94
Frequency (Hz)	0.129	0.150	0.157	0.157	0.157
	--	--	--	**T6L30D6**	**T8L22.5D6**
Efficiency (%)	--	--	--	75.06	72.86
Frequency (Hz)	--	--	--	0.164	0.164

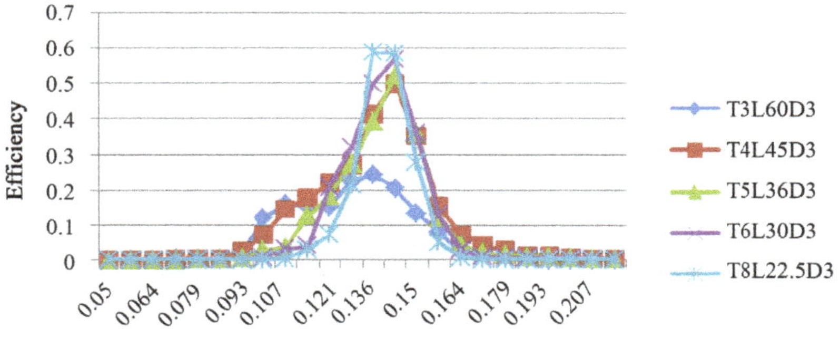

Figure 5. Test set two: efficiency curves for attenuators with 3 m diameter (T = tube number, L = tube length).

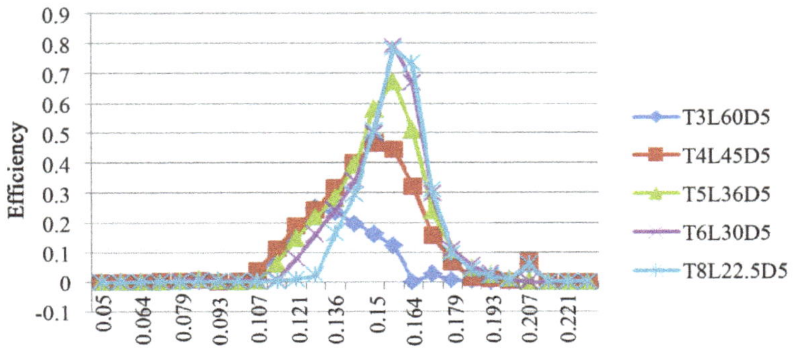

Figure 6. Test set two: efficiency curves for attenuators with 5 m diameter (T = tube number, L = tube length).

30 m tube length, and a 5 m tube diameter, is the highest performer with an efficiency of 78.905% and is used for Test Set Three. As shown in **Table 6** and **Figure 7** in all four cases for test set three, the maximum efficiency is greater than that of constant tube length model T6L3D5. The T6D5 SBS model (middle tubes longer than the end ones) is the highest performer, 94.72%, followed by the T6D5 Increasing model, 85.41%, which in turn is closely followed by the T6D5 BSB model, 85.10%; the T6D5 Decreasing model was only slightly more efficient than the T6L30D5 model. The two models with the highest efficiency occur at the same frequency of

Table 6. Test set three: maximum efficiencies and corresponding frequency for attenuators at different tube arrangements.

Model number	T6L36D5	T6D5 increasing	T6D5 decreasing	T6D5 SBS	T6D5 BSB
Efficiency (%)	78.90	85.41	79.81	94.72	85.10
Frequency (Hz)	0.157	0.164	0.171	0.164	0.171

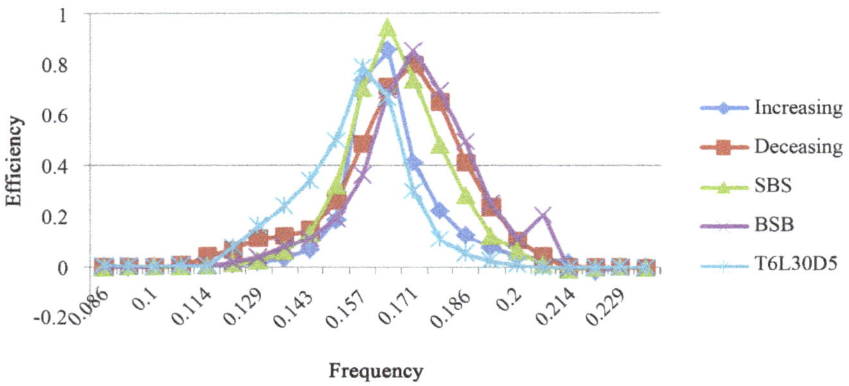

Figure 7. Test set three: T6D5 series of efficiency curves.

0.164 Hz and the two with the lower efficiencies occur at the same frequency of 0.171 Hz, yet all four models experienced peak efficiencies at higher frequencies compared to the model T6L3D5 with consistent tube lengths, $f = 0.156$ Hz.

While the short end tubes of the T6D5 SBS model have a low damping value, the middle tubes act as sort of vertical mooring structure that allow for these shorter end tubes to generate large vertical elevation which increase the angular rotation at the hinges, resulting in larger power generation and higher efficiency. The model T6D5 BSB operates on an opposing principle; there is an even balance of high damping and low elevation with low damping and high elevation. The large end tubes create a large amount of damping and experience slight elevation, while the smaller tubes connected to them are able to increase their elevation and angular rotation, resulting in high power generation and efficiency values. The model T6D5 Decreasing has its largest tubes at the head of the structure; this causes a greater amount of wave damping, leaving less wave height available for energy extraction at later hinges. The very opposite of this principle leads to the model T6D5 Increasing absorbing a larger amount of energy, as the smaller front tubes create less wave damping and hence leave more of the wave to be extracted at later hinges.

Thusly, out of all three test sets, the best model simulation based solely on maximum efficiency would be the model T6D5 SBS, having six tubes, a diameter of five meters, and the tubes varying in size, with the middle tubes longer than the end tubes ($L1 < L2 < L3 = L4 > L5 > L6$). However, selecting a model based on overall maximum efficiency is not always the best decision. It is important to create the model with the characteristics of the environment in mind, specifically by knowing what frequency range is experienced at the proposed location.

3.5. Effect of Frequency and a Case of Application

Ocean locations have energy distributed over a range of wave heights and periods, or frequencies. Off the coast of Scotland where the actual Pelamis P2 model resides, the ocean state experiences a mean wave period between 8.1 s in winter and 6.3 s in summer which corresponds to frequencies of 0.123 Hz and 0.159 Hz, respectively, [26]. With this in mind, the optimal design of the attenuator for this site needs to capture the most energy over that entire range. While all models (diameters 3 m to 5 m) have their peak efficiency within that range, the efficiency of the three-meter diameter models is too small for these devices to be considered viable; when the diameter is increased the frequency is increased as well to be on the edge of the frequency range observed off the coast of Scotland. The six-meter diameter models are out of the range altogether. Therefore, the four-meter diameter models are the best fit for these constraints. Of these five models, the T3L60D4 and T4L45D4 models produced significantly less efficiencies and will be excluded from consideration. While it has the higher efficiency, the T8D22.5L4 model also has a narrower frequency range and it would not produce any viable energy during the winter months. The T8D22.5L4 mode is therefore excluded from consideration. Between the remaining models, T5L36D4 and T6L30D4, the control model T5L36D4 has the higher efficiency and it absorbs viable energy over greater frequency range that includes the frequencies off the coast of Scotland. By combining this with the results from Test Set Three, the optimal design of the attenuator that is being deployed off the coast of Scotland would be the model T5D4 SBS model, a device containing five tubes with four-meter diameters whose tube lengths increase then decrease ($L1 < L2 < L3 > L4 > L5$), given the restrictions of 180 m overall length and 1350 tons overall weight.

4. Conclusions

In this study, the wave energy converter, amultibody self-referencing attenuator that is similar to the actual Pelamis P2 is studied using numerical modeling to determine the optimal size and number of the attenuator structure. Three sets of tests are conducted by varying total tube number, each tube length and the order of each tube with different length. The analysis is based pm linearized hydrodynamic fluid waves loading on a floating body by employing three-dimensional radiation/diffraction theory in regular waves in the frequency domainand the conclusion is drawn as follows.

1) Given the restrictions of maintaining the overall length and weight of the attenuator structure, the efficiencies are highest with a smaller tube length (*i.e.*, larger number of tubes).

2) For a given tube length, there is an optimum tube diameter corresponding to the maximum efficiency.

3) Out of all three test sets, the best model simulation based solely on maximum efficiency of 94.74% would be the model T6D5 SBS with six tubes, a 5 m diameter, and the tubes varying in size, with the middle tubes longer than the end tubes ($L1 < L2 < L3 = L4 > L5 > L6$).

4) Given the restrictions of 180 m overall length and 1350 tons overall weight, the optimum design of the attenuator to be deployed off the coast of Scotland is the model T5D4 SBS with five tubes of four-meter diameters, whose tube lengths increase *then decrease* ($L1 < L2 < L3 > L4 > L5$).

The future work after this paper will be to credit the optimal design in the time-domain analysis by introducing nonlinearities and viscous effect with cost-effective in mind. It is expected that the extraction power and efficiency will be decreased by 3% - 5% after introducing nonlinearities and viscous effect. However, the results and conclusion should hold for irregular waves if the range of wave frequencies is appropriate.

Acknowledgements

The research project was partially supported by 2013-2014 Provost's Research Fellow program and Provost's Research Incentive Fund Summer 2014.

References

[1] Leybourne, M., Bahaj, A.S., Minns, N. and O'Nians, J. (2014) Preliminary Design of the OWEL Wave Energy Converter Pre-Commercial Demonstrator. *Renewable Energy*, **61**, 51-56. http://dx.doi.org/10.1016/j.renene.2012.08.019

[2] Ruehl, K., Brekken, T.A., Bosma, B. and Paasch, R. (2010) Large-Scale Ocean Wave Energy Plant Modeling. 2010 *IEEE Conference on Innovative Technologies for an Efficient and Reliable Electricity Supply*, Waltham, Massachusetts, 27-29 September 2010, 379-386. http://dx.doi.org/10.1109/citres.2010.5619775

[3] Babarita, A., Halsb, J., Muliawanb, M.J., Kurniawanb, A., Moanb, T. and Krokstadc, J. (2012) Numerical Benchmarking Study of a Selection of Wave Energy Converters. *Renewable Energy*, **41**, 44-63. http://dx.doi.org/10.1016/j.renene.2011.10.002

[4] Silva, D., Rusu, E. and Soares, C.G. (2013) Evaluation of Various Technologies for Wave Energy Conversion in the Portuguese Nearshore. *Energies*, **6**, 1344-1364. http://dx.doi.org/10.3390/en6031344

[5] Yu, Y.-H., Li, Y., Hallett, K. and Hotimsky, C. (2014) Design and Analysis for a Floating Oscillating Surge Wave Energy Converter. *ASME* 2014 33*rd International Conference on Ocean, Offshore and Arctic Engineering*, San Francisco, 8-13 June 2014, V09BT09A048. http://dx.doi.org/10.1115/omae2014-24511

[6] Cruz, J. (2008) Ocean Wave Energy. 1st Edition, Springer, Berlin. http://dx.doi.org/10.1007/978-3-540-74895-3

[7] Falcão, A.F.D.O. (2010) Wave Energy Utilization: A Review of the Technologies. *Renewable and Sustainable Energy*

Reviews, **12**, 899-918. http://dx.doi.org/10.1016/j.rser.2009.11.003

[8] Li, Y. and Yu, Y.-H. (2012) A Synthesis of Numerical Methods for Modeling Wave Energy Converter-Point Absorbers. *Renewable and Sustainable Energy Reviews*, **16**, 4352-4364. http://dx.doi.org/10.1016/j.rser.2011.11.008

[9] Folley, M., Babarit, A., Child, B., Forehand, D., O'Boyle, L., Silverthorne, K., Spinneken, J., Stratigaki, V. and Troch, P. (2012) A Review of Numerical Modelling of Wave Energy Converter Arrays. *ASME* 2012 31*st International Conference on Ocean, Offshore and Arctic Engineering*, Vol. 7: Ocean Space Utilization; Ocean Renewable Energy, Rio de Janeiro, 1-6 July 2012, 535-545. http://dx.doi.org/10.1115/omae2012-83807

[10] Pizer, D.J., Retzler, C.H. and Yemm, R.W. (2000) The OPD Pelamis: Experimental and Numerical Results from the Hydro-Dynamic Work Program. 4*th European Wave Energy Conference*, Aalborg, 4-6 December, 2000, 227-234.

[11] Retzler, C.H. and Pizer, D.J. (2001) The Hydrodynamics of the Pelamis Wave Energy Device: Experimental and Numerical Results. *Proceedings of OMAE*'01, 20*th International Conference on Offshore Mechanics and Arctic Engineering*, Rio de Janeiro, 3-8 June 2001, 31-35.

[12] Carcas, M.C. (2003) The OPD Pelamis WEC: Current Status and Onward Programme. *International Journal of Ambient Energy*, **24**, 21-28. http://dx.doi.org/10.1080/01430750.2003.9674899

[13] Henderson, R. (2008) Case Study: Pelamis. In: Cruz, J., Ed., *Ocean Wave Energy*, Springer, Berlin, 169-188.

[14] Yemm, R.W., Pizer, D.J., Retzler, C.H. and Henderson, R. (2012) Pelamis: Experience from Concept to Connection. *Philosophical Transactions of the Royal Society A*, **370**, 365-380. http://dx.doi.org/10.1098/rsta.2011.0312

[15] O'Connor, M., Lewis, T. and Dalton, G. (2013) Techno-Economic Performance of the Pelamis P1 and Wavestar at Different Ratings and Various Locations in Europe. *Renewable Energy*, **50**, 889-900. http://dx.doi.org/10.1016/j.renene.2012.08.009

[16] Dunnett, D. and Wallace, J. (2009) Electricity Generation from Wave Power in Canada. *Renewable Energy*, **34**, 179-195. http://dx.doi.org/10.1016/j.renene.2008.04.034

[17] Palha, A., Mendes, L., Fortes, C., Brito-Melo, A. and Sarmento, A. (2010) Modelling the Economic Impacts of 500 MW of Wave Power in Ireland. *Renewable Energy*, **25**, 62-77. http://dx.doi.org/10.1016/j.renene.2009.05.025

[18] Henderson, R. (2006) Design, Simulation, and Testing of a Novel Hydraulic Power Take-Off System for the Pelamis Wave Energy Converter. *Renewable Energy*, **31**, 271-283. http://dx.doi.org/10.1016/j.renene.2005.08.021

[19] ANSYS (2013) ANSYS AQWA. https://support.ansys.com/portal/site/AnsysCustomerPortal/template.fss?file=/prod_docu/15.0/Aqwa%20Users%20Manual.pdf

[20] EMEC10, Wave Data. http://www.emec.org.uk/facilities/live-data/wave-data/

[21] The Scottish Government (2010) 3.6 West of Shetland. http://www.scotland.gov.uk/Publications/2010/09/17095123/14

[22] Flanes, J. (2002) Ocean Waves and Oscillating Systems: Linear Interactions Including Wave-Energy Extraction. Cambridge University Press, Cambridge. http://dx.doi.org/10.1017/CBO9780511754630

[23] Mei, C., Stiassnie, M. and Yue, D. (2005) Theory and Application of Ocean Surface Waves, Part 1: Linear Aspects. World Scientific Publishing, Singapore.

[24] Pelamis Wave Power, Scottish Power Renewables at EMEC. http://www.pelamiswave.com/our-projects/project/2/ScottishPower-Renewables-at-EMEC

[25] Kim, S., Novak, D., Weems, K. and Chen, H. (2008) Slamming Impact Design Loads on Large High Speed Naval Craft. ABS Technical Papers.

[26] The Scottish Government (2008) Chapter 2 Physical Characteristics and Modeling of the Marine Environment. http://www.scotland.gov.uk/Publications/2008/04/03093608/13

4

Evaluation of the Power Generation Capacity of Hydrokinetic Generator Device Using Computational Analysis and Hydrodynamic Similitude

Oladapo S. Akinyemi[1], Terrence L. Chambers[1], Yucheng Liu[2]

[1]Department of Mechanical Engineering, University of Louisiana at Lafayette, Lafayette, USA
[2]Department of Mechanical Engineering, Mississippi State University, Starkville, USA
Email: osa4975@louisiana.edu

Abstract

This paper presents a similitude and computational analysis of the performance of a scaled-down model of a paddle wheel style hydrokinetic generator device used for generating power from the flow of a river. The paddle wheel dimensions used in this work are one-thirtieth scale of the full-size paddle wheel. The reason for simulating the scaled-down model was to prepare for the testing of a scaled-down physical prototype. Computational Fluid Dynamics using ANSYS Fluent 14.0 software was used for the computational analysis. The scaled-down dimensions were used in the simulations to predict the power that can be generated from the scaled size model of the paddle wheel, having carried out similitude analysis between the scaled down size and its full-size. The dimensionless parameters employed in achieving similitude are the Strouhal number, power coefficient, and pressure coefficient. The power estimation of the full-size was predicted from the scaled size of the paddle wheel based on the similitude analysis.

Keywords

Similitude, Paddle Wheel, CFD, Power Generation, Hydrokinetics

1. Introduction

The demand for energy coupled with maintaining its affordability and decreasing its pollution makes it imperative to look into ways in which affordable, clean and readily available energy sources can be harnessed.

Hydropower is an energy form obtainable from water and it is a clean form of energy source which is also renewable. Different forms of hydroelectric power generation have previously been developed, the most important forms being (a) Conventional hydroelectric dam, (b) Tidal power and (c) Ocean wave energy [1].

The above mentioned forms of hydroelectric power generation use either the pressure-head of water (dams) or oscillatory motion of the water (ocean waves and tides) for electricity generation, though some [2] [3] have actually looked into the run-off of rivers as a power generation means.

The paddle wheel is a hydrokinetic power generating device that uses the flow of a river (low-head) for its energy. The paddle wheel in this paper is designed such that the rotational effect created on it by the water current is transferred from the paddles onto a shaft in its midsection, and this shaft drives a generator in order to produce electric power.

Previous research work on the generation of electricity from a paddle-wheel was carried out using analytical and computational analysis on a full-size (pilot) paddle wheel [4]. Computational Fluid Dynamics (CFD) using ANSYS FLUENT13.0 version software was employed for the computational analysis in the previous work.

The aim of this paper is to generate a computational analysis result for a scaled down version of the paddle wheel, in order to have a comparison to results that will be generated in a laboratory test of the scaled size model, ensure similitude between the full-size and the scaled-down model, and also develop a mathematical relationship that can predict the power generation capacity of a full scale-sized paddle wheel from a scaled-down size of the paddle wheel.

2. Problem Statement

Over the last 22 years, global electricity production has doubled and electricity demand is rising rapidly around the world as economic development spreads to emerging economies [5]. Not only has electricity demand increased significantly, it is the fastest growing end-use of energy [6]. According to the International Energy Agency, the world needs to invest $48 trillion between now and 2035 to meet global energy demands [7]. The bulk of this will be used in offsetting the declining base oil and gas production, replace aging power plants and fund projects to meet the growing needs of emerging economies.

Energy Overview

There are two main sources from which energy is generated. These are renewable and non-renewable energy sources.

Renewable energy: It is generally defined as the energy that comes from resources which are replenished over time. Renewable energy sources are much sought after due to their low carbon dioxide emission [8], thus providing eco-efficient solution for developed and developing countries. The European Union is said to be generating 71% of its electricity from renewable energy sources [9].

The different forms of renewable energy sources include; solar energy, wind power, hydropower, biomass, and geothermal energy. This study looks into a means of generating energy from hydropower.

3. Hydropower

Hydropower can simply be referred to as the power captured from water as it flows. This movement could be from potential to kinetic energy, the process of which is achieved by gravity [10]. Hydropower is at present; the primary large-scale renewable alternative to fossil fuel and nuclear in the generation of electricity. It provides approximately 19% of the electricity produced worldwide [11]. Studies show that a total of 66 countries obtain at least half of their electricity from hydropower, with large economies like Brazil, Canada, and Norway having 97%, 62%, and 99% of national electricity production, respectively [12]. Worldwide, only about a third of the economically feasible potential of hydropower has been developed.

Unlike other renewable energy technologies, the multiplicity of wave, tidal, and hydrokinetic power devices show that the power generation potential from world's ocean and rivers is considerable [13]. Many devices are currently being explored and tested in order to harness the kinetic energy in the ocean and rivers. Examples of some hydrokinetic energy devices are point absorbers, hydrokinetic turbines, and paddle wheels.

3.1. Point Absorber

One form of Wave Energy Converter (WEC), developed in 2011, is used when the horizontal size of the device is much smaller than the typical wavelength [14]. The point absorber, either floating or submerged, transforms the vertical motion of the ocean waves into rotational or linear motion, which is then used to drive an electrical generator by means of a power take off (PTO) system.

Initially, high speed rotary machines, such as hydraulic turbines, were being used by point absorbers to extract energy from ocean waves, but in recent years, linear generators have been proposed in several marine applications as a well-suited technology [15] [16].

Disadvantages of the point absorber approach include: the complexity of the drive mechanism, difficulty of mooring in an offshore environment, and its low efficiency when not in resonance with the incident wave regime.

3.2. Hydrokinetic Turbines

These types of turbines are fully submerged in water and make use of the kinetic flow of the water in driving the turbine blades and hence generate electricity. The power generation capacity of hydrokinetic turbines is in the range of 30 - 50 kW for a water speed of 3 - 4 m/s [17]. A major disadvantage of these turbines is the damage to the blades of the turbines as a result of debris that may be in the water flow path which consequently stops the power production of the turbine [18]. Typical examples of the hydrokinetic turbine are the submerged windmill turbine [19] and transverse horizontal axis turbine [17] [20].

3.3. Paddle Wheel

The paddle wheel has been used for many years on ships or vessels to produce thrust, when powered by an engine. However, the paddle wheel took a cue from the water wheel in the generation of power. The water wheel, initially used to lift water and irrigate fields and later a means to generate mechanical power for milling, attracted scientific interest as an important energy generating device. The water wheel was utilized well into the last century where about 7554 wheels were in operation 1927 in Germany [21], and in Switzerland about 7000 small scale hydropower stations were in operation till 1924 [22]. Despite the variety of designs of paddle wheels proposed by Watkins [23], Rogers [24], and Winus [25], very few modern researchers have looked into paddle wheel as a power generation device.

Recent development of hydrokinetic devices have created renewed interest in the use of these devices for power generation and paddle wheels are more advantageous in that they may be more resistive to debris that may be in the water flow [18].

This study is a furtherance of a research on paddle wheel as a device for power generation. Liu and Penyanmi [4] carried out empirical and computational analysis on a paddle wheel making use of the full-size design in the analysis. This device is required to be tested in a laboratory in order to validate the empirical and computational results obtained. In order to achieve this feat, a scaled-down size of the paddle wheel is required, since it would be expensive, time consuming, and unwise to carry out experiment on a full-size prototype.

The aim of this study is to carry out CFD analysis on the scaled-down model. The behavior of the full-size prototype must be similar to that of the scaled-down model; hence similitude must be achieved between these two sizes, and the power generation capacity of the full-size should be approximately predicted from the performance of the scaled-down model.

4. Similitude

Similitude is used to describe model tests and is used to transfer model test results to the real application. It makes use of the laws of similarity. The most important thing in scaling is to achieve similarity between the scaled model and its test conditions to that of the real application and its test conditions.

There are three basic types of similitude in hydraulic problems namely: geometric similarity (similarity of shape), kinematic similarity (similarity of motion), and dynamic similarity (similarity of forces). These similarities should be fulfilled to ensure hydrodynamic similarity between a scaled model and its full-size.

Dynamic similarity is often used as a catch-all, because it implies that geometric and kinematic similitudes have already been met. Dimensionless numbers play an important role in establishing dynamic similarity, and these numbers are obtained from ratios of relevant forces that act in fluid dynamics. The dimensionless numbers often come across in fluid dynamics are [26]: Reynolds number, Froude number, Euler number, Cauchy number, Weber number, and Strouhal number.

4.1. Similitude for Turbomachinery

Dimensionless analysis plays an important role in the development and utilization of turbomachinery, as it has enabled the test of relatively small turbomachines. Dimensionless coefficients can be obtained through various parameters involved in the application of turbomachinery, and the coefficients form the basis of similitude in turbomachinery.

Parameters of significant consideration in turbomachines include: Power, rotational speed, flow rate, pressure change across the blade, density of fluid, viscosity of fluid, and the outer diameter of the machine [26]. Using these parameters, various dimensionless groupings can be obtained.

4.2. Scale Effects

Scale effects result from distortions introduced by a non-dominant force in a hydrodynamic analysis [27]. They occur when one or more dimensionless parameters differ between a model and its real world prototype. Scale effects are often small, but not always negligible altogether. Some forces become more dominant in the model than in the full size prototype, and this distorts the results [28].

5. Analysis of the Scaled Model of Paddle Wheel

For the paddle wheel in a river scenario, the parameters of significant consideration are: Power ($\dot{W}$), velocity of fluid (V), rotational speed of paddles (ω), pressure change across paddle blades (ΔP), gravitational acceleration (g), density of fluid (ρ), viscosity of fluid (μ), and characteristic length of paddle wheel (wetted depth) (l).

$$\dot{W} = f\left(V, \omega, \Delta P, g, \rho, \mu, l\right)$$

Using ρ, v and l as the repeating variables, a set of dimensionless groupings obtained are:

$$C_{\dot{W}} = \frac{\dot{W}}{\rho V^3 l^2}$$

$$\mathrm{Fr} = \frac{V^2}{gl}$$

$$\mathrm{Re} = \frac{\rho V l}{\mu}$$

$$\mathrm{Eu} = \frac{\Delta P}{\rho V^2}$$

$$\mathrm{St} = \frac{l\omega}{V}$$

The Euler number is shown to be a function of the Froude and Reynolds number in most practical hydraulic model tests [27], thus satisfying the Euler number approximately satisfies the Froude and Reynolds numbers. Hence the dimensionless grouping for the paddle wheel model reduces to:

$$f\left(C_{\dot{W}}, \mathrm{Eu}, \mathrm{St}\right) = 0$$

Making these numbers constant for both the model and its real application ensures similitude of the paddle wheel.

The Euler number $\frac{\Delta P}{\rho V^2}$ can be written in force form as $\frac{F}{\rho V^2 A}$ and this has to be constant for both full scale prototype (shown with the subscript f) and model (shown with the subscript m),

$$\frac{F_f}{\rho A_f V_f^2} = \frac{F_m}{\rho A_m V_m^2} \tag{1}$$

And if the scaling factor, λ, is defined as the characteristic length of the full-size prototype (l_f) over the characteristic length of the model (l_m), we get the equations shown in (2) below.

$$F_m = \frac{F_f}{\lambda},\ \frac{A_f}{A_m} = \lambda^2,\ \frac{l_f}{l_m} = \lambda \tag{2}$$

If ρ is constant,

$$V_m = V_f\sqrt{\lambda} \tag{3}$$

If ρ is not constant,

$$V_m = V_f\sqrt{\lambda}\sqrt{\frac{\rho_f}{\rho_m}} \tag{4}$$

From the Strouhal number, St:

$$\frac{l_f\omega_f}{V_f} = \frac{l_m\omega_m}{V_m}$$
$$\omega_m = \sqrt{\lambda^3}\,\omega_f \tag{5}$$

From the power coefficient, C_W:

$$\frac{P_f}{\rho_f V_f^3 l_f^2} = \frac{P_m}{\rho_m V_m^3 l_m^2}$$

If ρ is constant,

$$\frac{P_f}{V_f^3} = \frac{P_m l_f^2}{V_m^3 l_m^2}$$
$$\frac{P_f}{V_f^3} = \frac{\lambda^2 P_m}{V_m^3} \tag{6}$$

The Euler number is also a form of the pressure coefficient, given as:

$$\frac{\Delta P}{\rho V^2} = \text{Constant}$$

$$\frac{\Delta P_f}{\rho_f V_f^2} = \frac{\Delta P_m}{\rho_m V_m^2}$$

If ρ is constant,

$$\frac{\Delta P_f}{V_f^2} = \frac{\Delta P_m}{V_m^2} \tag{7}$$

If the paddle wheel parameters satisfy these coefficients, the model and full scale of the paddle wheel are in mechanical similitude.

Based on the above analysis of the model of the paddle wheel's water velocity and angular velocity, and assuming that density is a constant; the empirical analysis of the model of the paddle wheel was evaluated using the equations from the analysis of the full scale size [4], and the result is as shown in **Figure 1**.

5.1. Computational Model

Computational analysis was used to estimate the power generation capacity of the paddle wheel, and the soft ware package ANSYS FLUENT [29] was used in modeling, analyzing, and simulating the paddle wheel in water scenario. The input variables to the computational analysis were based on the results obtained from the similitude analysis.

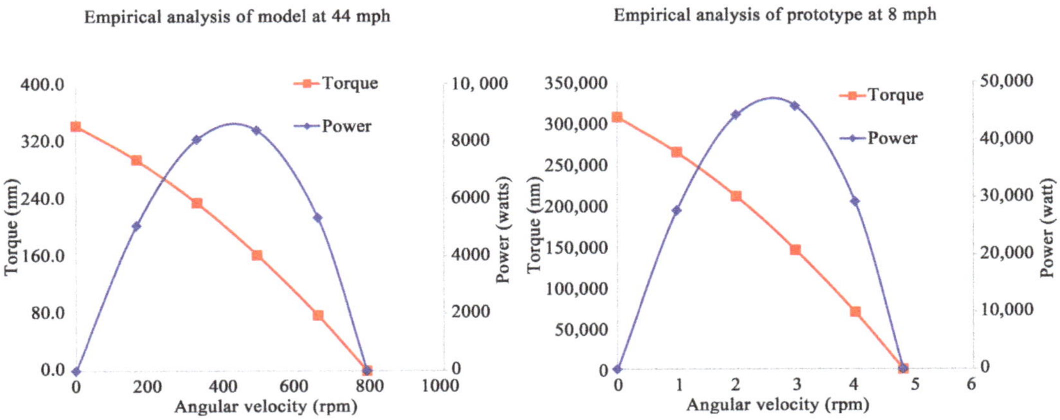

Figure 1. Empirical analysis of model and prototype.

5.1.1. Geometry

The geometry of the paddle wheel and the associated moving water was designed in the Design-Modeler, 3D Computer Aided Design (CAD) software available in the ANSYS workbench. In the computational environment, the physical domain is considered to be only fluid as there is no heat or mass transfer between the paddle wheel and water [4]. The air above the water line was not considered in the simulation as this area was taken as a pressure outlet. The geometry of the paddle wheel under water is shown in **Figure 2(a)**.

5.1.2. Meshing

Mesh generation is one of the most important steps during the pre-processing stage after the definition of the domain geometry. CFD requires the subdivision of the domain into smaller, non-overlapping subdomains in order to solve the flow physics within the domain geometry that has been created. This results in the generation of a grid of cells (elements or control volumes) overlaying the whole domain geometry [30]. The meshing for the entire body of the model was carried out using automatic (quadrilateral dominant) mesh method. The mesh of the paddle wheel is shown in **Figure 2(b)** having a number of cells of 4668.

5.1.3. CFD Analysis

Water velocities for the scaled-down model of 22, 27, 33, 44, and 55 mph were used. These velocities were obtained from the similitude analysis based on 4, 5, 6, 8, and 10 mph applied to the full-size model. Thus the scaled-up velocities were used for the CFD analysis to find the power generation capacity of the scaled-down model. At each velocity, six angular velocities (resulting from the Strouhal number analysis) from 0 to a maximum angular velocity (where the maximum angular velocity was defined as being when the net generated torque on the wheel becomes zero), were applied to the paddle wheel for the simulation. In carrying out the analysis, some important algorithms, theoretical models, and other settings were used as shown in **Table 1**.

5.2. Power Prediction from Scaled Model

The essence of building a model is to be able to save cost in experimenting, and also predict the performance of the full-scale device from the results obtained from the model, by scaling up these results. Hence, the power and torque obtained from the simulation of the model of the paddle wheel was scaled up using the power coefficient as given in Equation (6).

$$\frac{P_f}{V_f^3} = \frac{\lambda^2 P_m}{V_m^3},$$

From (3) $V_m = V_f\sqrt{\lambda}$

$$P_f = \sqrt{\lambda} P_m \quad (8)$$

Similarly $P = T\omega$

From (7), $T_f \omega_f = \sqrt{\lambda} T_m \omega_m$

But from (4) $\omega_m = \sqrt{\lambda^3}\,\omega_p$

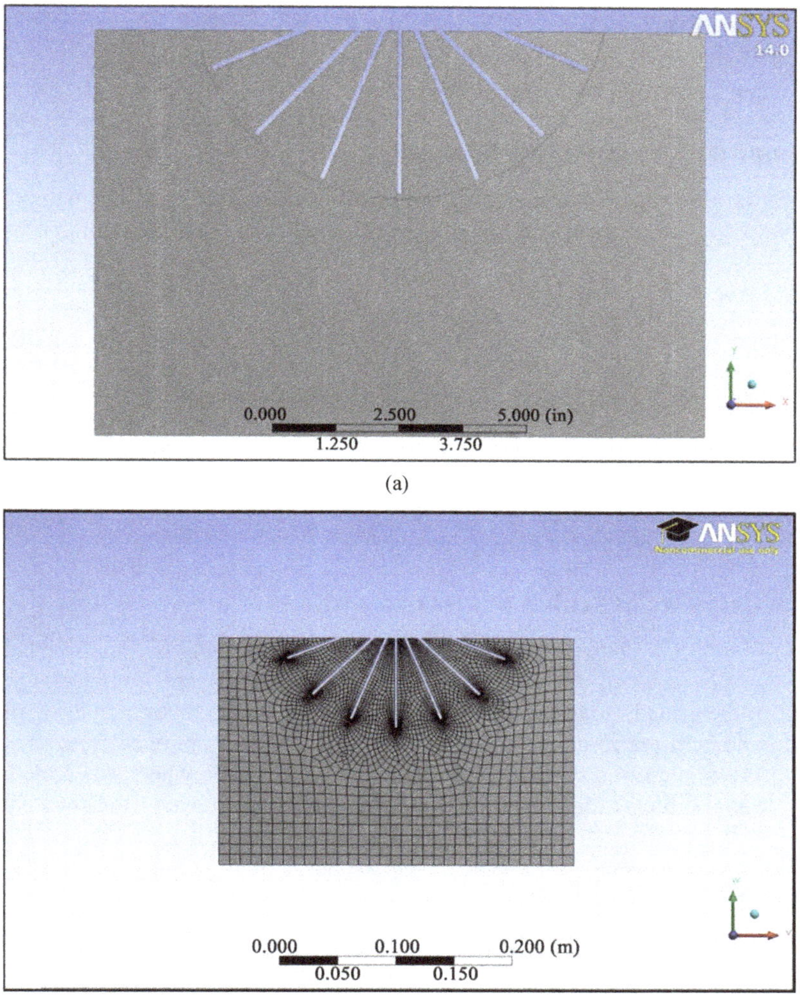

Figure 2. (a) Geometry and (b) meshing of paddle wheel under water.

Table 1. CFD analysis settings [4].

Solver Type	Pressure based, double precision, steady state, 2D
Viscous model	k-ε realizable with standard wall function
Fluid	Water with density of 998 kg/m^3
Reference frame	Rotational for paddle wheel zone
Pressure-velocity coupling	Coupled
Gradient Discretization	Least square cell based
Pressure Discretization	Standard
Momentum Discretization	Second-order upwind
Turbulent kinetic energy Discretization	Second-order upwind
Turbulent dissipation energy Discretization	Second-order upwind
Convergence criteria	1×10^{-4}
Solution initialization	Standard initialization with absolute relativity
Boundary condition	Velocity inlet (22, 27, 33, 44, 55 mph) Pressure outlet 0 l b/ft^2

Therefore

$$T_f = \lambda^2 T_m \tag{9}$$

6. Results

From the analysis of similitude in (3), the velocities to be applied on the scaled down size of the paddle wheel for laboratory test and simulation are given in **Table 2**.

6.1. Power Generated from Paddle Wheel

The power generation capacity of the scaled model of the paddle wheel was estimated from simulations at the evaluated water velocities and corresponding angular velocities, and the results obtained are shown in **Figure 3**.

6.2. Similarity Law

To ensure similitude between the full-scale size and the scaled-model of the paddle wheel, dimensionless groupings power coefficient, $C_{\dot{W}} = \frac{\dot{W}}{\rho V^3 l^2}$ and pressure-coefficient, $\frac{\Delta P}{\rho V^2}$ = constant should be the same for both scale. This was evaluated, using the results shown above, and assuming the same fluid was used in the full-size analysis and the scaled model analysis *i.e.* density is constant, and is shown in **Table 3**. The results in **Table 3** show that the Power Coefficient and the Pressure Coefficient are very similar, leading to the conclusion that similitude was achieved to a large extent during the analysis.

6.3. Prediction Capacity of Scaled Model of Paddle Wheel

Another essence of similitude, other than dimensionless groupings being constant for both full-sizes and scaled models, is to predict the result of the variable of interest for the full size from scaling up the results obtained from the analysis of the scaled model. Equation 8 was used for the prediction analysis of the power estimate of the full-size prototype, and the result obtained was compared to that obtained from the analysis carried out by Liu and Penyanmi [4] on the full-size model. The full-size of the paddle wheel was scaled down to a one-thirtieth of its dimension to make the scaling factor λ = 30. The predicted power from the scaled model is shown in **Table 4**.

Table 2. Velocities of scaled model of paddle wheel.

Velocity on full-size (mph)	Velocity on scaled size (mph)
4	22
5	27
6	33
8	44
10	55

Table 3. Justification of similitude.

	Full-size Prototype	Scale model
Power coefficient	0.91	0.95
Pressure coefficient	1003	1026

Table 4. Power prediction from model.

Water Velocity	Model Power	Predicted Power	Full-size Power [21]
22	1037.4	5682.1	5478.0
27	2091.9	11,457.8	11,186.1
33	3580.0	19,608.5	19,368.7
44	8024.1	43,949.8	44,024.7
55	16,227.9	88,883.9	86,116.3

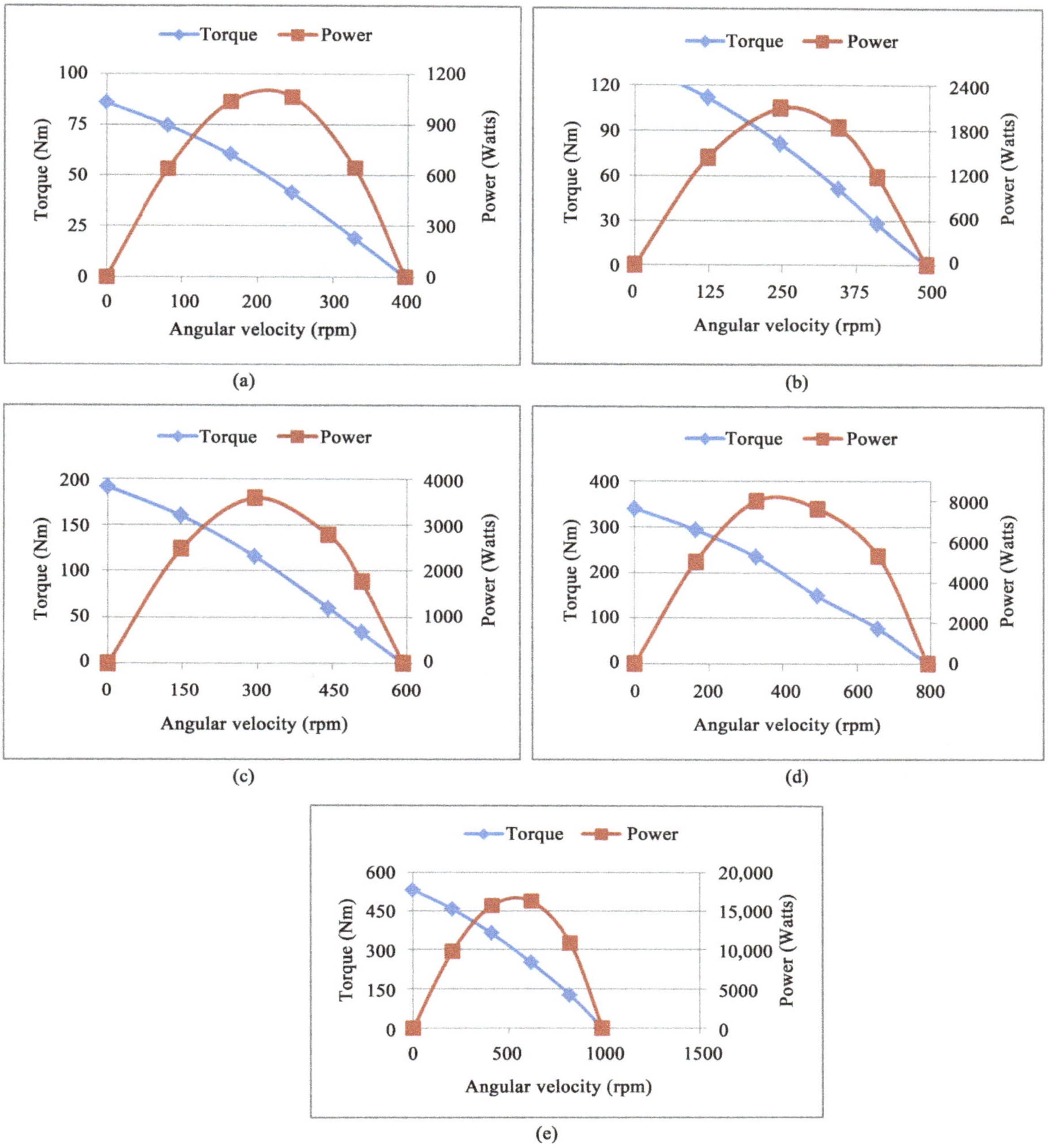

Figure 3. CFD result of model at water velocities (a) 22 mph (b) 27 mph (c) 33 mph (d) 44 mph (e) 55 mph.

7. Discussion

Similitude analysis was used to obtain the velocities that would be required to be applied on the scaled-down size of the paddle wheel as seen in **Table 2**, from which the corresponding angular velocities used for the simulation were obtained. The velocities obtained were based on the premise that the working fluid for both the full-size and scaled down size of the paddle wheel is water, hence the density is constant.

Figure 4 shows the pressure and velocity distributions of water around the paddle wheel and it was observed that the four blades to the left of the wheel contribute significantly to the power generation, while the remaining three blades merely assist in the continuous rotation of the wheel. Further work could be done to optimize the number of blades used.

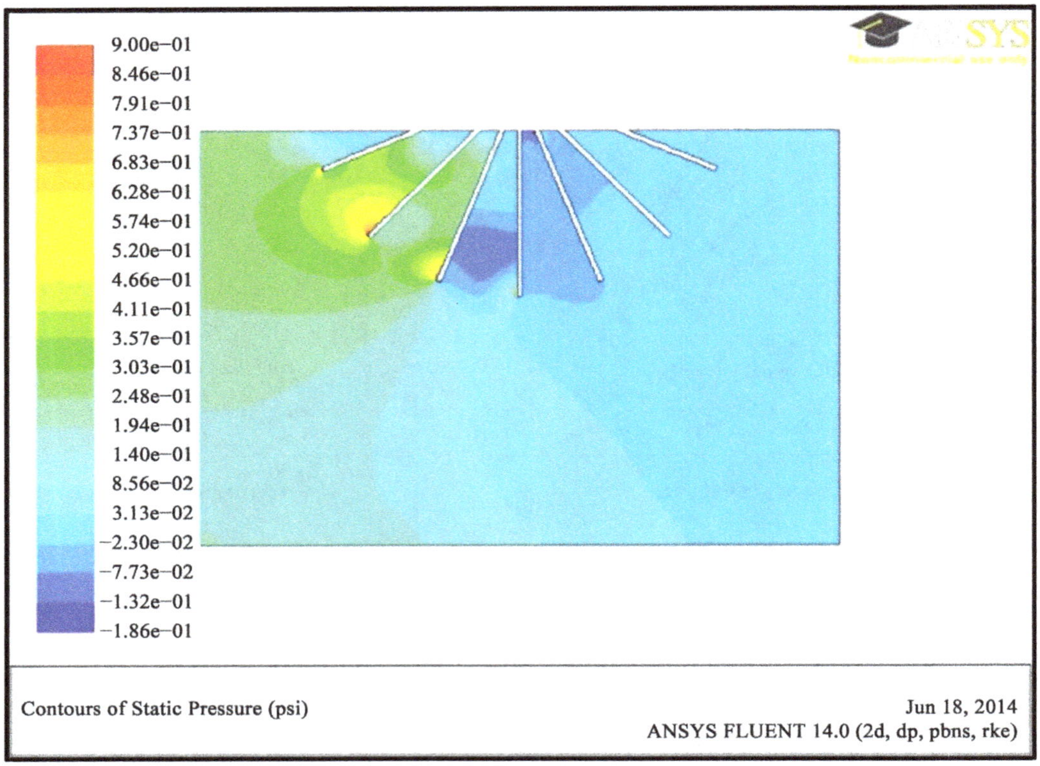

(a)

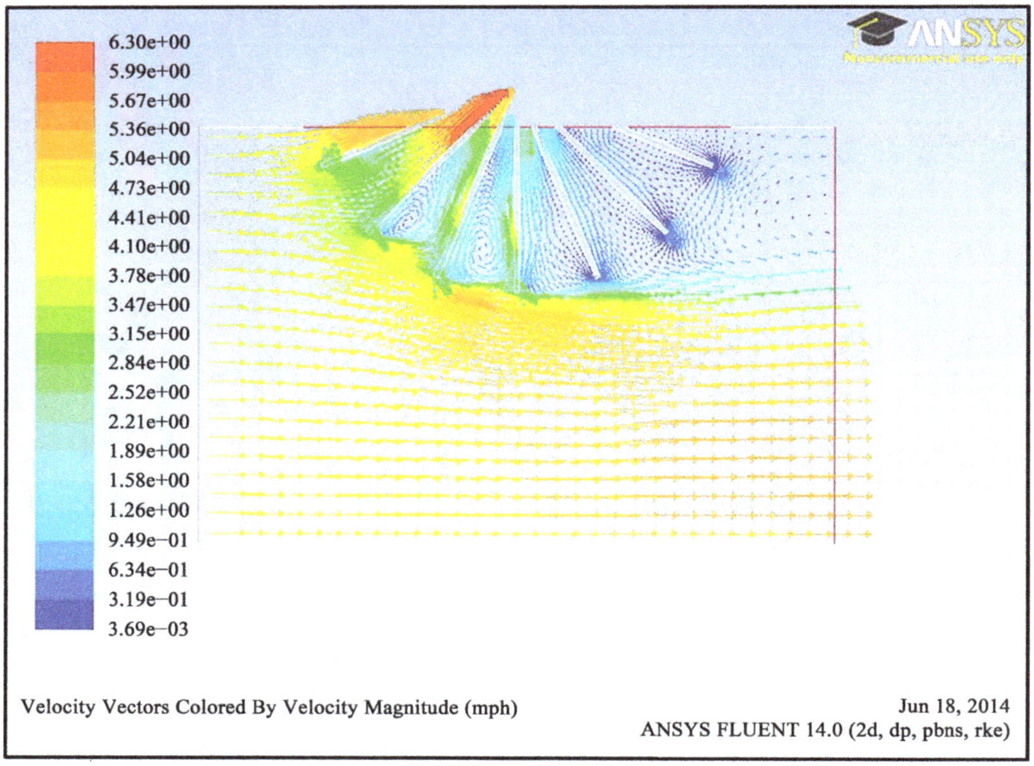

(b)

Figure 4. (a) Static pressure (b) Velocity vector distribution of water around paddle wheel.

For similitude to be justified, an evaluation of the power-coefficient and the pressure-coefficient was carried out on the full-size and the scaled model of the paddle wheel and the result as seen in **Table 3** shows that these coefficients are close. The resulting disparities are as a result of scaled effects which occur when non-dominant forces on the full-size become dominant in the scaled model. In this instance, the gravitational force on the model is reduced which was dominant in the full size while turbulence is increased which was less dominant, and thus brings about an increase in the power coefficient in the model compared to that in the full size.

To further justify similitude, the result of power estimated from the simulation of the scaled model of the paddle wheel was scaled up based on the equation relating the model to the full-size for power prediction, it can be observed that the model sufficiently predicts the power that can be obtained from the full-size prototype, as **Table 4** clearly depicts.

In reality, the velocities and angular velocities (up to 55 mph and 988 rpm) used in the simulation of the model of the paddle wheel will be highly difficult to achieve in the laboratory test of the wheel. It is recommended that a different, higher density, fluid be used for the laboratory test of the model. Based on the similitude analysis, the density of the fluid to be used for the model needs to be denser than water, as this will help reduce the operating velocity and make the laboratory test practicable.

8. Conclusion

The present work carried out CFD analysis on the scaled-down model of the paddle wheel. It demonstrated how similitude can be applied to a paddle wheel for the relevance of laboratory test or experimentation. It shows the relevant dimensionless groupings to the paddle wheel application and gives equations for predicting power and torque for hydrodynamic problems related to the paddle wheel. The velocity of water applied to the scaled model of the paddle wheel is very high, and this velocity is highly impractical to be achieved for the laboratory test of an actual scaled down prototype of the paddle wheel. The results obtained from this analysis will be compared to the laboratory test that will be carried out on the scaled down size of the paddle wheel at a future date.

9. Future Work

It is proposed that a liquid of higher density than water be used for the laboratory test in order to reduce the velocity of liquid that will be applied on the scaled model of paddle wheel for the laboratory test. This velocity can be obtained by the application of (4).

The paddle wheel in this study can be redesigned such that the blades do not run solidly from the axle edge, as this causes a wake zone behind each blade hence increasing the drag force that limits the rotation of the wheel and hence the power production capacity. The proposed design will have a section of the blade near the tip of the wheel and an open space between the axle and the blade. This will help eliminate wake zones behind the blades and improve power generation.

The power generation capacity can also be improved by adding a plate (bottom-fin) to the design of the wheel. This plate is placed beneath the paddle wheel at an inclined angle, and it directs water towards blades that the flow may not come in contact with. This helps to increase the number of blades that contribute to power generation.

References

[1] Carrasco, F. (2011) Introduction to Hydropower. World Technologies Edition, New Delhi.

[2] Sharma, H. and Singh, J. (2013) Run Off River Plant: Status and Prospects. *International Journal of Innovative Technology and Exploring Engineering* (*IJITEE*), **3**, 210-213.

[3] Parish, O. (2002) Small Hydro Power: Technology and Current Status. *Renewable and Sustainable Energy Reviews*, **6**, 537-556. http://dx.doi.org/10.1016/S1364-0321(02)00006-0

[4] Liu, Y. and Penyami, Y.F. (2012) Evaluation of Paddle Wheels in Generating Hydroelectric Power. *ASME International Mechanical Engineering Congress and Exposition*, **6**, 675-684. http://dx.doi.org/10.1115/imece2012-85121

[5] Yüksel, I. (2007) Hydropower in Turkey for a Clean and Sustainable Energy Future. *Renewable and Sustainable Energy Reviews*, **12**, 1622-1640.

[6] Yüksel, I. (2009) Dams and Hydropower for Sustainable Development. *Energy Sources*, *Part B*: *Economics*, *Planning*, *and Policy*, **4**, 100-110.

[7] International Energy Agency (IEA) (2014) World Energy Investment Outlook, World Needs $48 Trillion in Investment to Meet Its Energy Need to 2035. Press Release London.

[8] Ahmed, S., MdIslam, T., Karim, A. and Karim, N.M. (2014) Exploitation of Renewable Energy for Sustainable Development and Overcoming Power Crisis in Bangladesh. *Renewable Energy*, **72**, 223-235. http://dx.doi.org/10.1016/j.renene.2014.07.003

[9] "Global status report" (2012) REN21. Renewable Energy Policy Network for the 21st Century. http://www.ren21.net/wp-content/uploads/2015/06/2012KFen.pdf

[10] Ed Hiserodt (2007) The Other Renewables. *The New American*, **23**, 25-27.

[11] IEA (2010) Renewable Energy Essentials: Hydropower. http://www.iea.org/publications/freepublications/publication/hydropower_essentials.pdf

[12] Hydropower and the World's Energy Future. The Role of Hydropower in Bringing Clean, Renewable Energy to the World. (2000). www.ieahydro.org/reports/Hydrofut.pdf

[13] Musial, W. (2008) Status of Wave and Tidal Power Technologies for the United States. Technical Report NREL/TP-500-43240.

[14] Silva, B., Adria, M.M., Alessandro, A., Giuseppe, P. and Renata, A. (2013) Modeling of a Point Absorber for Energy Conversion in Italian Seas.

[15] Polinder, H., Gardner, F. and Vriesema, B. (2000) Linear PM Generator for the Wave Energy Conversion in the AWS. *Proceedings of the* 14*th International Conference on Electrical Machines* (*ICEM*), Espoo, 28-30 August 2000, 309-313.

[16] Mueller, M.A. (2002) Electrical Generators for Direct Drive Wave Energy Converters. *IEE Proceedings—Generation, Transmission and Distribution*, **149**, 446-456. http://dx.doi.org/10.1049/ip-gtd:20020394

[17] Nenana Hydrokinetic (RivGen) Power Systems. (2011). http://energy-alaska.wikidot.com/nenana-hydrokinetic-turbine

[18] Anyi, M. and Kirke, B. (2010) Evaluation of Small Axial Floe Hydrokinetic Turbines for Remote Communities. *Energy for Sustainable Development*, **14**, 110-116. http://dx.doi.org/10.1016/j.esd.2010.02.003

[19] Could Wind Farms of the Future Be Underwater? (2014). http://www.fut-science.com/wind-farms-future-underwater-company/

[20] Kumar, A. and Saini, R.P. (2014) Development of Hydrokinetic Power Generation System: A Review. *International Journal of Engineering Science and Advanced Technology*, **4**, 464-477.

[21] Muller, G., Kauppert, K. and Mach, R. (2002) Back to the Future. International Water Power and Dam Construction, 30-33.

[22] Klunne, W. (2003) Micro and Small Hydropower for Africa. ESI Africa Issue 4. http://renewables4africa.net/klunne/publications/hydropower_africa_esi.pdf

[23] Watkins, A.G. (1918) Water Power Plant with Vertical and Horizontal Surface Conveyor. US Patent No. 1280617.

[24] Rogers, E.R. (1981) Water Wheel Electric Generation Device. US Patent No. 4268757.

[25] Winius, H.C. (2011) Paddle Wheel Electricity Generator. US Patent No. 7969034.

[26] Heller, V. (2012) Model-Prototype Similarity. 4*th CoastLab Teaching School*, *Wave and Tidal Energy*, Porto, 17-20 January 2012.

[27] Chanson, H. (2004) The Hydraulics of Open Channel Flow: An Introduction. 2nd Edition, Butterworth-Heinemann, London, 258.

[28] Heller, V., Pfister, M. and Chanson, H. (2011) Scale Effects in Physical Hydraulic Engineering Models. *Journal of Hydraulic Research*, **49**, 293-306. http://dx.doi.org/10.1080/00221686.2011.578914

[29] Simulation Driven Product Development. ANSYS Fluent 14.0, ANSYS Inc. (2012).

[30] Tu, J., Yeoh, G.H. and Liu, C. (2008) Computational Fluid Dynamics: A Practical Approach. Butterworth-Heinemann, London, 35-37.

5

A Strategy for PMU Placement Considering the Resiliency of Measurement System

Jyoti Paudel, Xufeng Xu, Karthikeyan Balasubramaniam, Elham B. Makram

Electrical and Computer Engineering Department, Clemson University, Clemson, USA
Email: jpaudel@clemson.edu, xufengx@clemson.edu, bbalasu@clemson.edu, makram@clemson.edu

Abstract

This paper aims to find strategic locations for additional Phasor Measurement Units (PMUs) installation while considering resiliency of existing PMU measurement system. A virtual attack agent is modeled based on an optimization framework. The virtual attack agent targets to minimize observability of power system by coordinated attack on a subset of critical PMUs. A planner agent is then introduced which analyzes the attack pattern of virtual attack agent. The goal of the planner agent is to mitigate the vulnerability posed by the virtual attack agent by placing additional PMUs at strategic locations. The ensuing problem is formulated as an optimization problem. The proposed framework is applied on 14, 30, 57 and 118 bus test systems, including a large 2383 node western polish test system to demonstrate the feasibility of proposed approach for large systems.

Keywords

Observability, Optimization Model, PMU Placement, Resiliency, Scenario Technology, Virtual Attacker

1. Introduction

PMUs play a significant role in wide area monitoring and control. PMUs are capable of measuring node voltages and line currents as phasors. The measured quantities are time stamped based on global positioning satellite (GPS) signal. The time stamp allows analysis of measurement data that is geographically dispersed. Physical properties of power network enable computing the voltage and current phasors across the entire network by installing PMUs at only a subset of nodes.

The PMUs placement in strategic locations has been the vital research topic for PMU application and various methodologies have been introduced by power engineers all across the world [1]. The researchers have approached the PMUs placement problem using two methods: (i) Heuristic approach, (ii) Mathematical approach.

1) Heuristic approaches. They have been widely adopted in this area. Simulated annealing is used in [2] [3] to find the placement location based on desired depth of unobservability. Reference [4] solves the PMU placement problem using recursive Tabu search. Though the algorithm used for this approach gives satisfactory results for larger bus systems but no robust contingency is considered. Literature [5] addresses on N-1 PMU failure and solves the PMU placement problem using differential evolution. Immunity genetic algorithm is used in [6] to investigate the PMU placement. This approach is relatively time consuming and is not preferable for large bus systems. Binary Particle Swam Optimization (PSO) is another optimization approach that is enormously used in this field. In [7] a simple PMU placement has been implemented using BPSO but the algorithm does not consider details regarding PMU vulnerability. Since all the techniques are discussed in heuristic approach, being iterative in nature requires time for convergence and also the convergence fully depends on the initial guess.

2) Mathematical approaches. Mathematical approach has been gaining popularity from recent years. They are easy to apply in the situation where a definite solution is required. They are based on formulae derived from mathematical calculations. Integer linear programming is a common approach as presented in [8], in which a general formulation for PMU placement using conventional and without conventional measurement is taken into consideration. Contingency constrained optimal PMU placement using exhaustive search approach is proposed in [9]. This literature has taken several zero-injection buses in account for PMU placement considering single PMU loss and measurement channel limitation. Mixed Integer Linear Programming is used in [10] which considers zero injection and branch flow measurements in order to maximize the measurement redundancy and reduce the number of PMUs. However, the approach in [10] requires almost twice the amount of PMUs to obtain full system observability under contingency operation than at normal operating conditions.

Due to the critical nature of power systems, complete observability of all nodes at all times is required. However, the networked PMUs might be rendered out of service by natural disasters such as hurricanes or PMUs can be intentionally taken down by malicious attacks. Enough attention should be given to PMU vulnerability while placing PMUs in the system. The concept of economically deploying PMUs considering resiliency of existing system post attack is missing in the above literatures. Hence, this paper highlights a considerable interest in improving PMU redundancy at minimum cost. In order to ascertain a subset of nodes which are most likely to be attacked, a virtual attack agent is modeled. The aim of the virtual attack agent is to reduce system observability to a minimum while carrying out a coordinated attack on a subset of PMU installation nodes. This virtual attack is used by the operator agent to identify a set of critical nodes whose redundancy needs to be increased. The planner agent then finds strategic locations to place additional PMUs in order to increase redundancy of critical nodes while minimizing incurred cost.

This paper is organized as follows: Section 2 introduces two agents including the attacker and the planner to design a framework for classifying critical PMUs and planning scheme. Section 3 establishes a mathematical model to corporate on their objectives: the attacker aims to disable critical PMUs while the planner tries to design remedial measures. In Section 4, the model is applied to different standard test systems. Finally, Section 5 summaries the paper.

2. Agent Based PMU Placement Framework

An uncertainty constraint PMU placement problem can be expressed in three different agent based stages:

- *Attacker*: A virtual attack agent is introduced whose goal is to take down a set of installed PMUs to reduce system observability. Uncertain events like intentional attacks are an important aspect that needs to be considered while making PMU placement decision. Due to geographical span of interconnected power systems planning a coordinated attack on all of the installed PMUs is improbable. Hence, the virtual attack agent will carry out coordinated attacks on a subset of installed PMUs that are deemed critical. Here, the set of critical PMUs are the ones which when taken out of service minimizes system observability. Cardinality of the critical set is assumed to vary depending on the resources available to virtual attack agent.
- *Operator*: At this stage, the operator has to take corrective measures to mitigate the possible damage caused by the attacker. The operator agent identifies a set of critical nodes based on virtual attack agents attack plan. The operator agent then relays the corrective measure, which in this case is to increase the redundancy of critical nodes, to the planner agent.
- *Planner*: The task of planner is to deploy additional PMUs to increase redundancy of critical nodes at minimum cost.

Schematic representation of the three cyclic stages is shown in **Figure 1**. The schematic is cyclic in nature because of the nature of the problem, where the virtual attack agent comes up with strategies to minimize system observability given a set of PMU locations. The operator and planner agents then mitigate the effect of virtual attack agent by placing additional PMUs at strategic locations. The virtual attack agent then starts a new cycle with the new set of PMU installation locations.

Each undesired PMU outage caused by the virtual attack agent is an optimization scenario for the operator. These undesired outages can be single, double or multiple based on virtual attack agent's resources. Let P be the number of PMUs deployed into the system and Ψ be the scenario which corresponds to the number PMUs to be attacked by the attacker. The total scenario can be represented as combinatorial number ${}_P C_\Psi$ as:

$${}_P C_\psi = \frac{P!}{\psi!(P-\psi)!} \tag{1}$$

Since there are hundreds of thousands of possible attack scenarios, it is impossible to enumerate all scenarios for large systems due to computational burden. Instead, by adopting the approach in (2) a worst case scenario can be obtained.

$$\psi = \eta\% \times P \tag{2}$$

where $\eta \in [0, 100]$—representing the percentage of installed PMUs that are attacked. As a worst-case scenario, an assumption has been made that the attacker can attack up to 50% of the total deployed PMUs. Depending upon η value, a set of attacked PMUs $\Psi = \{\Psi_1, \Psi_2, \cdots, \Psi_z\}$ is obtained from the optimization problem and this set is named as critical PMUs. The programming framework for the agent based PMU placement is shown in **Figure 2**.

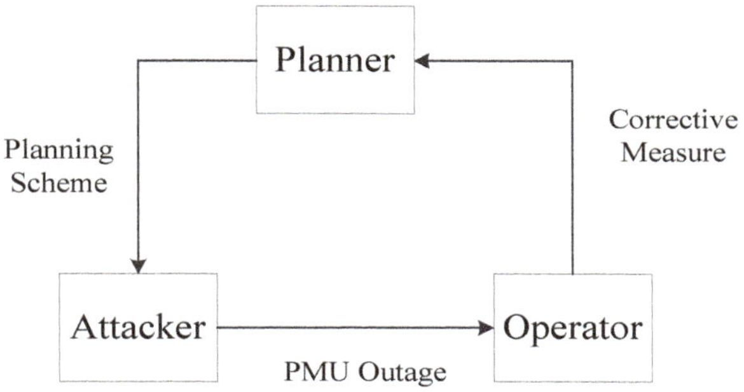

Figure 1. Relationship between three agents in PMU placement.

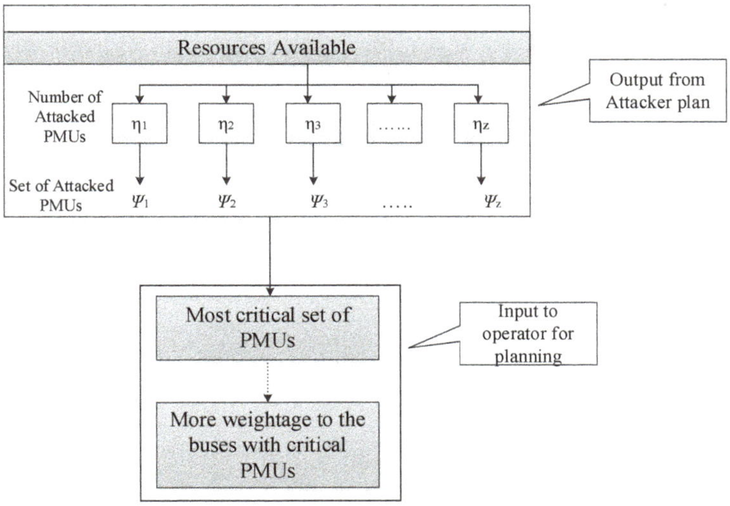

Figure 2. PMU placement framework.

3. Mathematical Formulation

Development of agent models as an optimization problem is discussed in this section. The initial deployment locations for PMUs, which act as the starting point for the proposed agent based framework are obtained using optimal PMU placement algorithm from [8].

3.1. Virtual Attack Agent

The objective of virtual attack agent is to attack a subset of installed PMUs in the system such that the system bus observability is minimized. The attack agent is modeled using binary integer programming.

The mathematical formulations for attacker's objective is as follows:

$$\min \sum_{k=l+1}^{m} \xi_k \tag{3}$$

S.t.

$$\left(\sum A_i\right)\xi_k \geq A_i x_i \tag{4}$$

$$\sum_{p=1}^{l} x(p) = \left(\sum x\right) - \psi_z \tag{5}$$

$$\xi_k \in \{0,1\} \quad \text{and} \quad x_p, x_i \in \{0,1\} \tag{6}$$

The objective function (3) ξ_k is the decision variable that tends to give the observability of each bus in terms of binary variable. If the bus is observable by PMUs remaining in the system after the coordinated attack by virtual attack agent then ξ_k will take the value of 1 and if the bus is not observable by any of the PMUs then ξ_k will take the value '0'. In general, observability of a bus can be 0 in which case the bus is not observable or observability can be a positive number which means the bus is observable.

$$\xi_i = \begin{cases} 1 & \text{if } A_i \cdot x_p > 0 \\ 0 & \text{otherwise} \end{cases} \tag{7}$$

Since the available PMUs were placed based on system network topology, it becomes necessary to define a network connectivity matrix A.

Elements in matrix A are defined as follows:

$$A_{ij} = \begin{cases} 1 & \text{if } i = j \text{ or } i \text{ and } j \text{ are adjacent} \\ 0 & \text{otherwise} \end{cases} \tag{8}$$

In constraint (4), x_i is an auxiliary binary variable of PMU placement. If the PMU is present at the ith bus then x_i is regarded as 1 otherwise 0. Before the attack, the observability of the ith bus denoted by left-hand side of (4) should be equal to the product of connectivity matrix of bus i and PMU placement variable x_i. Since the attacker already know the exact location of the PMUs, the attacker agent tries to enumerate all the possibilities to destroy or damage the PMU which are critical. This procedure is presented in (5). The word 'critical' defines those set of PMUs whose installation in the system increases the system observability. Post attack the variable x_i is zero for the disabled or attacked PMU. In this case, the constraint (4) will act as inequality constraint because the observability of the bus at right hand side will be greater than left hand side. The connectivity matrix is always fixed as long as all the transmission lines in the system are in service. The variable x_p is the PMU placement variable post attack. Depending upon the auxiliary variable x_p, the attacker performs all combinatorial number and checks the observability of each bus one by one. Those combination sets where the observability of bus shows the maximum number, the attacker tries to attack on those particular sets of PMUs. Constraint (4) helps the attacker to judge the most attractive set of PMUs to act on.

Computational complexity of this optimization model increases substantially when dealing with large number of system buses. From (4), the total number of inequality constraints is equal to the number of system buses N and the equality constraint (5) is split into two sections, one for the set of the buses where PMUs were installed and other for the set of buses where PMUs were not installed. Therefore the total number of constraints is $N + 1 + 1$. Similarly the total numbers of variables are twice the number of system buses M. This is because the first half $M/2$ denotes the auxiliary variable of PMU placement post attack and the other half $M/2$ denotes the bus observability.

3.2. Operator Agent

The responsibility of the operator is to identify vulnerable nodes based on the behavior of virtual attack agent. Vulnerable nodes in this context are a set of critical buses whose observability is compromised by the virtual attack agent. Critical buses are the buses include critical PMU installation buses and buses that are observable by critical PMUs.

The number of PMUs attacked by virtual attack agent is a percentage of the total number of installed PMUs. Since, larger systems have larger number of installed PMUs, the number of critical buses also tends to increase with system size. Since various sets of PMUs were obtained depending upon the availability of attacker's resources. Now, with the concern of PMU's and their installation cost, from those several sets of classified critical PMUs, the planner has to choose only the most repeated PMUs among all sets of critical PMUs. To obtain this, following formulation is used.

$$R = \psi_1 \cup \psi_2 \tag{9}$$

$$S = R \cap \psi_3 \cap \cdots \cap \psi_Z \tag{10}$$

where $S = \{s_1, s_2, \cdots, s_w\}$, denotes set of critical PMUs in (10).

The critical buses are those buses that are observable from the set of critical PMUs.

$$W_c^B = \theta\left(A_i \middle| s_w\right) \tag{11}$$

$$W_c^B = \left\{w_{c,1}^B, w_{c,2}^B, \cdots, w_{c,f}^B\right\} \tag{12}$$

where W_c^B is represented for set of critical buses obtained from each critical PMUs s_w and θ is the index of buses which are adjacent to critical PMU located buses.

3.3. Planner Agent

The objective of the planner agent is to install additional PMUs in strategic locations to mitigate the vulnerability posed by virtual attack agent. The optimal PMU placement considering the critical PMUs is as follows:

$$\min \sum_{i=1}^{N} c_i x_i' \tag{13}$$

Subject to

$$\left(A_i \middle| \overline{w_{c,f}^B}\right) x_i' \ge b_i \tag{14}$$

$$\left(A_i \middle| w_{c,f}^B\right) x_i' \ge b_i' \tag{15}$$

$$A_{eq} X' = \sum_{i=1}^{N} x_i \tag{16}$$

$$c_i = \begin{bmatrix} 1 & 1 & \cdots & 1 \end{bmatrix}_{1\times N} \tag{17}$$

$$x_i' \in \{0,1\},\ \forall i = \{1, 2, \cdots, N\} \tag{18}$$

The objective function (13) implies that minimum number of PMU is placed in the system and x_i' is the new decision variable for PMU placement for this particular model. It is defined same as x_i as described earlier in attacker's model. In this model, b_i is observability constraint for non critical buses and is considered equivalent to one. Whereas for critical buses, the observability constraint b_i' is considered as two. Therefore constraints (14) and (15) describes that each non-critical bus w_c^B and critical buses w_c^B must be observable by at least one PMU and two PMUs respectively. Equality constraint (16) represents that original PMUs has to be placed in the same location. Thus, under any uncertain events or attacks, all the buses are still observable and with higher redundancy with additional number of PMUs in the system. There are N number of variables and $2N$ number of constraints. In this proposed model, the restrictions on number of additional PMUs are not implemented. However, this optimization model has the potential to optimize the fixed amount of additional PMUs just by adding a new constraint such that the summation of decision variable is equal to a constant number.

4. Discussion and Result

The performance of proposed model is tested on 14, 30, 57 and 118 IEEE test bus systems including large power system 2383 bus Western Polish system [11] [12]. All the testified cases are implemented on 1.70 GHz processor with 6 GB of RAM using CPLEX12.6.2 Solver [13]. The optimization is executed in MATLAB environment.

4.1. Critical PMUs

The number of critical PMUs depends upon the size of the system and the system topology. The set of PMUs that poses a higher influence in increasing the system bus observability are shown in the **Table 1**. The critical PMUs are obtained based upon the resources available to the attacker. The percentage shown in the **Table 1** indicates that the attacker has ability to damage certain percentage of the total deployed PMUs in the system. For a small system like14 bus system, only 4 PMUs are needed in the system for full observability before attack. 10% of 4 PMUs being a negligible number, 20% and 50% of total placed PMUs is considered for execution. $N_{\min}$ is the number of attacked PMUs. Similarly, **Table 1** demonstrates all the critical PMUs for different IEEE systems.

To further analyze strictly critical PMUs, only one set of PMUs per system is evaluated. The PMUs that happens to be critical for more than twice among the differentiated level of resources availability are only considered as most critical PMUs. **Figure 3** shows all such single set of most critical PMUs for 14, 30, 57 and 118 IEEE bus systems only. The model was further tested for larger power systems like IEEE 300 and 2383 Western Polish system. For the larger system, the most critical PMU buses are shown in Table II. The critical PMUs for larger systems are selected based on 10% of total installed PMUs. Since the numbers of PMUs installed in IEEE 300 and 2383 Western Polish system outnumbered to smaller system, PMU installed buses are not shown in the described **Table 2**.

4.2. Planning Scheme for PMU Placement

The PMU placement planning scheme is presented in this section. The goal of the planning scheme is to place additional PMUs in order to mitigate the loss of observability in the event of an attack. From the previous section, the set of critical buses with respect to loss of observability was obtained. The planner agent uses this

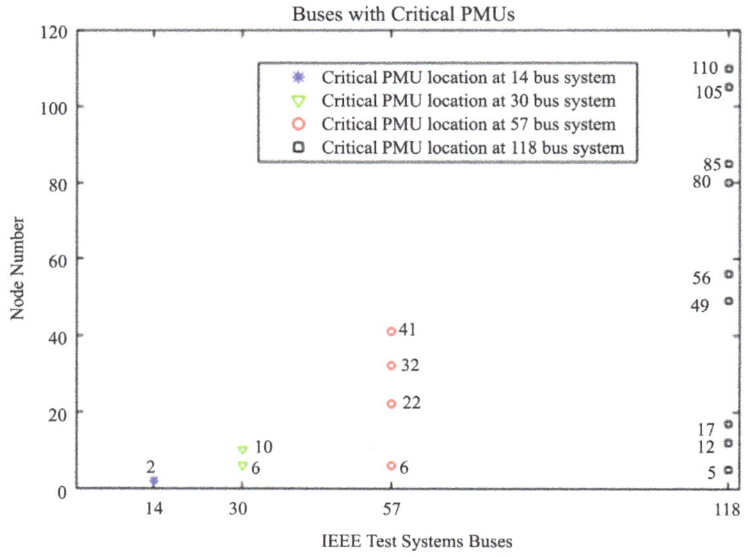

Figure 3. Critical PMUs in different test systems.

information to obtain PMU placement scheme for installing additional PMUs with least cost to increase redundancy of critical nodes. For the most critical buses as shown in **Table 3**, the measurement redundancy was set to 2 *i.e.* the most critical buses must be observable by at least 2 PMUs. The resultant optimal numbers of PMUs are shown in **Table 4**. For IEEE 14 bus system, two additional PMUs are required to increase redundancy of five critical buses. Similarly for IEEE 30 bus system four additional PMUs are required to increase redundancy of 12 critical buses. Since the original PMU deployment was shown in **Table 1**, **Table 4** shows PMU locations only for additional PMUs. Due to space limitation, location of additional PMUs for larger systems are not tabulated but are rather summarized as follows. For IEEE 300 bus system, 30 additional PMUs are required. While the 2383 bus polish system required 252 PMUs in addition to the originally placed 746 PMUs to obtain full bus system observability and increased redundancy at critical buses.

Table 1. Critical PMUs depending upon IEEE test systems.

IEEE System	PMU location	Resources available to the attacker									
		10%		20%		30%		40%		50%	
		N_{min}	Ψ_1	N_{min}	Ψ_2	N_{min}	Ψ_3	N_{min}	Ψ_4	N_{min}	Ψ_5
14	2, 7, 10, 13	_	_	1	2	_	_	_	_	2	2, 13
30	1, 2, 6, 10, 11, 12, 15, 19, 25, 29	1	10	2	6, 10	3	6, 10, 25	4	6, 10, 12, 15	5	6, 10, 12, 15, 19
57	2, 6, 12, 19, 22, 25, 27, 32, 36, 39, 41, 45, 46, 49, 51, 52, 55	2	6, 41	3	6, 32, 41	5	6, 22, 32, 41, 46	7	6, 12, 22, 32, 41, 49, 55	9	6, 12, 22, 32, 36, 39, 41, 49, 55
118	1, 5, 9, 12, 15, 17, 21, 25, 28, 34, 37, 40, 45, 49, 52, 56, 62, 64, 68, 70, 71, 76, 77, 80, 85, 87, 91, 94, 101, 105, 110, 114	3	56, 105, 110	6	49, 56, 80, 85, 105, 110	10	5, 12, 17, 49, 56, 80, 85, 105, 110	13	5, 12, 15, 17, 34, 37, 40, 49, 56, 80, 85, 105, 110	16	5, 12, 15, 17, 34, 37, 40, 45, 49, 56, 62, 80, 85, 94, 105, 110

Table 2. Critical buses with PMUs on Larger systems.

IEEE Test System	Total installed PMUs	Selected critical PMU buses
300 bus system	87	315 109 112 190 268 269 270 272
2383 polish	746	6 18 29 133 246 309 310 321 322 353 354 361 365 366 374 425 456 494 511 525 526 527 546 556 613 644 645 679 694 717 750 754 755 796 797 870 923 944 978 979 1050 1096 1120 1138 1190 1201 1212 1213 1216 1217 1245 1483 1504 1524 1647 1664 1669 1680 1761 1822 1882 1883 1885 1919 1920 2112 2113 2166 2195 2196 2235 2258 2261 2274 2323

Table 3. Critical buses for different test systems.

IEEE Test System	Critical Buses
14	1 2 3 4 5
30	2 4 6 7 8 9 10 17 20 21 22 28
57	4 5 6 7 8 11 21 22 23 31 32 33 34 38 41 42 43 56
118	2 3 4 5 6 7 8 11 12 14 15 16 17 18 30 31 42 45 47 48 49 50 51 54 55 56 57 58 59 66 69 77 79 80 81 83 84 85 86 88 89 96 97 98 99 103 104 105 106 107 108 109 110 111 112 113 117

Table 4. Comparison of No. of Optimal PMUs under normal condition and uncertainty events.

IEEE Test System	No. of Optimal PMUs		% of Additional PMUs compared with original placement	Additional PMU Placement Location considering critical buses
	Normal Operating Condition	More weight age to critical PMUs		
14	4	6	50%	1 4
30	10	14	40%	5 8 12 16 22
57	17	26	53%	4 7 11 21 23 30 33 34 42
118	32	51	59%	4 6 8 18 32 46 54 57 58 78 83 88 96 100 106 108 111 112 117

5. Conclusion

This paper proposes a planning approach for optimal PMU placement making the system more resilient to PMU failure. The likelihood of undesired events is analyzed by creating a virtual attack agent which intends to damage some of the critical PMUs in the system. Operator agent is used to obtain a subset of buses that are critical based on the attack pattern of virtual attack agent. Simulation results illustrate the ability of the planner agent to place additional PMUs at strategic locations to increase the redundancy of critical buses. The developed framework was tested on several test systems including a 2383 bus western polish system and optimal results were obtained in all cases. Future work will consider the account of zero-injection measurement and branch flow measurement for more economical solution.

References

[1] Manousakis, N.M., Korres, G.N. and Georgilakis, P.S. (2012) Taxonomy of PMU Placement Methodologies. *IEEE Transactions on Power Systems*, **27**, 1070-1077. http://dx.doi.org/10.1109/TPWRS.2011.2179816

[2] Zhang, J., Welch, G., Bishop, G. and Huang, Z. (2010) Optimal PMU Placement Evaluation for Power System Dynamic State Estimation. *IEEE PES, Innovative Smart Grid Technologies Conference Europe* (*ISGT Europe*), Gothenburg, 11-13 October 2010, 1-7.

[3] Nuqui, R.F. and Phadke, A.G. (2005) Phasor Measurement Unit Placement Techniques for Complete and Incomplete Observability. *IEEE Transactions on Power Delivery*, **20**, 2381-2388. http://dx.doi.org/10.1109/TPWRD.2005.855457

[4] Koutsoukis, N.C., Manousakis, N.M., Georgilakis, P.S. and Korres, G.N. (2013) Numerical Observability Method for Optimal Phasor Measurement Units Placement Using Recursive Tabu Search Method. *IET Generation, Transmission & Distribution*, **7**, 347-356. http://dx.doi.org/10.1049/iet-gtd.2012.0377

[5] Peng, C., Sun, H. and Guo, J. (2010) Multi-Objective Optimal PMU Placement Using a Non-Dominated Sorting Differential Evolution Algorithm. *International Journal of Electrical Power & Energy Systems*, **32**, 886-892. http://dx.doi.org/10.1016/j.ijepes.2010.01.024

[6] Aminifar, F., Lucas, C., Khodaei, A. and Fotuhi-Firuzabad, M. (2009) Optimal Placement of Phasor Measurement Units Using Immunity Genetic Algorithm. *IEEE Transactions on Power Delivery*, **24**, 1014-1020. http://dx.doi.org/10.1109/TPWRD.2009.2014030

[7] Peppanen, J.,Alquthami, T., Molina, D. and Harley, R. (2012) Optimal PMU Placement with Binary PSO. *IEEE Energy Conversion Congress and Exposition* (*ECCE*), Raleigh, NC, 15-20 September 2012, 1475-1482.

[8] Gou, B. (2008) Generalized Integer Linear Programming Formulation for Optimal PMU Placement. *IEEE Transactions on Power Systems*, **23**, 1099- 1104. http://dx.doi.org/10.1109/TPWRS.2008.926475

[9] Azizi, S., Dobakhshari, A.S., Nezam Sarmadi, S.A. and Ranjbar, A.M. (2012) Optimal PMU Placement by an Equivalent Linear Formulation for Exhaustive Search. *IEEE Transactions on Smart Grid*, **3**, 174-182. http://dx.doi.org/10.1109/TSG.2011.2167163

[10] Esmaili, M., Gharani, K. and Shayanfar, H.A. (2013) Redundant Observability PMU Placement in the Presence of Flow Measurements Considering Contingencies. *IEEE Transactions on Power Systems*, **28**, 3765-3773.

[11] Christie, R. (1993) Power System Test Archive. https://www.ee.washington.edu/research/pstca

[12] MatPower. http://www.pserc.cornell.edu//matpower/

[13] The ILOG CPLEX Website, 2015. http://www.ilog.com/products/cplex

6

An Approach to Assess the Resiliency of Electric Power Grids

Navin Shenoy, R. Ramakumar

School of Electrical and Computer Engineering, Oklahoma State University, Stillwater, USA
Email: navin.shenoy@okstate.edu

Abstract

Modern electric power grids face a variety of new challenges and there is an urgent need to improve grid resilience more than ever before. The best approach would be to focus primarily on the grid intelligence rather than implementing redundant preventive measures. This paper presents the foundation for an intelligent operational strategy so as to enable the grid to assess its current dynamic state instantaneously. Traditional forms of real-time power system security assessment consist mainly of methods based on power flow analyses and hence, are static in nature. For dynamic security assessment, it is necessary to carry out time-domain simulations (TDS) that are computationally too involved to be performed in real-time. The paper employs machine learning (ML) techniques for real-time assessment of grid resiliency. ML techniques have the capability to organize large amounts of data gathered from such time-domain simulations and thereby extract useful information in order to better assess the system security instantaneously. Further, this paper develops an approach to show that a few operating points of the system called as landmark points contain enough information to capture the nonlinear dynamics present in the system. The proposed approach shows improvement in comparison to the case without landmark points.

Keywords

Grid Resilience, Machine Learning, Smart Grids, Time-domain Analysis, Dynamic Security Assessment

1. Introduction

In the wake of new vulnerabilities such as those arising from severe weather events and cyber-attacks, current electric grids can no longer be allowed to operate as they did in the past. It is becoming increasingly difficult to analyze different combinations of contingencies under changing scenarios. Grid resilience and improved situational awareness will form the basis of future electric grids in order to tackle these new challenges. The most cost effective way to meet such stringent requirements is through intelligent operation of the grid by employing data driven models that are both informational and analytical in nature. The key attribute involved here is the ability to assess the current state of the power system in real-time in terms of its security. Power system security is defined as its ability to survive imminent disturbances (contingencies) without interruption of customer service. Historically, it has been recognized that for a power system to be secure, it must be stable against all types of disturbances [1] [2]. Hence, stability analysis is an important component that can facilitate the assessment of power system security and thus, its resiliency.

Security in terms of operational requirements implies that following a sudden disturbance, power system would be secure if and only if: 1) it could survive the transient swings and reach an acceptable steady state condition, and 2) there are no limit violations in the new steady state condition. The first requirement can be met by carrying out time-domain simulations in order to investigate the instability phenomena such as loss of synchronism or voltage collapse in the post-contingency transient phase. The second requirement is met by using power-flow based methods in order to assess the new steady state condition for voltage and current limit violations.

Time-domain simulations (TDS) are computationally involved and too complex to be performed in real-time. Therefore, for many years in the past, the electric utility industry's framework for real-time security assessment mainly consisted of solution methods that would meet only the second requirement stated earlier. Such a type of real-time security analysis is prevalent even today and is commonly referred to as "Static Security Assessment (SSA)". On the other hand, a "Dynamic Security Assessment (DSA)" procedure would strive to meet both the requirements (as stated earlier) in real-time in order to assess power system security.

Different forms of DSA practices have existed in North America since the late 1980s [3]. Modern DSA implementations are able to complete a computation cycle within 5 - 20 minutes after a real-time snapshot (base case) of the system is available [4]. Real-time snapshots are provided by existing SCADA-based state estimators every few seconds or minutes depending on the size of the system [5]-[9]. Thus, these modern DSA implementations can be termed as "near real-time" and not "real-time". However, the latest PMU-based data collection technology can provide much better snapshots wherein the measurements are transmitted to the main control center at rates as fast as 60 samples/second [10]. Thus, DSA implementations of the future will be required to handle large amounts of data and complete the computation cycles much faster in order to assess the system security in true "real-time". Mathematically, such an instantaneous assessment would be possible only if grid resilience against any contingency can be expressed as a function of the state estimator output. In other words, input to the data-driven models must consist of only steady-state (static) quantities namely bus voltages and bus angles derived from power-flow based methods.

Machine learning (ML) techniques have the ability to assimilate and reason with knowledge the way human brain does. Such techniques are primarily driven by data that could be in the form of various power system parameters such as [11]-[13]: voltage, current, power, frequency, power angles etc. ML techniques can capture the nonlinear dynamics of power systems by extracting useful information from such large amounts of data. DSA tools employing such ML techniques will have the ability to determine stability limits in real-time. Such sophisticated tools will be able to analyze the current and future dynamics of power systems without carrying out extensive time-domain simulations. Additionally, these tools would also benefit the system operators by providing them with real-time information on trends in system security, thereby facilitating faster decision-making during crucial times. Also, as the entry of renewable energy systems further increases grid complexity, it is possible to extend the proposed work in order to accommodate online training, thereby resulting in a smart tool that can very effectively assess the system security in real-time.

This paper presents a framework that would enable implementation of such powerful machine learning techniques for real-time assessment of grid resilience. A standard IEEE 14-bus system is used in this paper for simulation purposes [14]. Firstly, a set of multiple steady-state operating points is generated by performing a SSA on the base case. Secondly, a TDS is performed on each operating point to assess the grid resilience against a specific contingency, thus generating a dataset for this work. The paper highlights the importance of selecting a few cases as landmark points in the operational space under consideration. Further, it presents a procedure to select the best landmark points in order to improve the prediction accuracy on the original dataset, thereby enhancing the ability to assess grid resilience instantaneously.

2. Static Security Assessment (SSA)

Static security assessment (SSA) provides a mathematical framework to compute stability limits for individual buses and lines based on power flow based methods. This involves checking for steady state voltage violations at every bus in the system. Power-Voltage (PV) curves are plotted for each bus by systematically loading the base case of the power system under consideration. This is achieved by means of an algorithm called as "Continuation Power Flow (CPF)" [15].

CPF is a "case worsening" procedure where the power system is loaded in steps as follows:

$$\begin{aligned} P_G &= \lambda P_{G0} \\ P_L &= \lambda P_{L0} \\ Q_L &= \lambda Q_{L0} \end{aligned} \tag{1}$$

where P_{G0}, P_{L0}, Q_{L0} are the base case generator and load powers (in per-unit) and λ is the loading parameter (in per-unit). CPF facilitates plotting of voltage curves as a function of loading parameter λ, for each bus.

As stated earlier, such a framework can be used to generate a dataset consisting of multiple steady-state operating points. For an n-bus system, every such operating point can be represented by a feature vector x of dimension 2n consisting of n bus voltages and n bus angles as features. A set S containing such objects is given by,

$$\left\{ x \in \Re^{2n} : x = \left[V_1, \cdots, V_n, \delta_1, \cdots, \delta_n \right]^{\mathrm{T}} \right\} \tag{2}$$

where V_i's are bus voltages (in per unit) and δ_i's are bus angles (in degrees).

SSA is performed on the standard IEEE 14-bus system for the following voltage stability criteria at each bus: $V_{\max} = 1.2$ pu and $V_{\min} = 0.8$ pu. Generators are represented by machine models along with automatic voltage regulators and turbine governors. A CPF routine is performed for each line outage of this power system. Thus, a maximum loading parameter $\lambda_{\max i}$ is calculated for each line outage i, taking voltage stability criteria into account. The set represented by Equation (2) is generated only for values of λ given by,

$$1 \le \lambda \le \lambda_{\max i}, \forall i \tag{3}$$

It has to be noted that these $\lambda_{\max i}$ values account for only steady-state voltage violations and hence, do not provide any information about dynamic system security. In order to account for dynamic stability, time-domain simulations are performed for each operating point, as described in the next section. All routines are carried out using the PSAT toolbox for Matlab [16]. **Figure 1** shows V-λ curves for a particular line outage.

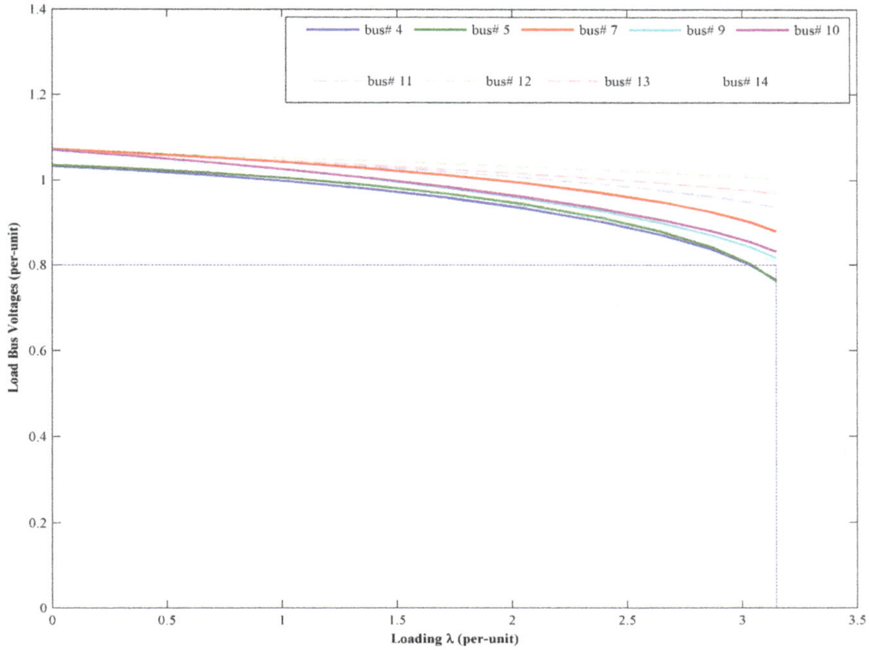

Figure 1. V-λ curves of PQ buses for outage of line#16 (bus 2 to bus 4), bus 5 voltage reaches $V_{\min}$ at $\lambda = 3.1487$.

3. Dynamic Security Assessment (DSA)

The goal of a DSA is to classify different cases based on their dynamic security severity. Dynamic security depends on the time responses of various system variables for the contingency under consideration. As mentioned earlier, it is not possible to perform computationally intensive time-domain simulations in real-time. Nonetheless, machine learning techniques have the ability to extract information from offline time-domain simulations. Subsequently, such useful information can be used to predict dynamic system security for new configurations in order to avoid lengthy time-domain simulations. To implement such an application, detailed time-domain simulations are required to be conducted for different operating points. Thus, a database, on which ML techniques can operate, needs to be generated in offline mode.

The database is generated in the form of a feature matrix X and an output vector y. Each row of the feature matrix X represents a steady state operating point in the form of object $x \in \Re^{2n}$ from set S as defined in equation (2). Matrix X contains total number of 'm' such objects and hence, its size is ($m \times 2n$). A time-domain simulation for a specific contingency is performed on each of these m objects. These simulations are tagged as "stable" or "unstable" depending on the time responses of system variables. Output vector y is a binary column vector with m rows wherein each row represents whether the corresponding TDS is stable(1) or unstable(0). For the IEEE 14-bus test system considered in this paper, a load disturbance of 0.2 per-unit (increase) is applied to every steady state operating point generated in the previous section. Stability is decided based on the average values of voltage violations over the entire simulation period (V_{max} = 1.2 pu and V_{min} = 0.8 pu). **Figure 2(a)** and **Figure 2(b)** show voltage dynamics at all buses for a stable and unstable case respectively.

Essentially, DSA is a mapping between each object x and its resiliency against the contingency under consideration, expressed by function f such that,

$$f(x) = \begin{cases} 1, & \text{if TDS is stable} \\ 0, & \text{if TDS is unstable} \end{cases} \tag{4}$$

The next section describes the application of machine learning techniques in order to arrive at this unknown function f.

4. Application of Machine Learning Techniques

Machine learning techniques can be applied to the database as generated in the previous section in the form of feature matrix X (size $m \times 2n$) and output vector y (size $m \times 1$). Each row i of matrix X is in the form of object $x \in \Re^{2n}$ from set S as defined in Equation (2) and is referred to as the ith training example: $x^{(i)}$. Similarly, the ith row from vector y represents the output of the ith training example and is represented by a bit $y^{(i)}$ (either 0 or 1). Therefore, we have,

$x^{(i)}$ = ith training example

$y^{(i)}$ = output (stability) of the ith training example

For "2n" features and "m" training examples, matrix X and vector y are given as follows,

$$X = \begin{bmatrix} \cdots & x^{(1)^T} & \cdots \\ \cdots & x^{(2)^T} & \cdots \\ \cdots & \vdots & \cdots \\ \cdots & x^{(m)^T} & \cdots \end{bmatrix}_{m\times 2n} \qquad y = \begin{bmatrix} y^{(1)} \\ y^{(2)} \\ \vdots \\ y^{(m)} \end{bmatrix}_{m\times 1} \tag{5}$$

Next, a prediction/hypothesis function h in terms of parameter vector θ (column vector) of size $2n$ is proposed as follows,

$$h_\theta(x) = g\left(\theta^T x\right) \tag{6}$$

where x is any training example vector and g depends on the machine learning algorithm being employed.

The cost function J for machine learning algorithms is generally of the form [17],

$$J(\theta) = \frac{1}{2m}\sum_{i=1}^{m}\left[h_\theta\left(x^{(i)}\right) - y^{(i)}\right]^2 \tag{7}$$

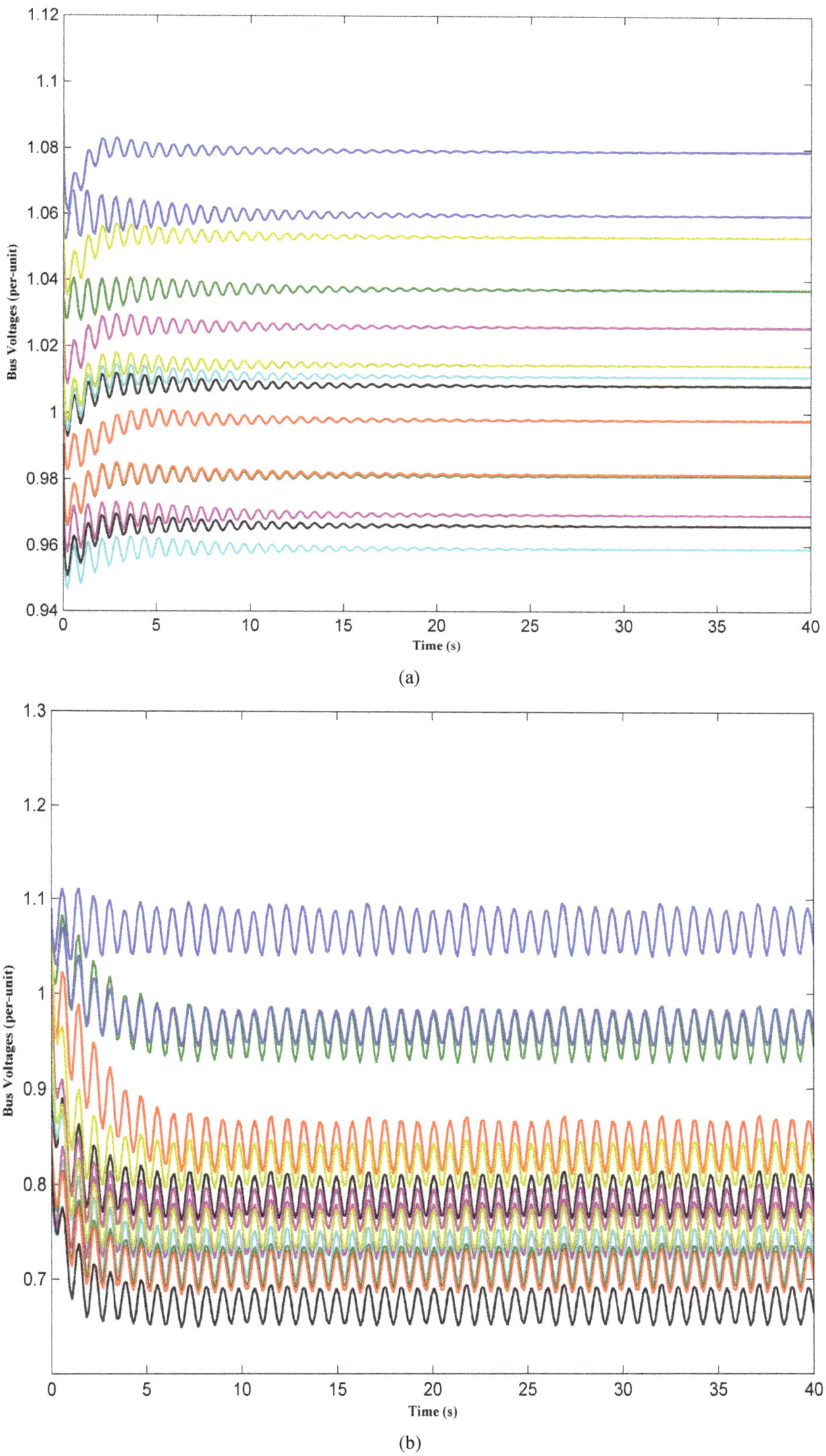

Figure 2. (a) Bus voltages (stable case); (b) Bus voltages (unstable case).

The above cost function is the mean of the sum of squared errors in predicting the outputs of m training examples. Such a cost function can be minimized by using analytical method or batch gradient descent method. The optimal parameter vector θ thus derived can be used for predicting the stability of future cases in real-time.

The problem presented in this paper is to classify a TDS as stable (1) or unstable (0). For such classification problems, logistic regression can be used, in which case functions g and J are given as follows [18],

$$g(z) = \frac{1}{1+e^{-z}} \tag{8}$$

and

$$J(\theta) = -\frac{1}{m}\sum_{i=1}^{m}\left[y^{(i)}\log\left(h_\theta\left(x^{(i)}\right)\right) + \left(1-y^{(i)}\right)\log\left(1-h_\theta\left(x^{(i)}\right)\right)\right] \tag{9}$$

The function $g(z)$ given in equation (8) is a sigmoid function and its value lies between 0 and 1. For classification purposes, TDS cases for which $g(z)$ is greater than 0.5 can be considered as stable and the rest as unstable. At this point, it should be noted that function h given in equation (6) approximates the unknown function f of the previous section, when the parameter θ is optimal. The approximated function f_{apprx} can be given by,

$$f_{apprx}(x) = \begin{cases} 1, & \text{if } h_\theta(x) \geq 0.5 \\ 0, & \text{if } h_\theta(x) < 0.5 \end{cases} \tag{10}$$

In order to test the algorithm, the 14-bus dataset represented by matrix X and vector y (as generated in the previous section) can be divided into a training set (75%) and a test set (25%), which is a normal practice in ML domain. We may also delete the constant feature columns from X such as those containing PV bus voltages and reference angles, since such constant feature values do not add any valuable information. Therefore, an original matrix X with 22 columns (features) is used in this paper. **Figure 3(a)** and **Figure 3(b)** show the learning curves for the training and test sets respectively. Learning curves are plotted by varying the number of examples m in the training set. As highlighted in these figures, the average prediction errors on the training and test sets are calculated as 1.245% and 2.599% respectively. For m objects, prediction error is the percentage of examples that are classified incorrectly by the function f_{apprx} given in Equation (10) and it is calculated as follows,

$$\%\ \text{Error} = \frac{100}{m}\sum_{i=1}^{m} err\left(f_{apprx}\left(x^{(i)}\right), y^{(i)}\right)$$

where (11)

$$err\left(f_{apprx}\left(x^{(i)}\right), y^{(i)}\right) = \begin{cases} 1, & \text{if } f_{apprx}\left(x^{(i)}\right) = 1 \ \& \ y^{(i)} = 0 \ \text{ or} \\ & \text{if } f_{apprx}\left(x^{(i)}\right) = 0 \ \& \ y^{(i)} = 1 \\ 0, & \text{otherwise} \end{cases}$$

The next section of this paper introduces the concept of "landmark points" and "linear kernel". Further, this paper presents a strategy to select best landmark points in order to improve the prediction accuracy.

5. Landmark Points and Linear Kernel

The concept of selecting landmark points gains importance from the fact that a few training examples may contain the most relevant information about the inherent dynamics present in the dataset [19]. This section investigates the possibility of selecting such landmark points within the operational space under consideration in order to improve prediction accuracy without compromising computational efficiency. Essentially, these landmark points are 2n-dimensional objects belonging to the same set S given by Equation (2).

In order to demonstrate the effectiveness of this concept, L number of landmark points are drawn at random from the rows of matrix X and then, every (training example, landmark) pair is compared using a linear kernel [20]. A linear kernel measures the similarity between training example $x^{(i)}$ and landmark $l^{(j)}$ using the dot product and is given by,

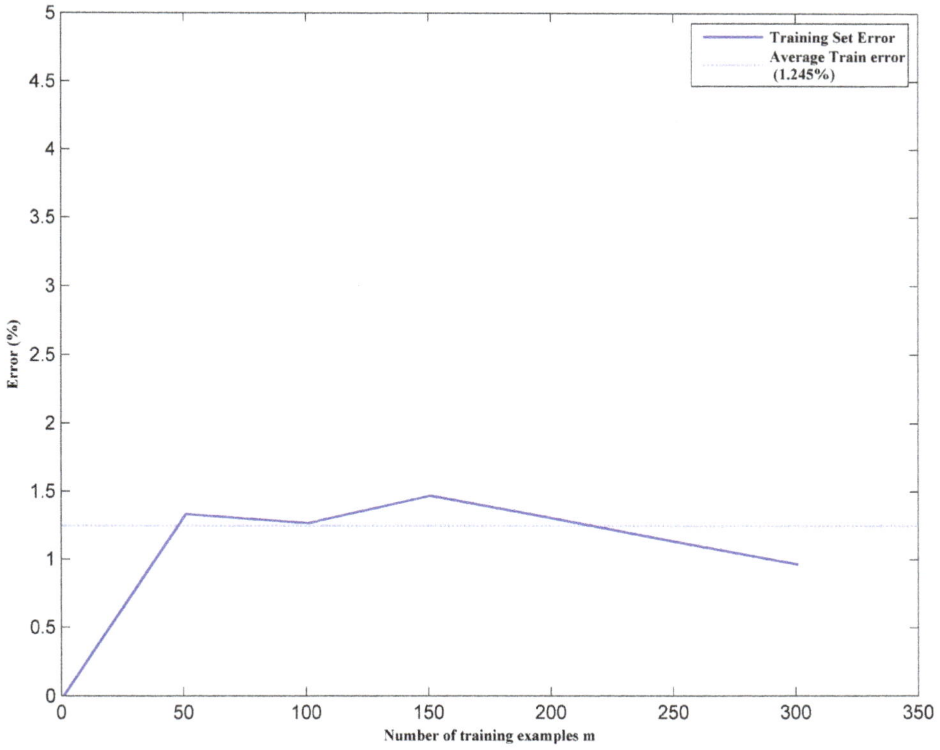

(a)

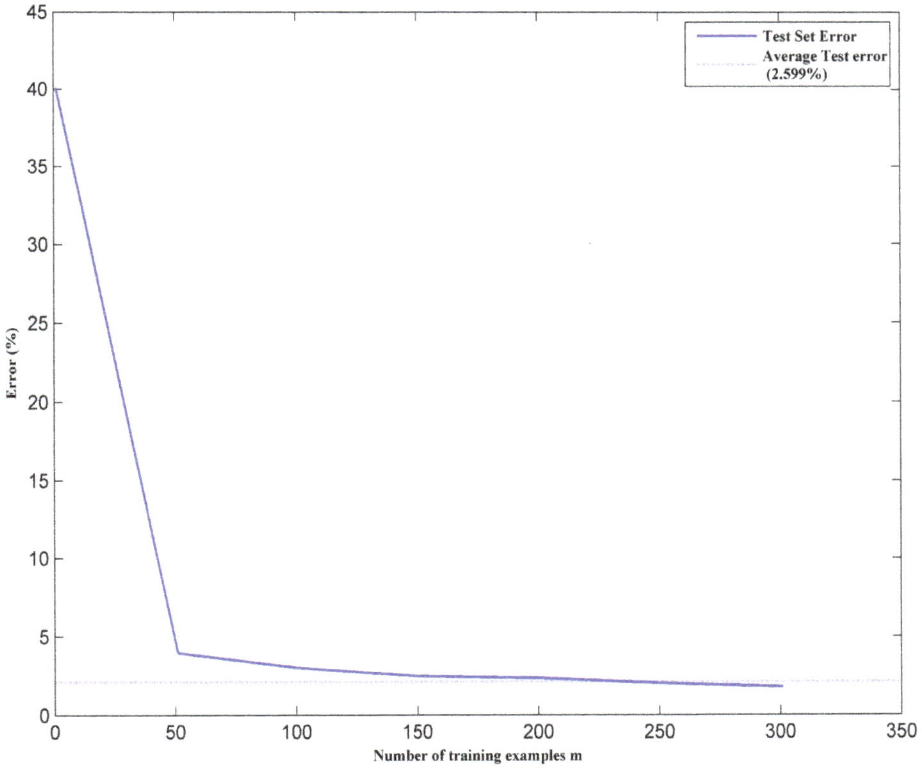

(b)

Figure 3. (a) Learning curve (training set); (b) Learning curve (test set).

$$sim\left(x^{(i)}, l^{(j)}\right) = x^{(i)\mathrm{T}} l^{(j)} \tag{12}$$

Similarity is calculated between all training examples i: $1 < i < m$ and landmark points j: $1 < j < L$. The original feature matrix X (size $m \times 2n$) gets transformed into a new matrix X'_{random} (size $m \times L$) which can be now used for training and testing purposes. Computational efficiency is maintained by enforcing the following constraint,

$$L \leq 2n \tag{13}$$

As shown in **Figure 4(a)** and **Figure 4(b)**, prediction errors on the training and test sets decrease as the number of landmark points increase. However, it should be noted that such a random selection of landmark points does not guarantee better performance when compared with the average training and test set errors calculated in the previous section.

6. Strategy to Select Best Landmark Points

Choosing the most appropriate set of landmark points for a given dataset is not an easy task. In this section, the k-means algorithm is used to derive better landmark points as compared to the random ones selected in the previous section [21]. Using k-means algorithm, centroids can be calculated for any feature matrix X. A total number of L such centroids are generated from X for use as landmark points and then, using linear kernel a new matrix $X'_{\text{centroids}}$ is formed like in the previous section.

In an attempt to find the best landmark points, the original matrix X is divided into 2 matrices X_{stable} and X_{unstable} consisting of only stable and unstable cases respectively. Using k-means, a total number of L centroids are generated for each of these matrices separately and again using linear kernel, two new matrices X'_{stable} and X'_{unstable} are formed.

The strategy for selecting best landmark points can be stated as follows,

- Select L random examples from original matrix X as landmarks and generate X'_{random}
- Select L centroids from original matrix X as landmarks and generate $X'_{\text{centroids}}$
- Select L centroids from X_{stable} as landmarks and generate X'_{stable}
- Select L centroids from X_{unstable} as landmarks and generate X'_{unstable}
- Plot learning curves using X'_{random}, $X'_{\text{centroids}}$, X'_{stable}, X'_{unstable}
- Compare the training and test set errors and select the best L landmarks

Figure 5(a) and **Figure 5(b)** show the learning curves for each of the above matrices with L = 22 landmarks for the IEEE 14 bus dataset generated earlier. From these figures we can conclude that centroids selected from the unstable cases are the best landmark points for this dataset. Moreover, it has to be noted that computational efficiency is not compromised since the total number of landmarks used here (L = 22) is not greater than the total number of columns in the original matrix X (=22). **Figure 6(a)** and **Figure 6(b)** again compare the learning with increasing number of land marks for the case of random landmarks against landmarks selected as centroids from only unstable cases. **Figure 7(a)** and **Figure 7(b)** plot the learning curves for the original matrix X (without any landmarks) and X'_{unstable} (best landmarks). These plots confirm that when best landmarks are employed, prediction accuracy improves on both, the training set and the test set.

7. Concluding Remarks

The ability to assess the current state of the power system instantaneously is the key attribute needed for enhanced grid resilience. Electric power entities carry out large number of offline studies on power system models of different sizes, thus generating tons of data. Machine learning techniques can be employed to use such huge databases in order to learn the inherent non-linear relationships that exist among different power system parameters. Such useful information can be later used online for real-time security analysis.

This paper presents a framework to apply machine learning techniques for real-time assessment of the grid resilience against any contingency with respect to its static and dynamic stability using offline databases. Further, this paper demonstrates a strategy to select best landmark points in order to improve prediction accuracy without compromising computational efficiency. Moreover, ML algorithms are easily scalable and hence, the proposed approach can be extended for analyzing grid resilience against multiple contingencies. Metrics for grid resilience can be developed based on such multi-contingency analyses. With large-scale penetration of renewable energy

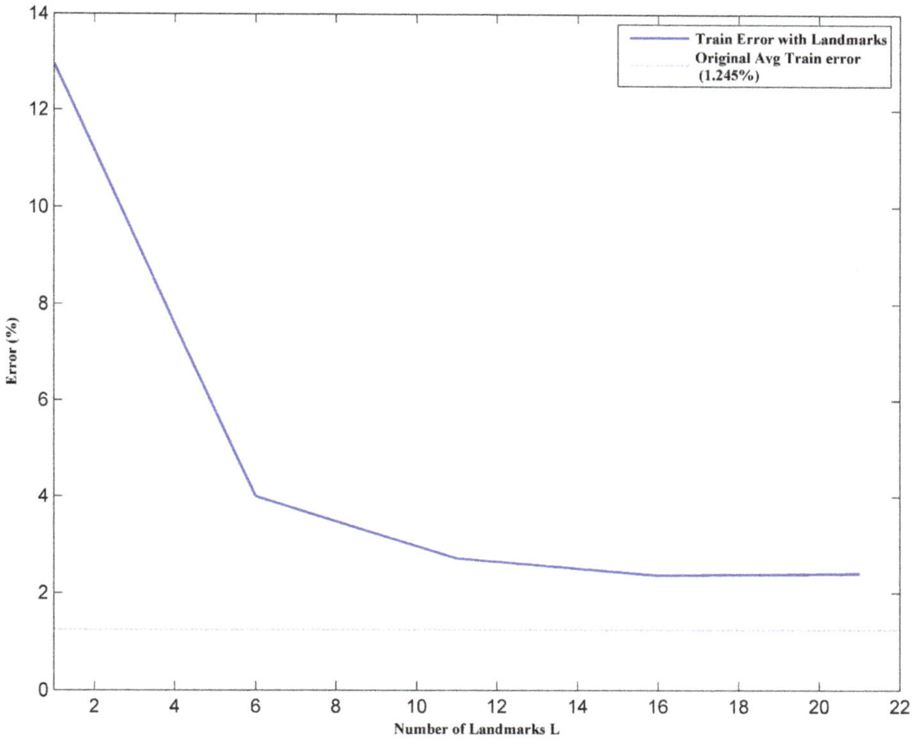

(a)

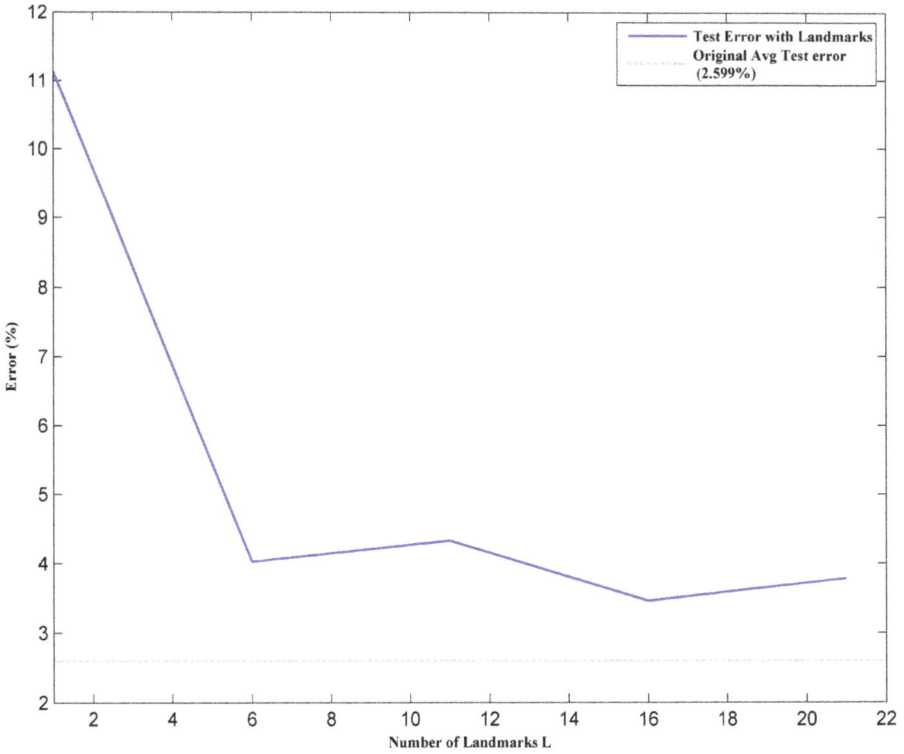

(b)

Figure 4. (a) % Error vs num. of landmarks (training set); (b) % Error vs num. of landmarks (test set).

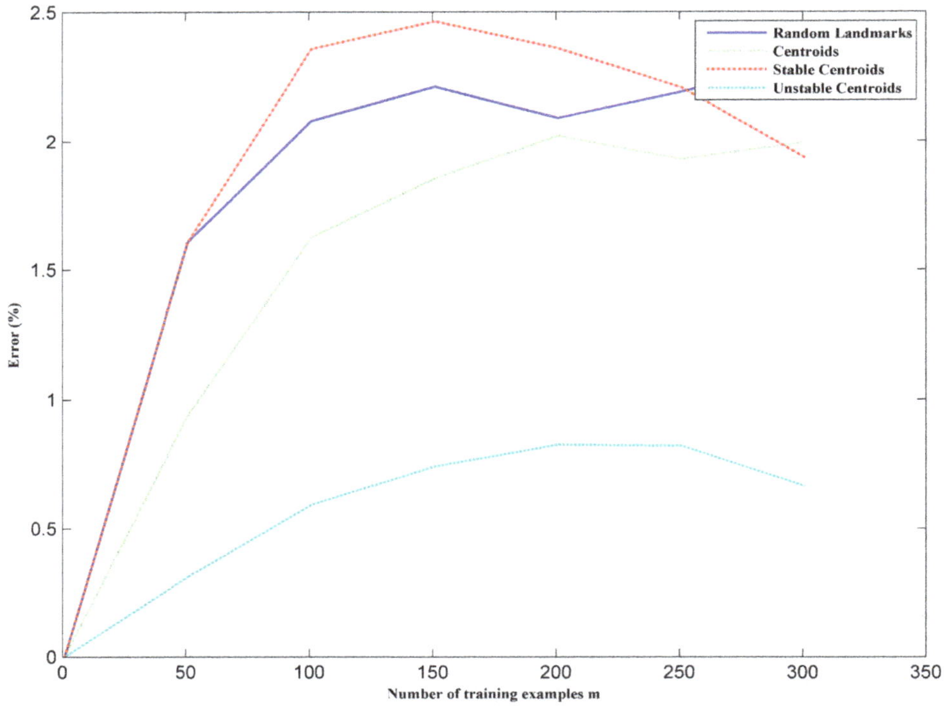

(a)

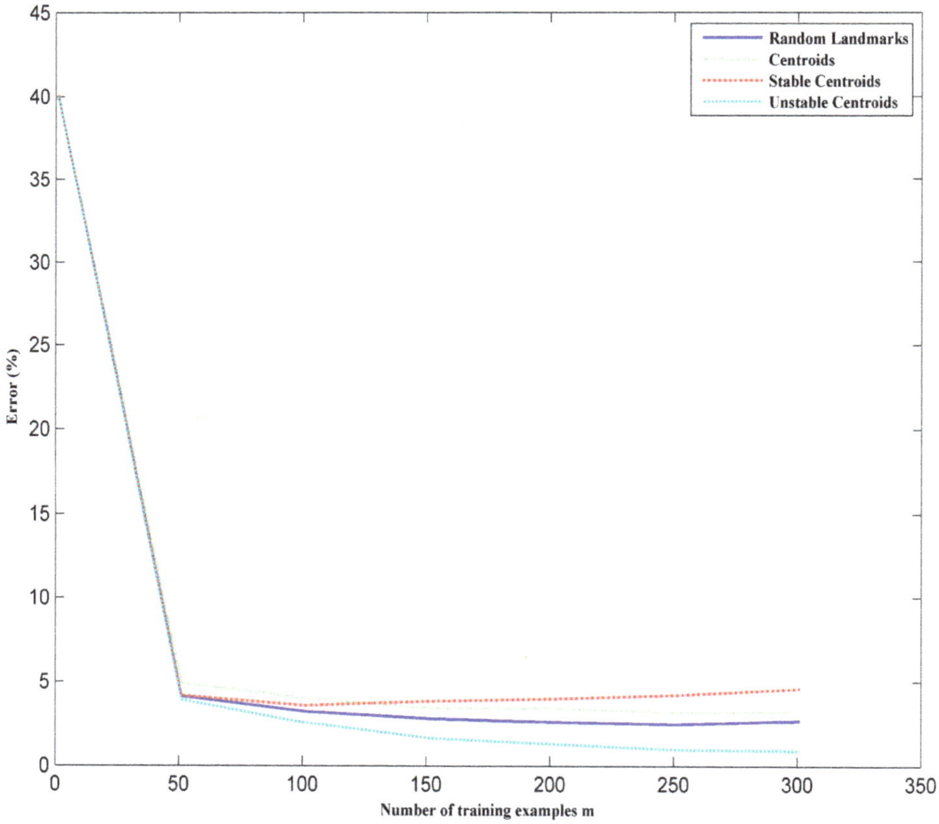

(b)

Figure 5. (a) Learning curves (training set); (b) Learning curves (test set).

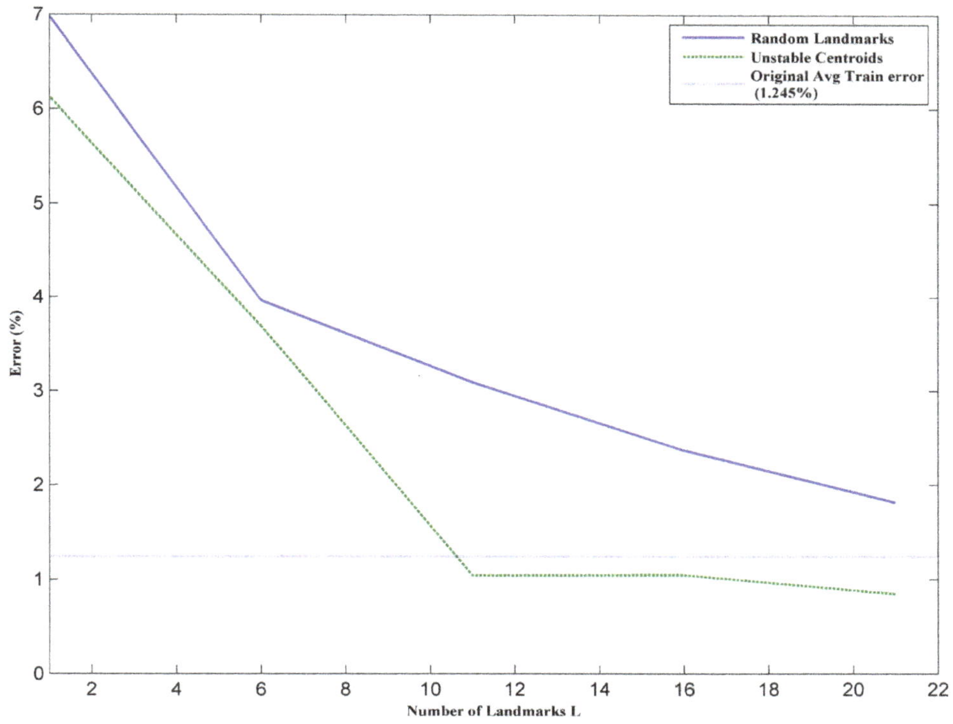

(a)

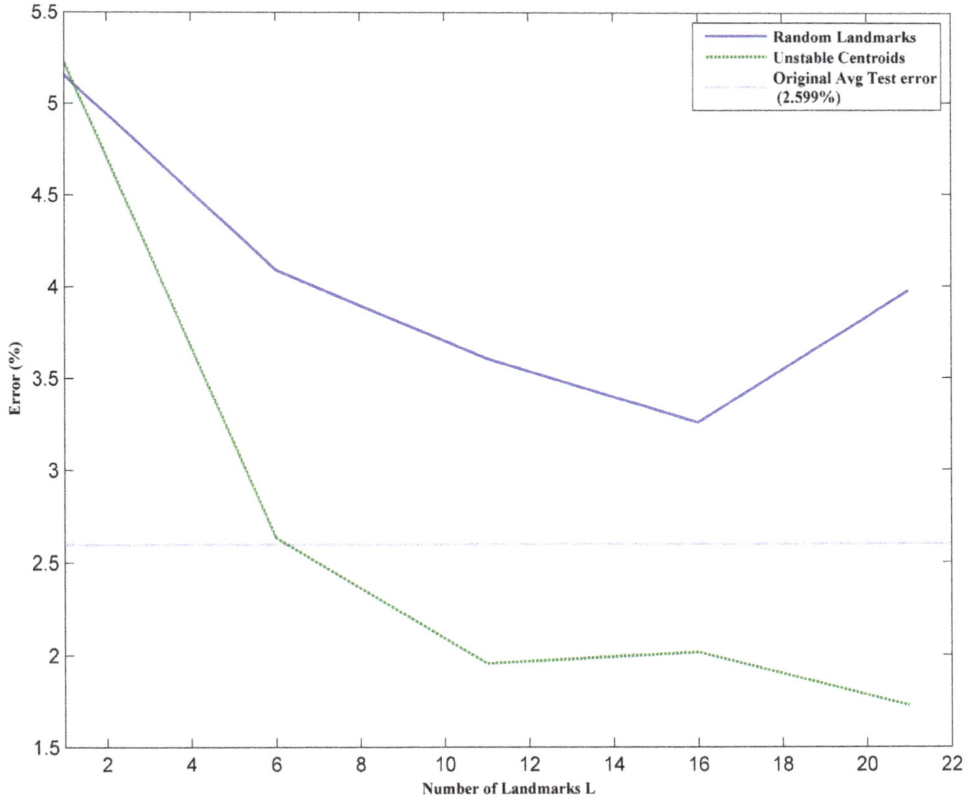

(b)

Figure 6. (a) % Error vs num. of landmarks (training set); (b) % Error vs num. of landmarks (test set).

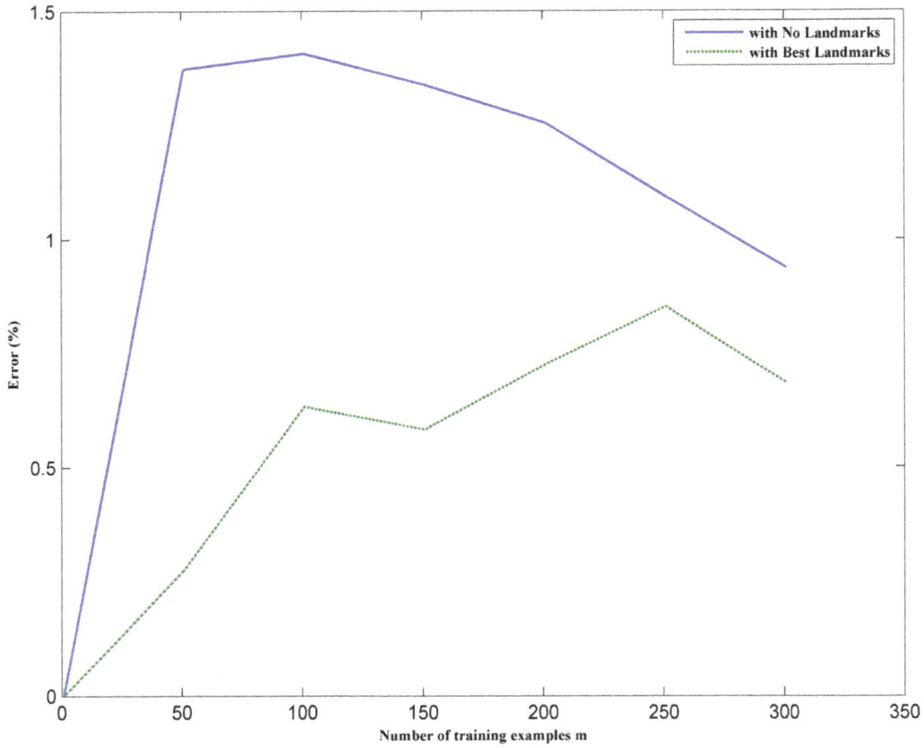

(a)

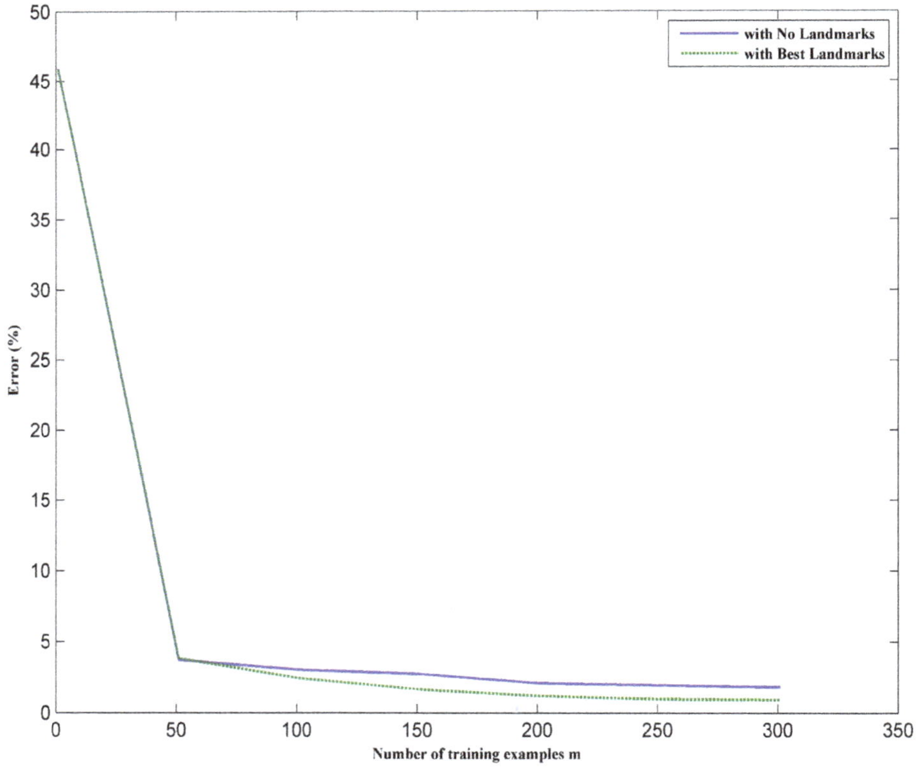

(b)

Figure 7. (a) Learning curves (training set); (b) Learning curves (test set).

in to the current grid and emergence of microgrids, future grid applications would require real-time training in order to extract useful information on a continuous basis. Machine learning techniques can accommodate such complex requirements posed by the continually changing electric grid and hence, would definitely play an important role in realizing next-gen real-time applications.

Acknowledgements

This work was supported by the OSU Engineering Energy Laboratory and the PSO/Albrecht Naeter Professorship in the School of Electrical and Computer Engineering.

References

[1] Kundur, P., *et al.* (2004) Definition and Classification of Power System Stability IEEE/CIGRE Joint Task Force on Stability Terms and Definitions. *IEEE Transactions on Power Systems*, **19**, 1387-1401. http://dx.doi.org/10.1109/TPWRS.2004.825981

[2] Wang, L. and Morison, K. (2006) Implementation of Online Security Assessment. *IEEE Power and Energy Magazine*, **4**, 46-59. http://dx.doi.org/10.1109/MPAE.2006.1687817

[3] Fouad, A., Aboytes, F. and Carvalho, V.F. (1988) Dynamic Security Assessment Practices in North America. *IEEE Transactions on Power Systems*, **3**, 1310-1321. http://dx.doi.org/10.1109/59.14597

[4] Grigsby, L.L. (2012) Power System Stability and Control. 3rd Edition, CRC Press, Boca Raton. http://dx.doi.org/10.1201/b12113

[5] Jardim, J., Neto, C. and dos Santos, M.G. (2006) Brazilian System Operator Online Security Assessment System. *IEEE Power Systems Conference and Exposition*, Minneapolis, 25-29 July 2010, 7-12.

[6] Tong, J. and Wang, L. (2006) Design of a DSA Tool for Real Time System Operations. *International Conference on Power System Technology*, Chongqing, 22-26 October 2006, 1-5. http://dx.doi.org/10.1109/icpst.2006.321419

[7] Savulescu, S.C. (2009) Real-Time Stability Assessment in Modern Power System Control Centers. John Wiley & Sons, Hoboken. http://dx.doi.org/10.1002/9780470423912

[8] Chiang, H.-D., Tong, J. and Tada, Y. (2010) On-Line Transient Stability Screening of 14,000-Bus Models Using TEPCO-BCU: Evaluations and Methods. *IEEE Power and Energy Society General Meeting*, Minneapolis, 25-29 July 2010, 1-8.

[9] Yao, Z. and Atanackovic, D. (2010) Issues on Security Region Search by Online DSA. *IEEE Power and Energy Society General Meeting*, Minneapolis, 25-29 July 2010, 1-4.

[10] Ekanayake, J., Jenkins, N., Liyanage, K., Wu, J. and Yokoyama, A. (2012) Smart grid: Technology and Applications. John Wiley & Sons, Hoboken. http://dx.doi.org/10.1002/9781119968696

[11] Ongsakul, W. and Dieu, V.N. (2013) Artificial Intelligence in Power System Optimization. CRC Press, Hoboken.

[12] Warwick, K., Ekwue, A. and Aggarwal, R. (1997) Artificial Intelligence Techniques in Power Systems. IEE Press, London.

[13] Song, Y.-H., Johns, A. and Aggarwal, R. (1996) Computational Intelligence Applications to Power Systems, Vol. 15. Springer Science & Business Media, Berlin, Heidelberg.

[14] Power Systems Test Case Archive. http://www.ee.washington.edu/research/pstca/

[15] Ajjarapu, V. and Christy, C. (1992) The Continuation Power Flow: A Tool for Steady State Voltage Stability Analysis. *IEEE Transactions on Power Systems*, **7**, 416-423. http://dx.doi.org/10.1109/59.141737

[16] Milano, F. (n.d.) PSAT, Matlab-Based Power System Analysis Toolbox. http://faraday1.ucd.ie/psat.html

[17] Ng, A. (2015) Machine Learning. http://cs229.stanford.edu/

[18] Barber, D. (2012) Bayesian Reasoning and Machine Learning. Cambridge University Press, Cambridge.

[19] Tipping, M.E. (2001) Sparse Bayesian Learning and the Relevance Vector Machine. *The Journal of Machine Learning Research*, **1**, 211-244.

[20] Murphy, K.P. (2012) Machine Learning: A Probabilistic Perspective. MIT Press, Cambridge, MA.

[21] Smola, A. and Vishwanathan, S. (2008) Introduction to Machine Learning. Cambridge University Press, Cambridge, UK.

7

Coordination of Overcurrent Relay in Distributed System for Different Network Configuration

Niraj Kumar Choudhary, Soumya Ranjan Mohanty, Ravindra Kumar Singh

Electrical Engineering Department, Motilal Nehru National Institute of Technology Allahabad, Allahabad, India
Email: niraj@mnnit.ac.in, soumya@mnnit.ac.in, ravindraksingh@gmail.com

Abstract

This paper presents a study on protection coordination of overcurrent relays (OCRs) in a distributed system by considering its different operating modes. Two different case studies which are considered in present work for protection coordination include: (i) DG interfaced distribution system in grid connected mode and (ii) DG interfaced distribution system in islanded mode of operation. The proposed approach is tested on the Canadian urban benchmark distribution system consisting of 9 buses. On the occurrence of fault, level of fault current changes which in turn changes the operating time of various OCRs. Therefore, it is important to calculate and suggest method of the relay setting in order to minimize the operating time of relays and also to avoid its mal-operation. In this paper, the protection scheme is optimally designed by taking into account the above mentioned conditions. The operating time of relays can be decreased and, at the same time, coordination can be maintained by considering the optimum values of time dial setting (TDS). Genetic Algorithm (GA) has been used for determining the optimum values of TDS and hence operating time.

Keywords

Distributed Generator (DG), Overcurrent Relays (OCRs), Time Dial Setting (TDS), Genetic Algorithm (GA), Grid Connected Mode, Islanded Mode

1. Introduction

The increase in load demand is forcing the utility to use non-conventional energy resources like photovoltaic, wind energy, biomass etc. These energy resources are an alternative which are used to decrease stress on the traditional utility grid. The distributed energy also known as decentralized energy is generated with the help of small distributed energy resources (DER) or distributed generators (DGs). These distributed generators are generally integrated to the low/medium voltage level distribution network as the range of voltage which can be generated has some limitation. Based on the interfacing medium DGs can be classified into two types: (i) rotating machine based DGs and (ii) electronically interfaced DGs. The DGs should be properly integrated by considering its impact on the performance of electric power distribution system. Integration of DG has several impacts on the performance of power system. The main impact is in terms of bidirectional power flow and change in short circuit current level etc. [1].

The protection system should be fast and capable enough to isolate the faulty part of the network in case the fault strikes. In the conventional distribution system, power flow is unidirectional *i.e.* from substation towards the load whereas the presence of another source causes the bidirectional flow of power [2]. Incorporation of DG in the distribution system has increased the complexity of protection coordination. The main changes due to DGs are bidirectional power flow and change in short circuit current level. Therefore, the existing coordination schemes may not be able to perform its function correctly [3]. The majority of protection schemes used in modern power system are based upon the short circuit current sensing capability [4] [5]. The main protection issues associated with the introduction of DERs to the distribution network include the blinding of protection, false sympathetic tripping, reclosure-fuse mis-coordination, lapse of inter fuse coordination and failed auto-reclosing [6] [7]. The most widely used form of protection in power system is overcurrent protection. Each relay in the power system should be properly coordinated with another relay protecting the adjacent equipment. In case the primary relay fails to clear the fault, the backup protection should initiate its operation after certain interval of time known as coordination time interval (CTI). If the relays are not properly coordinated, it may mal-operate. Thus over-current protection is one of the major concerns in power system protection.

Different optimization techniques for optimum coordination of OCRs have been proposed in literature. The optimized vale of time of operation of over-current relays can be calculated with the help of various optimization techniques. This section provides an extensive literature survey for some of the methods which have been used for the solution of protection coordination problem. The relay coordination is formed as mixed integer nonlinear programming and solved by the use of general algebraic modeling software [8] [9]. In [8], hybrid GA is used to solve the directional OCRs coordination problem for several network topologies. In [10], various linear programming problem techniques have been used for OCR coordination. In [11], sequential quadratic programming method has been used for optimizing all the settings of OCRs. Most recently some heuristic techniques with particle swarm optimization (PSO), evolutionary methods and harmony search algorithms have been used for solving the protection coordination problem of OCRs [12]-[14].

2. Impact of DG on Protection Coordination

In a distributed power system, DGs are used to generate power at low or medium voltage levels along with the utility grid. A distributed power system may contain different type of DG sources (electronically coupled distributed generators or rotating machine based distributed generators) depending upon the environmental/geographical conditions. The presence of multiple DERs may cause the increase in short circuit current flowing through the network. However, as the current rating of silicon device is limited, the fault current of electronically interfaced DGs should be restricted to a maximum of about two times their nominal current [15]. The higher current flowing through the relays affects the time dial setting (TDS) which in turn decreases its time of operation and also the coordination characteristics gets changed. Thus the established over-current protection technique remains no longer applicable for the DG integrated distribution systems [16]. The most important protection issues associated with the introduction of distributed energy resources in a distribution network includes blinding of protection and false/sympathetic tripping.

2.1. Blinding of Protection

The fault current measured by overcurrent relays in distributed power systems is lesser by an amount negatively contributed by DG connected to the system. This reduction in current may result in malfunction of overcurrent relays [17]. This undesirable condition may occur when DGs are connected anywhere between the feeding substation and fault location. Due to the contribution of the DER towards the fault current the current measured by the feeder relay decreases as compared to the situation when there is no DER connected to the network. In

Figure 1, for a fault F_2 the relay corresponding to circuit breaker CB_4 does not respond to the fault and comes under the effect of blinding. This may result in malfunction of the relay.

2.2. False/Sympathetic Tripping

False/Sympathetic tripping refers to a situation in which tripping occurs due to fault outside the zone of protection for a feeder embedded with DER. In this case, the DERs contribute to the fault via its feeder and the fault current flows upwards on the feeder. Thus, the non-directional relay of the healthy feeder may falsely detect a fault and may isolate the feeder, which is undesirable. Higher the short-circuit capacity; more adverse is its effect on the relay performance [18]. In **Figure 1**, for a fault F_1 circuit breaker CB_3 should operate but due to contribution of current I_{DG} from connected DG, circuit breaker CB_4 will operate which may lead towards unnecessary interruption of healthy feeders.

3. Methodology and Problem Formulation

The operating time of an OCR is inversely proportional to the short circuit current passing through it. The two parameters involved in the operating characteristics of relay represented by "Equation (1)" are its pick-up current (I_p) and time dial setting (TDS). In "Equation (1)", I_{SC} represents the short circuit current. The values of A and B decides the operating characteristics of the relays *i.e.* whether the relay has got normal inverse, very inverse or extremely inverse characteristics.

In this paper it has been assumed that the OCRs possess inverse definite characteristics and thus the values of A and B are taken to be 0.14 and 0.02 respectively [19] [20]. Therefore, the operating time of OCRs can be expressed as shown in "Equation (2)".

where, PSM is known as plug setting multiplier which can be determined for a known configuration after calculating the values of I_{SC} and I_p. The objective function denoted by T is the summation of coordination times of all relays, which is to be minimised and is expressed as following in "Equation (3)". Here t_{ii} indicates the operating time of primary relay i, for near end fault. Therefore the time of operation of individual relay is a function of TDS, which is represented by "Equation (4)". The value of C for each relay is a function of plug setting multiplier (PSM) which needs to be calculated for different fault location. The functional relationship of C with PSM is shown in "Equation (5)".

In "Equation (6)" C_i is the constant for i_{th} relay and its value for different fault location is to be calculated. The main objective is to minimize the operating time and to calculate the optimized value of $(TDS)_i$. The calculation of fault current and TDS is presented in Section V of this paper.

$$t = A\frac{TDS}{\left(\frac{I_{SC}}{I_p}\right)^B - 1} \tag{1}$$

$$t = A\frac{TDS}{(PSM)^B - 1} \tag{2}$$

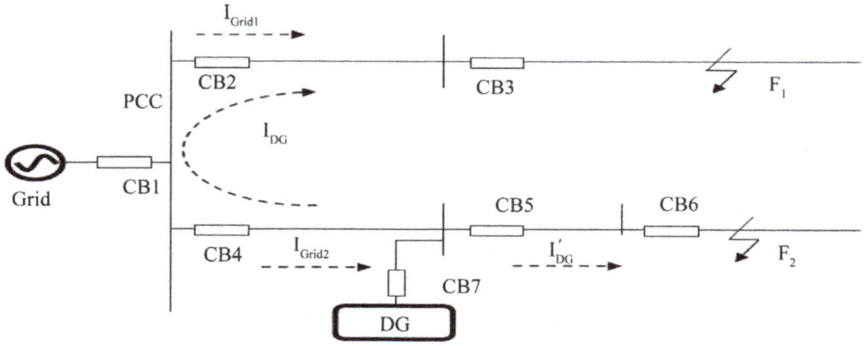

Figure 1. Operation of relay for different fault location.

$$\min T = \sum_{i=1}^{m} t_{i,i} \tag{3}$$

$$t = C(TDS) \tag{4}$$

$$C = \frac{A}{(PSM)^B - 1} \tag{5}$$

$$\min T = \sum_{i=1}^{m} C_i (TDS)_i \tag{6}$$

4. System Description and Simulation Setup

In **Figure 2**, a section of the Canadian urban benchmark distribution system is shown in which there are two radial feeders. The rating of each feeder is 8.7 MVA. The impedance of each feeder is $(0.1529 + j\,0.1406)$ Ω/km. These feeders are energized by utility through a transformer of 20 MVA, 115 kV/12.47kV. Four DGs are connected at different locations through transformers of 12.47 kV/480V voltage rating and having same power rating as that of DG. From simulation studies it has been observed that sixteen relays (R_1-R_{16}) are required for this system. The operating zone of each relay is to be identified in which the relays should be capable enough to operate on the occurrence of different types of fault.

Due to the presence of DG the distribution system no longer remains radial therefore each fault is associated with two primary relays, one from each side which in turn is associated with up to two backup relays. The study of protection coordination among the OCRs placed in this system has been studied for three cases: Distribution system without DG, DG integrated system in grid connected mode and DG integrated system in islanded mode of operation.

5. Simulation Results and Analysis

On the selected distribution system the study of protection coordination of OCRs are studied. Since directional overcurrent relays are used they will operate for the current flowing in a particular direction. If the current flows in reverse direction the relay will not operate. Now considering relay 13 (R13) as shown in **Figure 2**, which is at far end from the Grid. If fault occurs between bus 4 and bus 5 then relay 13 will act as primary relay and relay 11 acts as back up relay. The current flowing through the relay is 2600 A and current setting of relay is 200 A. Now PSM can be calculated by dividing fault current and current setting of relay and its calculated value is 13. From the inverse characteristics of overcurrent relay, operating time of relay is 2.8 sec. since there is no relay following the relay 13, therefore value of TDS should be small and here we have chosen 0.05. Thus the actual operating time of the relay is 0.14 sec. Now for the relay number 11, fault current is 1900 A and current setting is 400 A which results in PSM equal to 4.75. From the inverse time characteristics the operating time of relay is

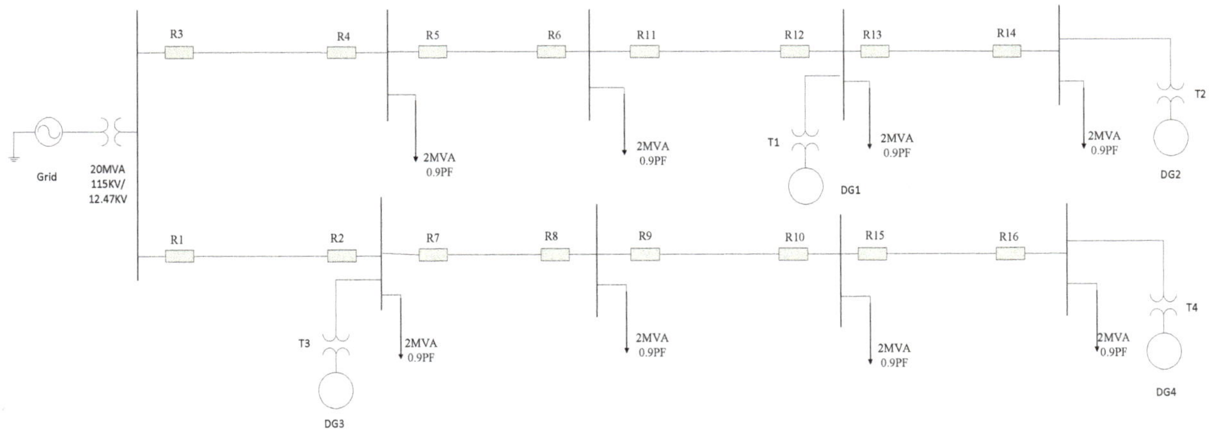

Figure 2. Canadian benchmark distribution system.

4.5 sec. But relay 11 is acting as backup relay and therefore grading margin of 0.3 sec is taken. Therefore required discrimination time of the relay is obtained by adding operating time of primary relay and grading margin. Therefore operating time of relay 11 is 0.44 sec. Hence required TDS is obtained by dividing required discrimination time and operating time obtained from inverse time characteristics. Thus the TMS of relay 11 is 0.097. Now if fault occurs between bus 3 and bus 4, relay 11 acts as primary relay for which the fault current is 2250 A and current setting is same *i.e.* 400 A. Therefore PSM is 5.625 and corresponding operating time from inverse time characteristic is 4 sec. But actual operating time is obtained by multiplying TMS of the relay and operating time obtained from the inverse time characteristics. Hence operating time of relay in this case is 0.388 sec. Following the same procedure TMS and operating time of relays can be calculated. This methodology has been used for solving the problem of protection coordination for the system configuration as following.

5.1. Grid Connected Mode of Operation

The 9-Bus electrical power distribution system operating in grid connected mode is shown in **Figure 2**. The prime objective of conducting this study is to investigate the impact of DG on fault current and how the coordination of OCRs gets altered. All the system parameters are explained in section IV of this paper.

From **Figure 3(a)** and **Figure 3(b)**, it can be observed that the level of current is higher as soon as the fault on feeder takes place. From the graph it is clear that the fault current through the feeder nearly become ten times the normal current. Due to this increased value of current the operating time of overcurrent relays should decrease compared to the condition when there is no fault in the system. **Table 1** shows the simulation results for the calculation of TDS and operating time for the various overcurrent relays connected to the distribution system.

Table 2 demonstrates the comparison of TDS of individual relays with and without optimization. For majority of relays available in the system, the optimized value of TDS is slightly lower than the un-optimized value. Thus the range of coordination for OCRs in the system increases and also the relay operating time will be more.

5.2. Islanded Mode of Operation

In this condition the overall distribution system is represented by two radial feeders which operate in isolation from the utility grid. The objective of conducting such study is to check the coordination of OCRs in islanded system under the influence of fault and also to find out the level of current which decides the time of operation of individual relays.

The islanded mode of operation of the system is represented as shown in **Figure 4**. From the diagram it is evident that in this condition the load demand is to be fulfilled by the DGs connected to the system. Islanded mode of operation is allowed when there is any fault on grid side and the connected DGs are capable enough to fulfill the load demand. The wave form of current in normal and faulty condition is shown in **Figure 5(a)** and **Figure 5(b)** respectively. The configuration and parameters of the distribution system are same as it is in case

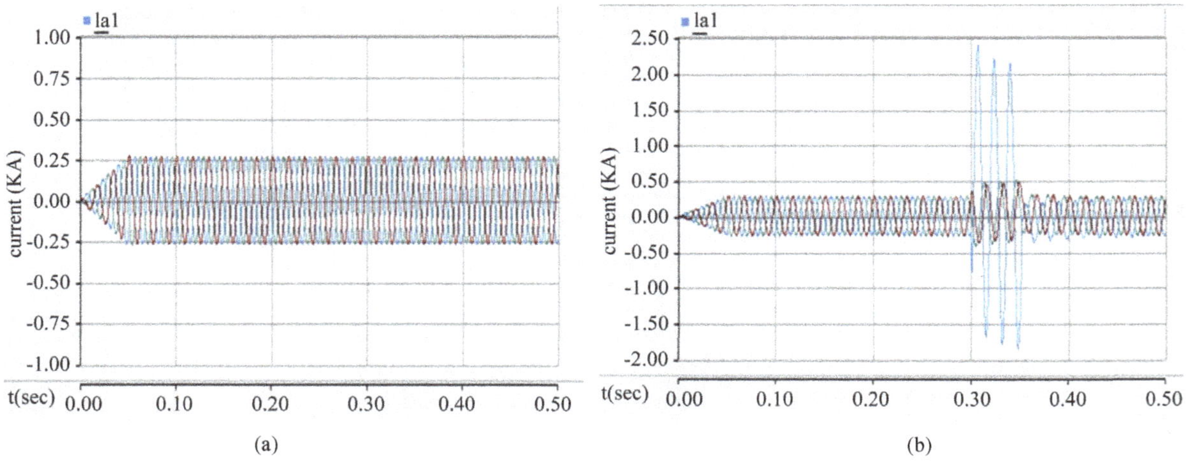

Figure 3. (a) Waveform of current in normal condition; (b) Waveform of current in faulty condition.

Table 1. Calculation of operating time and TDS in grid connected mode.

Relay No.	Operating Time and TDS		
	TDS	Near End Fault	Far End Fault
1	0.192	0.883	0.960
2	0.050	0.680	…
3	0.191	0.859	0.9468
4	0.050	1.000	…
5	0.132	0.646	0.688
6	0.144	0.993	1.300
7	0.150	0.660	0.707
8	0.032	0.544	0.980
9	0.110	0.407	0.440
10	0.084	0.633	0.844
11	0.097	0.388	0.440
12	0.182	1.019	1.293
13	0.050	0.140	…
14	0.231	1.155	1.319
15	0.050	0.140	…
16	0.173	0.761	0.933

Table 2. Optimized calculation of TDS and its comparison in grid connected mode.

Relay No.	Calculation of TDS by using GA and comparison with un-optimized value				
	TDS (without GA)	TDS (with GA)	Relay No.	TDS (without GA)	TDS (with GA)
1	0.192	0.1002	9	0.110	0.1300
2	0.050	0.0160	10	0.084	0.0270
3	0.191	0.1020	11	0.097	0.1180
4	0.050	0.0116	12	0.182	0.0340
5	0.132	0.0411	13	0.050	0.0751
6	0.144	0.0549	14	0.231	0.0826
7	0.150	0.0459	15	0.050	0.0729
8	0.032	0.0250	16	0.173	0.0954

of grid connected mode except that there is no utility grid and the system is completely operating in isolation.

Table 3 shows the simulation results for the calculation of TDS and operating time for the various overcurrent relays connected to the distribution system when operated in islanded mode. As compared to the grid connected mode as shown in **Table 1**, the values of TDS are larger in the islanded mode of operation. From the various simulation results it has been observed that the magnitude of short circuit current decreases in this mode of operation and hence the operating time of relay increases.

Table 4 demonstrates the comparison of TDS of individual relays with and without optimization in islanded mode of operation. Compared to the optimized values of TDS as obtained in case of grid connected mode which

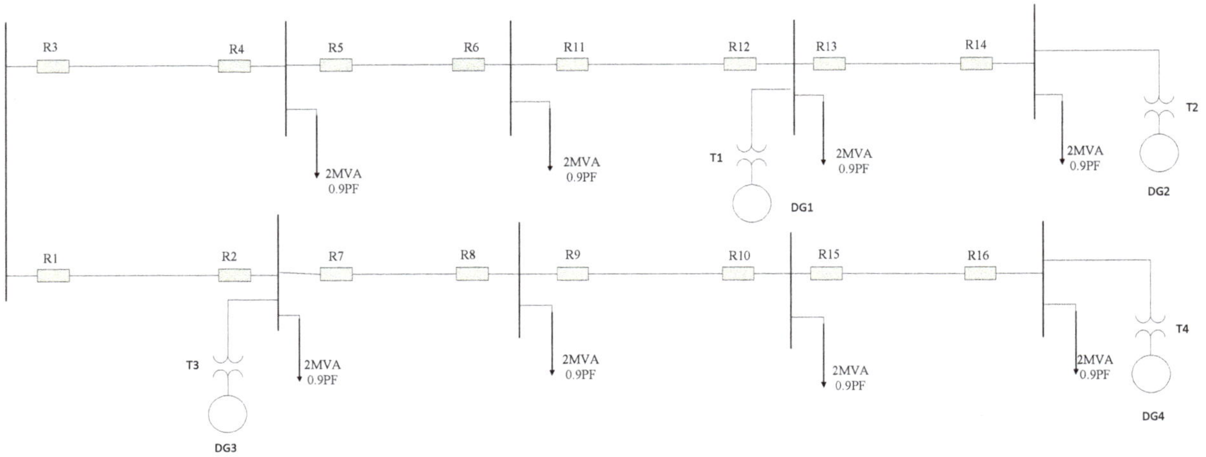

Figure 4. Canadian benchmark distribution system in islanded mode.

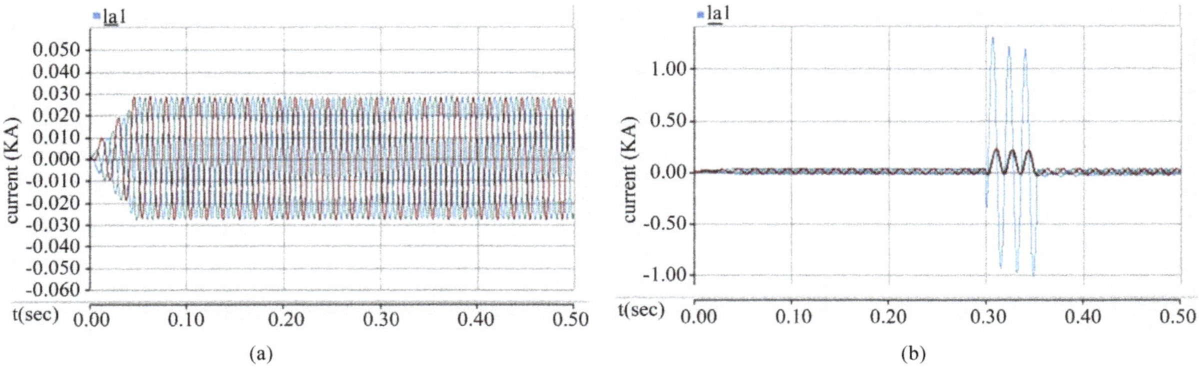

Figure 5. (a) Waveform of current in normal condition; (b) Waveform of current in faulty condition.

Table 3. Calculation of operating time and TDS in islanded mode of operation.

Relay No.	Operating Time and TDS		
	TDS	Near End Fault	Far End Fault
1	0.107	0.856	0.963
2	0.050	0.290	…
3	0.124	0.846	0.896
4	0.050	0.350	…
5	0.084	0.596	0.714
6	0.130	0.637	0.650
7	0.130	0.663	0.718
8	0.065	0.507	0.590
9	0.102	0.418	0.450
10	0.139	0.764	0.807
11	0.076	0.414	0.445
12	0.187	0.843	0.937
13	0.050	0.145	…
14	0.251	1.054	1.143
15	0.050	0.150	…
16	0.259	1.036	1.064

Table 4. Optimized calculation of TDS and its comparison in islanded mode of operation.

Relay No.	Calculation of TDS by using GA and comparison with un-optimized value				
	TDS (without GA)	TDS (with GA)	Relay No.	TDS (without GA)	TDS (with GA)
1	0.230	0.1189	9	0.115	0.1387
2	0.050	0.0147	10	0.079	0.0109
3	0.234	0.1189	11	0.115	0.1387
4	0.050	0.0147	12	0.050	0.0270
5	0.167	0.0480	13	0.050	0.0768
6	0.086	0.0450	14	0.051	0.0250
7	0.163	0.0480	15	0.050	0.0768
8	0.083	0.0431	16	0.063	0.0250

is shown in **Table 2**, the value of TDS increases. Thus the coordination margin for OCRs in the system increases and also the relays operating time will be more.

6. Conclusion

A comparative study for protection coordination of OCRs is presented for different network configuration in presence and absence of DG. The effect of DG penetration on the two main protection coordination problems *i.e.* blinding of protection and False/Sympathetic tripping is discussed in the first section of this paper which is supported with the simulation results. The novel idea used in this paper for the calculation of TDS is that both near and far end fault location has been considered. Whereas, in the second section, a comparative analysis for the calculation of TDS and operating time of relays has been presented. The range of protection coordination for the same system can be enhanced by using different types of fault current limiters (FCLs) at suitable location. In this paper, equivalent source based on DGs has been considered as a substitute of real distributed generators. Therefore, the effectiveness of the proposed method can be tested for electronically interfaced DERs (converter based DGs).

References

[1] So, C.W. and Lee. K.K. (2000) Overcurrent Relay Coordination by Evolutionary Programming. *Electric Power System Research*, **53**, 83-90. http://dx.doi.org/10.1016/S0378-7796(99)00052-8

[2] Chowdhury, A. and Koval, D. (2009) Power Distribution System Reliability: Practical Methods and Applications. Wiley-IEEE, Hoboken.

[3] Choudhary, N.K., Mohanty, S.R. and Singh, R.K. (2014) A Review on Microgrid Protection. 2014 *International Electrical Engineering Congress*, Chonburi, 19-21 March 2014, 1-4. http://dx.doi.org/10.1109/iEECON.2014.6925919

[4] Morren, J. and de Haan, S.W.H. (2008) Impact of Distributed Generation Units with Power Electronic Converters on Distribution Network Protection. *IET 9th International Conference on Developments in Power Systems Protection*, UK, 17-20 March 2008, 664-669. http://dx.doi.org/10.1049/cp:20080118

[5] Baran, M. and El-Barkabi, I. (2005) Fault Analysis on Distribution Feeders with Distributed Generation. *IEEE Transactions on Power Systems*, **20**, 1757-1764. http://dx.doi.org/10.1109/TPWRS.2005.857940

[6] Maki, K., Repo, S. and Jarventausta, P. (2008) Methods for Assessing the Protection Impacts of Distributed Generation in Network Planning Activities. *IET 9th International Conference on Developments in Power Systems Protection*, UK, March 2008, 484-489. http://dx.doi.org/10.1049/cp:20080085

[7] Kauhaniemi, K. and Knmpnlained, L. (2004) Impact of Distributed Generation on the Protection of Distribution Networks. 8*th IEE International Conference on Developments in Power System Protection*, **1**, 315-318. http://dx.doi.org/10.1049/cp:20040126

[8] Noghabi, A.S., Sadeh, J. and Mashhadi, H.R. (2009) Considering Different Network Topologies in Optimal Overcurrent Relay Coordination Using a Hybrid GA. *IEEE Transactions on Power Delivery*, **24**, 1857-1863. http://dx.doi.org/10.1109/TPWRD.2009.2029057

[9] Urdaneta, A.J., Ramon, N. and Jimenez, L.G.P. (1988) Optimal Coordination of Directional Relays in Interconnected Power System. *IEEE Transactions on Power Delivery*, **3**, 903-911. http://dx.doi.org/10.1109/61.193867

[10] Zeienldin, H., El-Saadany, E.F. and Salama, M.A. (2004) A Novel Problem Formulation for Directional Overcurrent Relay Coordination. 2004 *Large Engineering systems Conference on Power Engineering*, Halifax, 28-30 July 2004, 48-52. http://dx.doi.org/10.1109/lescpe.2004.1356265

[11] Birla, D., Maheshwari, R.P. and Gupta, H.O. (2006) A New Nonlinear Directional Overcurrent Relay Coordination Technique, and Banes and Boons of Near-End Faults Based Approach. *IEEE Transactions on Power Delivery*, **21**, 1176-1182. http://dx.doi.org/10.1109/TPWRD.2005.861325

[12] Mansour, M., Mekhamer, S. and El-Kharbawe, N.-S. (2007) A Modified Particle Swarm Optimizer for the Coordination of Directional Overcurrent Relays. *IEEE Transactions on Power Delivery*, **22**, 1400-1410. http://dx.doi.org/10.1109/TPWRD.2007.899259

[13] Bedekar, P.P. and Bhide, S.R. (2011) Optimum Coordination of Directional Overcurrent Relays Using the Hybrid GA-NLP Approach. *IEEE Transactions on Power Delivery*, **26**, 109-119. http://dx.doi.org/10.1109/TPWRD.2010.2080289

[14] Barzegari, M., Bathaee, S. and Alizadeh, M. (2010) Optimal Coordination of Directional Overcurrent Relays Using Harmony Search Algorithm. *9th International Conference on Environment and Electrical Engineering*, Prague, 16-19 May 2010, 321-324. http://dx.doi.org/10.1109/eeeic.2010.5489935

[15] Al-Nasseri, H., Redfern, M.A. and Li, F. (2006) A Voltage Based Protection for Micro-Grids Containing Power Electronic Converters. *IEEE Power Engineering Society General Meeting*, Montreal, 1-7.

[16] Miveh, M.R., Gandomkar, M., Mirsaeidi, S. and Nuri, M. (2011) Analysis of Single Line to Ground Fault Based on Zero Sequence Current in Microgrids. *ISCEE Conference*, Kermanshah.

[17] Hung, D.Q. and Mithulananthan, N. (2013) Multiple Distributed Generators Placement in Primary Distribution Networks for Loss Reduction. *IEEE Transactions on Industrial Electronics*, **60**, 1700-1708. http://dx.doi.org/10.1109/TIE.2011.2112316

[18] IEEE Recommended Practice for Protection and Coordination of Industrial and Commercial Power Systems. IEEE Std., 242-1986.

[19] Sortomme, E., Venkata, S. and Mitra, J. (2010) Microgrid Protection Using Communication-Assisted Digital Relays. *IEEE Transactions on Power Delivery*, **25**, 2789-2796. http://dx.doi.org/10.1109/TPWRD.2009.2035810

[20] Najy, W.K.A., Zeineldin, H.H. and Woon, W.L. (2013) Optimal Protection Coordination for Microgrids with Grid Connected and Islanded Capability. *IEEE Transactions on Industrial Electronics*, **60**, 1668-1677. http://dx.doi.org/10.1109/TIE.2012.2192893

8

Feasibility Study of a Hydro PV Hybrid System Operating at a Dam for Water Supply in Southern Brazil

Luis E. Teixeira[1], Johan Caux[1,2], Alexandre Beluco[1*], Ivo Bertoldo[3], José Antônio S. Louzada[1], Ricardo C. Eifler[4]

[1]Instituto de Pesquisas Hidráulicas, Universidade Federal do Rio Grande do Sul, Porto Alegre, Brazil
[2]Ecole Nationale Superieure de l'Energie, l'Eau et l'Environnement, Grenoble INP, Grenoble, France
[3]Companhia Riograndense de Saneamento, Santa Maria, Brazil
[4]Companhia Estadual de Energia Elétrica, Salto do Jacuí, Brazil
Email: luis.let@gmail.com, johan.caux@hotmail.com, *albeluco@iph.ufrgs.br, ivo.bertoldo@corsan.com.br, louzada@iph.ufrgs.br, ricardoce@ceee.com.br

Abstract

Dams for water supply usually represent an untapped hydroelectric potential. It is a small energetic potential, in most situations, usually requiring a particular solution to be viable. The use of pumps as power turbines often represents an alternative that enables the power generation in hydraulic structures already in operation, as is the case of dams in water supply systems. This potential can be exploited in conjunction with the implementation of PV modules on the water surface, installed on floating structures, both operating in a hydro PV hybrid system. The floating structure can also contribute to reducing the evaporation of water and providing a small increase in hydroelectric power available. This paper presents a pre-feasibility study for implementation of a hydroelectric power plant and PV modules on floating structures in the reservoir formed by the dam of Val de Serra, in southern Brazil. The dam is operated to provide drinking water to about 60% of the population of the city of Santa Maria, in the state of Rio Grande do Sul, in southern Brazil. The pre-feasibility study conducted with Homer software, version Legacy, indicated that the hydroelectric plant with a capacity of 227 kW can operate together with 60 kW of PV modules. This combination will result (in one of the configurations considered) in an initial cost of USD$ 1715.83 per kW installed and a cost of energy of USD$ 0.059/kWh.

Keywords

Hydro PV Hybrid System, PV Floating Cover, Water Supply Dam, Software Homer

*Corresponding author.

1. Introduction

Brazil is experiencing a time of uncertainty about the ability of the interconnected system to meet the demands of consumers for electricity. On the one hand, the pressures exerted by the economy, with an advertised and encouraged growth but not yet materialized in fact in all its possibilities. On the other hand, investments in the expansion and maintenance of installed capacity, faced with a scenario of little flexibility in achieving the necessary projects.

In this scenario, it is important to ensure a reasonable increase of installed capacity to the grid, especially when potential that are available and not used can be added to the system with simplified and fast work can be identified. Dams for water supply have reasonable storage capacity and are already in operation, making feasible the necessary investments for the implementation of micro and small scale hydropower plants.

The area flooded by the dam can also be utilized for the installation of a photovoltaic power plant. In many sites, the shading provided by the photovoltaic modules is seen as a negative point. In a dam for water supply, installing the photovoltaic modules would result in the reduction of water evaporation and a possible increase in the amount of water available for power generation, depending on local conditions.

A hybrid system constituted in this way, a hydro PV hybrid system, might seem unfeasible. A hydroelectric power plant implemented in an existing dam may represent a set of unsurpassable obstacles. The use of photovoltaic modules always seems to bump into excessively high initial costs. However, such a hybrid system has been studied for some time and these apparent difficulties can at least partially be overcome.

It's been over a decade; [1] described the ten years of a PV micro hydro hybrid system in Indonesia. Reference [2] also considered hydro PV hybrid systems for off grid rural electrification of Ethiopia. Reference [3] evaluated the feasibility of PV pico hydro systems for electrification in Cameroon. Reference [4] analyzed with Homer a PV micro hydro system that can be considered as a reference also for electrification of Cameroon.

Reference [5] does not consider hydro PV hybrid systems, but presents an interesting comparison between hydroelectric and PV ystems. Reference [6] evaluates the hydrogen production integrated with a hydro PV hybrid system in micro scale. Margeta and Glasnovic [7] evaluate a concept of hydro PV hybrid system in which photovoltaic modules create temporary flow rate for a hydroelectric power plant.

Referring to hydro PV hybrid systems, [8] present some general comments and evaluate some effects of possible complementarity between solar and hydroelectric resources. Subsequent works, [9] [10], explore the influence of energetic complementarity in time on the performance of hydro photovoltaic hybrid systems. A site with good complementarity in time between energy resources can lead to a hybrid system with less installed power and consequently lower costs.

Reference [11] evaluates the use of a PV system and a hydroelectric plant with reservoir to raise the capacity factor of the PV system. This idea has been explored in several research projects. Margeta and Glasnovic [12] evaluate the applicability of a concept in which photovoltaics and hydroelectric systems would be operated as hydro PV hybrid systems throughout Europe, based on energy storage in the hydraulic way.

Following this line of reasoning, leading to the natural growth of the use of photovoltaic modules, [13] evaluate the limit of solar penetration in interconnected systems. Reference [14] analyzes the growth of production of photovoltaic modules and presents interesting comments on how this growth should be accompanied with the appropriate incentives and regulations.

Glasnovic *et al.* [15] proposed a hydroelectric solar thermal hybrid system intending to overcome seasonal minimum energy availability, but would contribute mainly to raise the capacity factor of the solar thermal system. Glasnovic and Margeta [16] proposed a systematic alternative to obtaining energy supplies from systems with energy storage in hydroelectric reversible plants, in joint operation with other renewable resources.

This paper presents a pre-feasibility study of a hydro PV hybrid system to be installed at a dam for water supply in southern Brazil. The focus of the paper is to identify conditions mainly relating to components of photovoltaic installment, for which the system becomes feasible. The dam considered in the study is the dam of Val de Serra, in Santa Maria, in southern Brazil. This dam will be described in the next section.

In addition, the paper considers the installation of the photovoltaic modules in floating devices, possibly reducing evaporation and possibly increasing the amount of water available for generating power. The sections on power generation at a dam for water supply and the proposed hybrid system, below, discusses in more detail the adaptation of these hydraulic structures for power generation and the use of photovoltaic modules on floating structures.

2. Method of Analysis

The focus of this work is a dam used for water supply. This dam has been used only for water supply and the starting point of the work is the design of a power plant for harnessing the hydroelectric potential. The next section describes this dam and the following section discusses the power generation in water supply dams.

The feasibility study of a hydro PV hybrid system to be installed in this dam will be made with the use of the Homer software, Legacy version. This software can be learned with great ease and short time periods, with universal access at no cost. Section 5 presents the proposed system and Section 6 discusses the simulations with Homer.

It is worth noting that Homer software presents a space with optimal solutions to the problem being studied. However, a strong reason for its use is the way Homer performs the calculations and displays the results. Thus, the final solution may be chosen from the optimal solutions, and also considering the solutions that have not been selected as optimal solutions.

Finally, Section 7 then presents and discusses the results and section 8 reports the conclusions.

3. The Val de Serra Dam

The dam Rodolfo Costa e Silva, known as Val de Serra, was inaugurated on December 17, 1999. The reservoir is operated by the Companhia Riograndense de Saneamento, CORSAN, and is located on the boundary between the municipalities of Itaara [17] and São Martinho da Serra [18], in the center of the State of Rio Grande do Sul [19], the southernmost state of Brazil. The reservoir is responsible for supplying 60% of consumers in the urban area of the city of Santa Maria [20].

The reservoir covers an area [21] of 275 hectares and is located between the geographical coordinates: 29°29'01" to 29°30'56" South and 53°43'32" to 53°45'29" West (**Figure 1**). The dam [22] is built with roller compacted concrete, with a maximum height of 36.5 meters, maximum height above the river bed of 34.0 meters, crest length of 684 meters and energy dissipation with stepped spillway. **Figure 2** shows the structure of energy dissipation.

Figure 3 shows a view of the lake formed by the dam. The drainage area contributing to the reservoir is 49.4 km^2 and total useful volume is 23 million cubic meters. The reservoir is in the area of influence of a subtropical climate with an average annual temperature of 22°C, with maximum temperatures exceeding 30°C and minimum temperatures below 5°C. November is the least rainy month and June, September and October are the rainiest months.

The mean flow is regulated by a booster valve type Howell Bunger [23], with a nominal diameter of 600 mm. This valve has a fixed aperture to maintain constant flow equal to 1.25 m^3/s. The water level in the dam will vary over the year as that shown in **Figure 4**. These data were supplied by the utility company and were obtained with sensors installed on site with data acquisition every 5 minutes.

Figure 1. Image of the reservoir formed by the dam Val de Serra.

Figure 2. View of the structure of energy dissipation.

Figure 3. Partial view of the dam and lake.

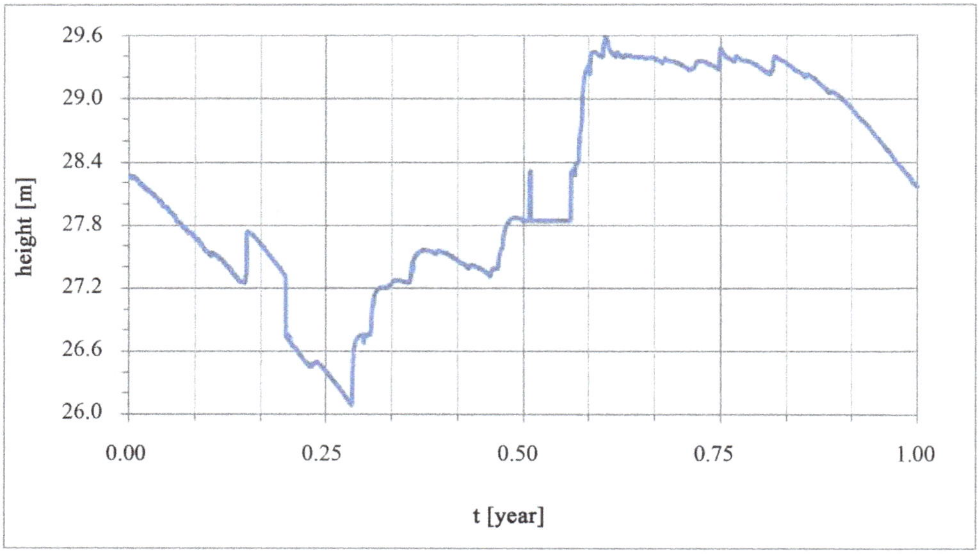

Figure 4. Levels in the dam Val de Serra during the year 2011.

4. Power Generation at a Dam for Water Supply

The water supply dams are usually designed with the sole purpose of reservation of water, mainly because of the usually small hydroelectric potential. Thus, the hydraulic structures are often designed to vent the necessary amounts of water, showing no appropriate structures for water adduction, machine room and returning water to the river. Moreover, they are operated in a simpler way, maintaining a constant flow, unlike hydroelectric plants.

The use of these structures in order to obtain energy supplies requires a solution with technical and economic commitment. The design of a conventional power plant will surely lead to infeasibility. It is necessary to devise an approach based on opportunity cost solution. That is why the use of pumps as turbines shown as the best solution for its simplicity and cost advantage over conventional turbines, without significant reduction in efficiency.

The adaptation of supply dams for power generation then require a change in design of water adduction, enabling installation of energy conversion machines where before there was an outlet valve. Moreover, it is necessary to design the installation of hydraulic machines and assembly of power conversion equipment, a living engine, a hydraulic structure that receives the water after its passage through hydraulic machine and then route the water to the riverbed.

The use of pumps as power turbines are already studied for several decades [24] [25] and is always considered as an alternative to electrification with lower costs [26] [27]. An idea that arose from the use of pumps as turbines for pressure recovery, still a subject of research [28], and that follows awakening interest for being an unconventional solution [29] [30].

5. The Proposed Hydro PV Hybrid System

Val de Serra has an untapped potential of about 260 kW, hardly feasible with a conventional hydroelectric project but feasible with a design based on the use of pumps as turbines. The area flooded by the dam would suggest the use of PV modules to compose a hybrid system. The modules should be installed on floating device and its use contribute to reducing the evaporation of water from the reservoir. The hybrid system thus obtained is shown schematically in **Figure 5**.

The hydro power plant to be installed will have three centrifugal pumps for operation at 16.2 m, 1125 m^3/h and 880 rpm, used as power turbines [31]. The engine room will be located at the base of the dam. The hydraulic machines will be installed at the end of the pipe that currently leads to the Howell Bunger water valve (cited above). These pumps will be connected by a shaft to electrical machines, which will be connected to the grid.

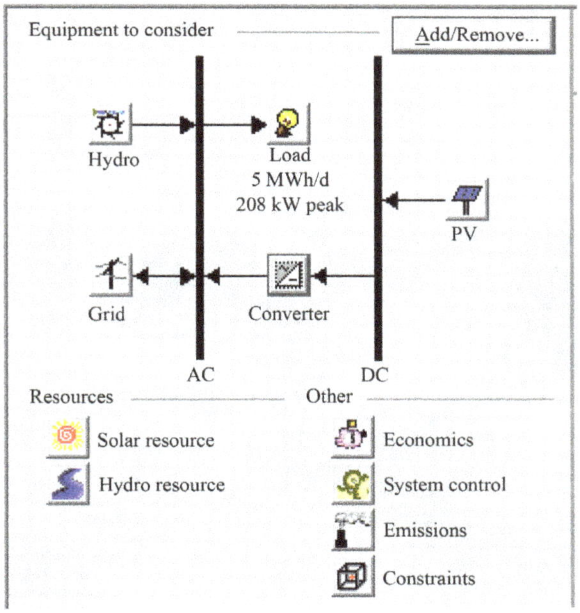

Figure 5. Schematic representation of hydro PV hybrid system to be adopted in Val de Serra.

The PV module assembly will be installed on floating structures, as recently suggested by [32] [33]. They propose and test a system with polyethylene floating modules that occupy an area that would not be used in a better way and that also contribute reducing evaporation. This study has not yet detailed the floating structures, having been restricted so far to the question of economic feasibility.

The reservoir has a relatively large area and full coverage would lead to a photovoltaic generator with an installed capacity of just over 100 MW. As the hydroelectric potential is much smaller, it was decided to consider in this study photovoltaic panels that add power comparable to the power of the proposed hydroelectric power plant. Thus, floating structures with installed capacity of 60 kW were considered in the simulations.

In addition to these components, there is still a connection to the grid and the converter. The connection to the grid is used together with the consumer load to simulate the power supply to be inserted in the interconnected system. The converter operates only as an inverter, allowing current flow from the dc bus to the ac bus. Energy storage was not considered in this study. The converter operates with an efficiency of 92% with a expected lifespan of 10 years.

6. Simulations with Homer

Homer [34] is a software for optimization of hybrid systems on micro and small scale. It was originally developed by National Renewable Energy Laboratory (NREL) and is available for universal access in its Legacy version. Homer simulates a system for power generation over the time period of 25 years at intervals of 60 minutes, presenting the results for a period of one year [35] [36].

PV modules with costs between US$ 4000/kW and USD$ 2000/kW were considered. The installation of floating structures, as suggested by [32] [33], raise the cost by 30%. **Figure 6** shows the solar radiation incident

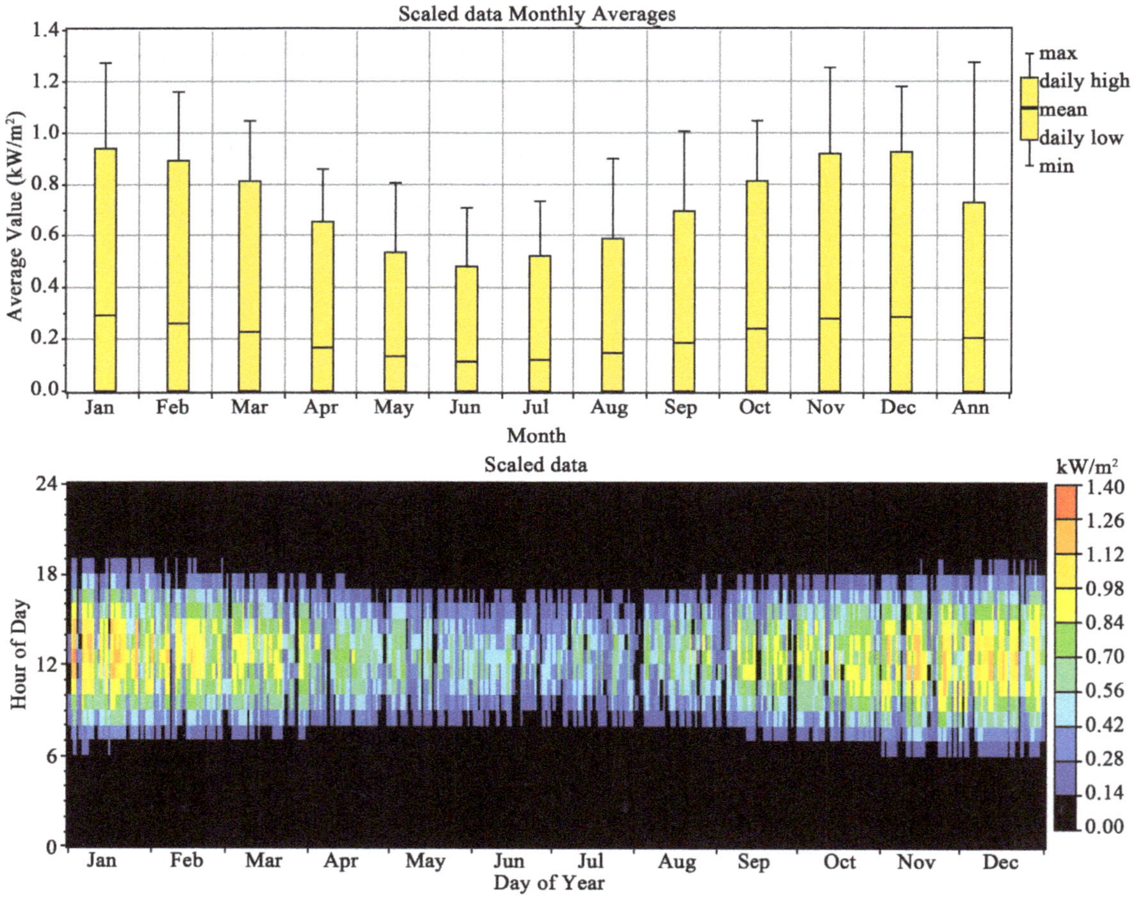

Figure 6. Solar radiation incident on a horizontal plane in Val de Serra.

on a horizontal plane on the site of Val de Serra. The first figure shows the minimum, average and maximum monthly values. The second figure shows the hourly distribution of the values throughout the year.

The operation of Val de Serra as a plant is a little different from a normal plant: height undergoes variation as the flow rate remains approximately constant. In fact, the flow also varies, as it consists in a reservoir being emptied, but this variation is small and was neglected. The hydroelectric plant will then have a flow of 1.25 m^3/s and a variable height (**Figure 4**). Thus, for simulation purposes, it was considered a height of 29.6 m and flow rates were then calculated again. **Figure 7** then shows the values used for the flow rate.

The data shown in **Figure 6** and **Figure 7** suggest a complementarity in time equal to 0.16 between hydro-power and solar energy. The map presented by [37] suggests a much greater complementarity, between 0.8 and 1.0. This difference is due to the consumption profile of the population served with the water stored in the reservoir. This case study clearly shows the influence of a reservoir that has no duties with power generation on energetic complementarity.

The design flow for sizing equipment was 1118.25 liters per second, the flow rate with 95% permanency, with equipment having efficiency of 80%. Implementation costs of the hydroelectric power plant were estimated at US$ 240,000. The life time was estimated at 25 years. The operation and maintenance as well as replacement costs of a plant equipped with pumps as turbines are approximately equal to those of a traditional plant.

The installation of floating structures alter evaporation of the water stored in the reservoir. Evaluation of evaporation was performed by traditional methods [38], with the search for time series data obtained by any instrument near the reservoir region. **Figure 8** shows the annual series of evaporation obtained for this work [39]. In the case of PV modules on 100% of the flooded area, values of **Figure 8** should be added cumulatively to the data of **Figure 7**. The results would be very similar to the values of **Figure 7**, with about only 3% more energy.

Simulations with the system of **Figure 5**, with the PV modules assembled on floating structures installed over the flooded surface of the reservoir, were performed. A set of 2,160,000 simulations were performed, with 300 values for optimization variables and 1440 values for sensitivity variables, repeated five times. Approximately 35% of the simulations result in feasible solutions or unfeasible solutions for technical issues; the other solutions correspond to non-viable combinations of the variables considered in the simulations.

The optimization variables considered were the following: 0 kW, 60 kW, 120 kW, 180 kW and 240 kW for PV array capacity; 0 kW, 60 kW, 120 kW, 180 kW and 240 kW for converter capacity and 0 kW, 60 kW, 120 kW, 180 kW, 240 kW and 300 kW for grid sales capacity.

The sensitivity inputs were the following: 6000 kWh/d, 6500 kWh/d, 7000 kWh/d, 7500 kWh/d and 8000 kWh/d for AC Load; 0 kW, 30 kW, 60 kW or 90 kW for grid capacity; 0.0%, 2.0%, 4.0%, 6.0%, 8.0% and 10.0% for maximum annual capacity shortage.

The sensitivity inputs also included USD$ 0.160/kWh, USD$ 0.240/kWh, USD$ 0.360/kWh for off peak energy price and USD$ 0.080/kWh, USD$ 0.120/kWh, USD$ 0.180/kWh for off peak energy sellback price, these two linked; 0.85, 1.00, 1.15, 1.30 for PV capital cost multiplier, for PV replacement cost multiplier and

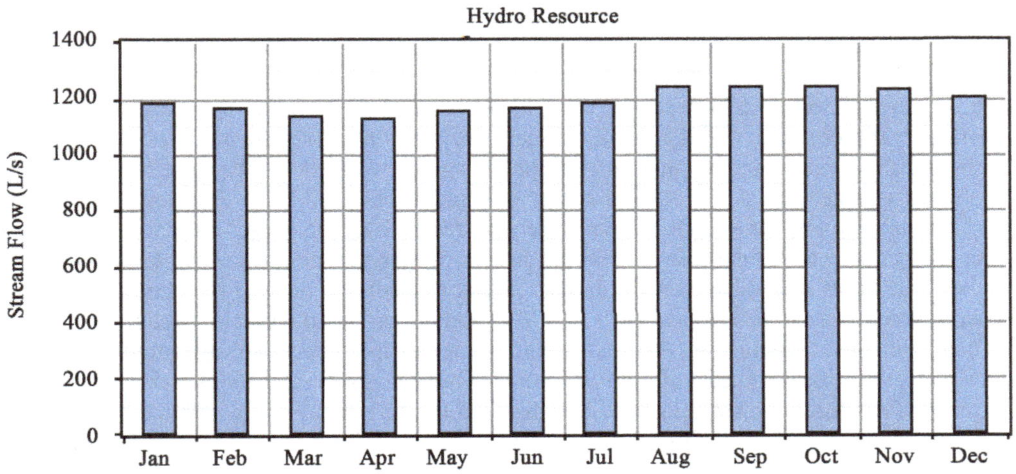

Figure 7. Monthly flow values for a height of 29.6 m equivalent to height variation (shown in **Figure 4**) for a flow of 1.25 m^3/s.

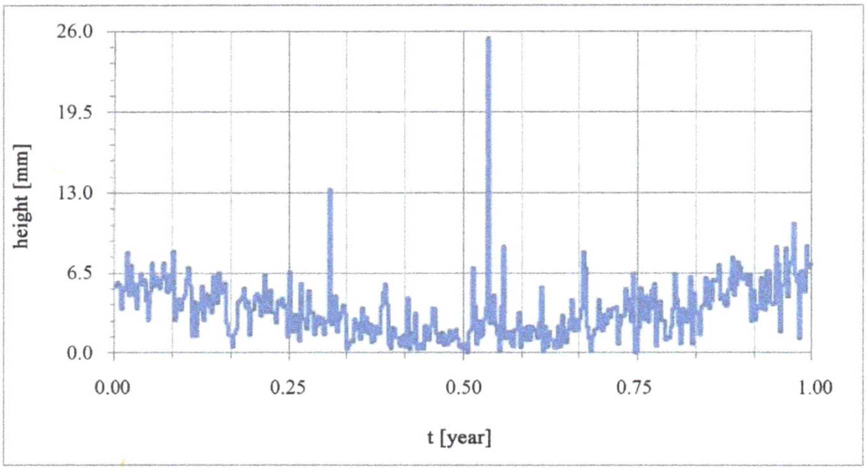

Figure 8. Annual daily values of evaporation obtained for the location of the dam considered in this work [39].

operation and maintenance cost multiplier, these three linked.

All these calculations were repeated for the values of USD$ 2000.00/kW, USD$ 2500.00/kW, USD$ 3000.00/kW, USD$ 3500.00/kW and USD$ 4000.00/kW of PV modules.

7. Results and Discussion

First, the optimization results provided by Homer should be analyzed. The simulations for lower energy costs obtained from the grid and higher PV modules acquisition costs always result solely in solutions composed of hydroelectric power plant and connection to the grid. The results including photovoltaic modules appear only for PV modules at USD$ 2500.00/kW and USD$ 2000.00/kW and at USD$ 0.24/kWh and USD$ 0.32/kWh. In a strict point of view, the optimal solutions provided by Homer does not include PV modules.

Figure 9 shows the optimization space for the PV capital cost multiplier as a function of the load, for the system of **Figure 5** without selling surplus energy to the grid, with power dispatch so there are no failures in the supply of energy to consumers, with PV modules acquired by USD$ 2500.00/kW. The results indicate two combinations of components: connection to the grid and hydroelectric plant with or without photovoltaic modules. The lowest demand and highest PV capital cost multiplier are met with hydro and grid while larger demands are met with hydro PV hybrid system, also connected to the grid.

Figure 10 and **Figure 11** show the optimization space for the PV capital cost multiplier as a function of the load for the same system without selling surplus energy to the grid, with power dispatch so there are no failures in the supply of energy to consumers, with photovoltaic modules acquired by USD$ 2000.00/kW. **Figure 12** shows the optimization space for the photovoltaic modules capital cost multiplier as a function of the load for this same system with excess energy sold to the grid. Along these figures, the same combinations of components appear with a clear increase in the number of solutions including photovoltaic modules.

The range of the photovoltaic capital cost multiplier values were chosen because the installation of floating structures represents an increase of 30% in costs (as suggested by [32] [33]). The results point out a change in slope of the separation line between the two different solutions in the optimization space.

It is obviously easier to reduce costs of floating structures than PV modules costs. This change in slope occurs close to the value of 30% corresponding to the increase in costs due to floating structures, suggesting a reduction in these structures costs can be decisive to include PV modules among the viable solutions.

This change in slope is defined by the relationship between capital costs and operating and maintenance costs. This change clearly presses the cost of the PV modules downward so that the system can be feasible. The smaller the number of PV modules, the greater will be the need for purchasing energy from the grid.

Figure 13 shows the system of **Figure 12** simulated with 8% failure in the power supply. In this case, the optimization space finally presents solutions including PV modules for the PV capital cost multiplier equal to 1.3. The solutions including PV modules appear nearest to the load of 7000 kWh/day, but the optimal solutions at the top of the chart show costs very close to the costs of the "non optimal" solutions, suggesting its feasibility.

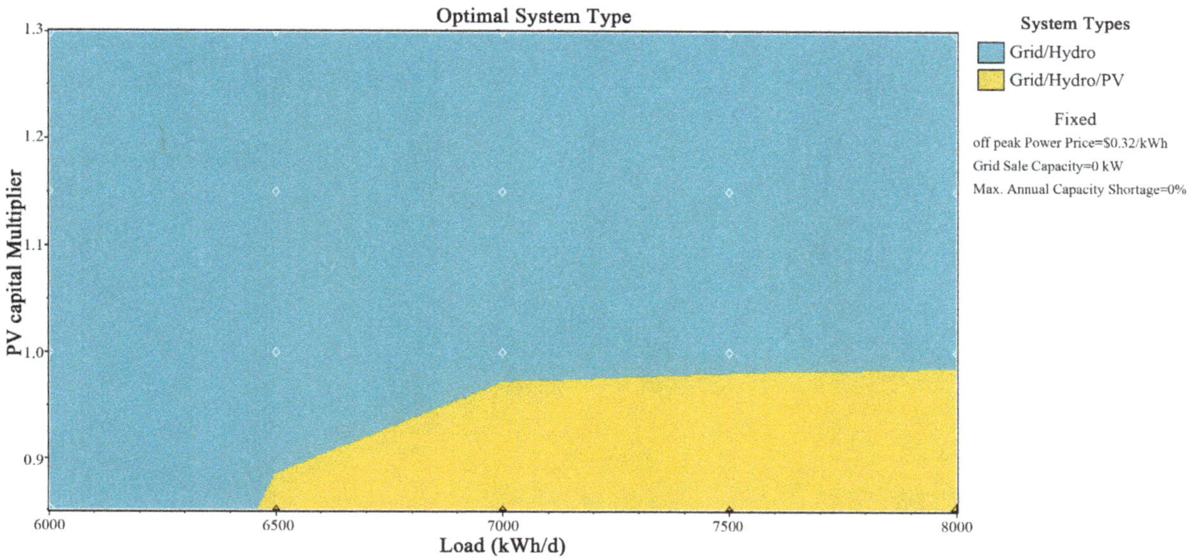

Figure 9. Optimization space for the system of **Figure 5**, with PV modules acquired by USD$ 2500/kW.

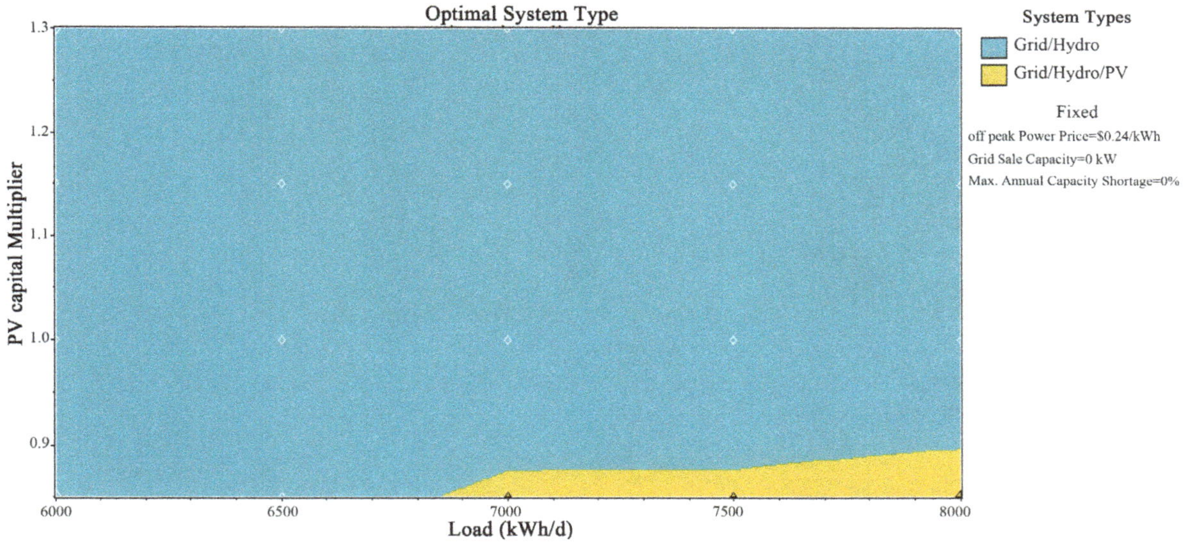

Figure 10. Optimization space for the system of **Figure 5**, with PV modules acquired by US$ 2000/kW.

In a second stage of analysis, "not so optimal" but still possibly appropriate solutions can be found among the other results provided by Homer. **Figure 14** shows the results provided by Homer to 7000 kWh per day, PV capital cost multiplier equal to 1.3, USD$ 0.32 per kWh from the grid and a grid sale capacity of 90 kW, corresponding to the PV capital costs equal to, from top to bottom, USD$ 2000/kW, USD$ 2500/kW, USD$ 3000/kW, USD$ 3500/kW and USD$ 4000/kW.

Solutions in the second lines in all five excerpts from Homer results screen, shown in **Figure 14**, indicate a combination of hydroelectric power plant of 227 kW and a photovoltaic generator with 60 kW. These solutions showed cost of energy with values ranging between USD$ 0.080/kWh and USD$ 0.090/kWh with a renewable fraction equal to 0.84.

Figure 15 shows initial costs, above, and cost of energy, below, as a function of PV capital cost for this system with 227 kW hydropower and 60 kW photovoltaic. Below, there is curves for power acquired from the grid at USD$ 0.32/kWh and USD$ 0.16/kWh. For power purchased from the grid at USD$ 0.16/kWh and PV modules at USD$ 3000/kW, closer to actual values, the initial cost will be USD$ 492,444 (or USD$ 1715.83/kW)

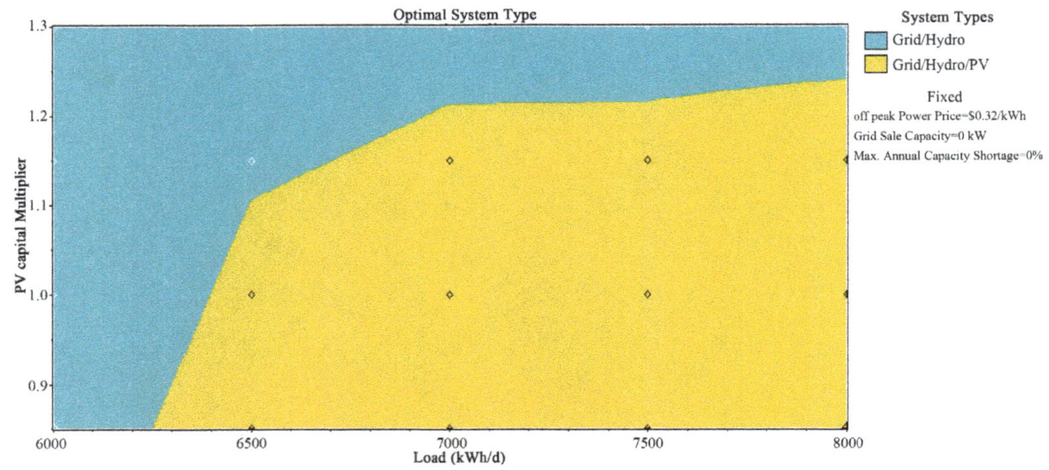

Figure 11. Optimization space for the system of **Figure 5**, with PV modules acquired by USD$ 2000/kW.

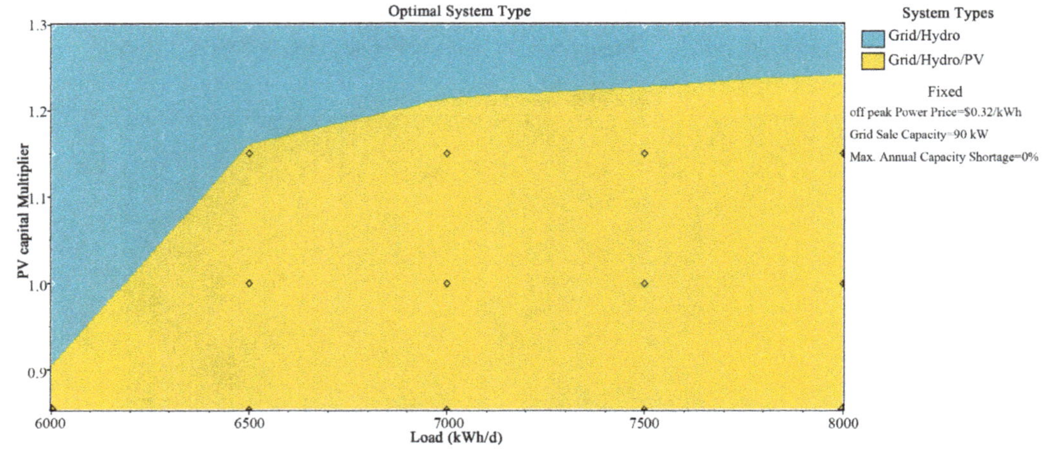

Figure 12. Optimization space for the system of **Figure 5**, with PV modules acquired by USD$ 2000/kW, with excess energy sold to the grid.

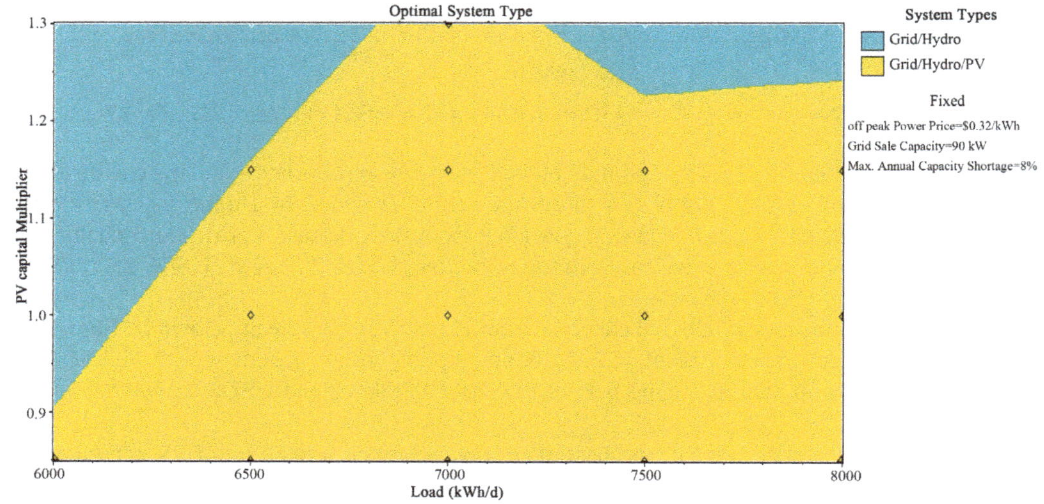

Figure 13. Optimization space for the system of **Figure 5**, with PV modules acquired by USD$ 2000/kW, with excess energy sold to the grid, with 8% failure in the power supply.

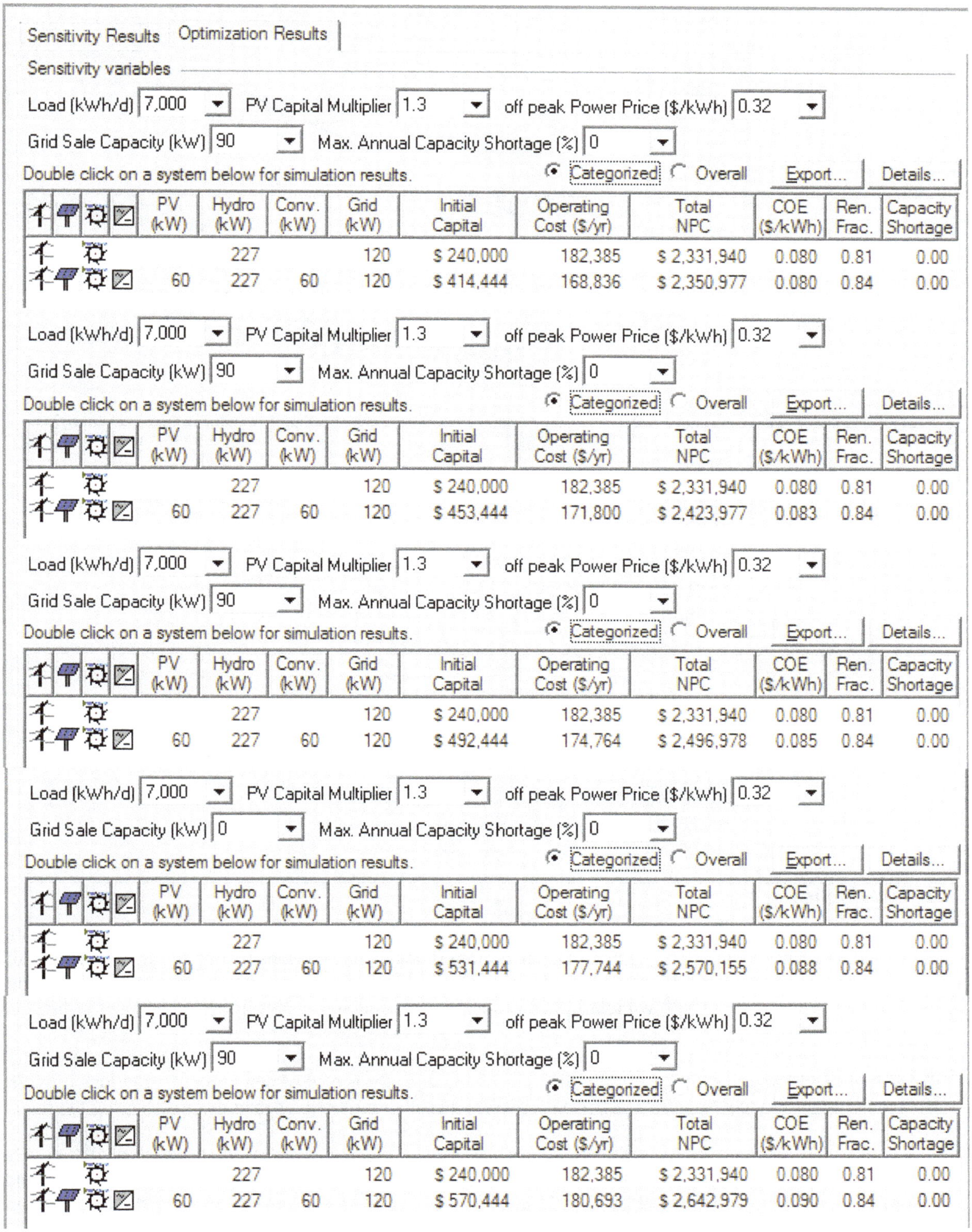

Sensitivity Results   Optimization Results

Sensitivity variables

Load (kWh/d) 7,000 ▾   PV Capital Multiplier 1.3 ▾   off peak Power Price ($/kWh) 0.32 ▾

Grid Sale Capacity (kW) 90 ▾   Max. Annual Capacity Shortage (%) 0 ▾

Double click on a system below for simulation results.   ⦿ Categorized ○ Overall   Export...   Details...

	PV (kW)	Hydro (kW)	Conv. (kW)	Grid (kW)	Initial Capital	Operating Cost ($/yr)	Total NPC	COE ($/kWh)	Ren. Frac.	Capacity Shortage
		227		120	$ 240,000	182,385	$ 2,331,940	0.080	0.81	0.00
	60	227	60	120	$ 414,444	168,836	$ 2,350,977	0.080	0.84	0.00

Load (kWh/d) 7,000 ▾   PV Capital Multiplier 1.3 ▾   off peak Power Price ($/kWh) 0.32 ▾

Grid Sale Capacity (kW) 90 ▾   Max. Annual Capacity Shortage (%) 0 ▾

Double click on a system below for simulation results.   ⦿ Categorized ○ Overall   Export...   Details...

	PV (kW)	Hydro (kW)	Conv. (kW)	Grid (kW)	Initial Capital	Operating Cost ($/yr)	Total NPC	COE ($/kWh)	Ren. Frac.	Capacity Shortage
		227		120	$ 240,000	182,385	$ 2,331,940	0.080	0.81	0.00
	60	227	60	120	$ 453,444	171,800	$ 2,423,977	0.083	0.84	0.00

Load (kWh/d) 7,000 ▾   PV Capital Multiplier 1.3 ▾   off peak Power Price ($/kWh) 0.32 ▾

Grid Sale Capacity (kW) 90 ▾   Max. Annual Capacity Shortage (%) 0 ▾

Double click on a system below for simulation results.   ⦿ Categorized ○ Overall   Export...   Details...

	PV (kW)	Hydro (kW)	Conv. (kW)	Grid (kW)	Initial Capital	Operating Cost ($/yr)	Total NPC	COE ($/kWh)	Ren. Frac.	Capacity Shortage
		227		120	$ 240,000	182,385	$ 2,331,940	0.080	0.81	0.00
	60	227	60	120	$ 492,444	174,764	$ 2,496,978	0.085	0.84	0.00

Load (kWh/d) 7,000 ▾   PV Capital Multiplier 1.3 ▾   off peak Power Price ($/kWh) 0.32 ▾

Grid Sale Capacity (kW) 0 ▾   Max. Annual Capacity Shortage (%) 0 ▾

Double click on a system below for simulation results.   ⦿ Categorized ○ Overall   Export...   Details...

	PV (kW)	Hydro (kW)	Conv. (kW)	Grid (kW)	Initial Capital	Operating Cost ($/yr)	Total NPC	COE ($/kWh)	Ren. Frac.	Capacity Shortage
		227		120	$ 240,000	182,385	$ 2,331,940	0.080	0.81	0.00
	60	227	60	120	$ 531,444	177,744	$ 2,570,155	0.088	0.84	0.00

Load (kWh/d) 7,000 ▾   PV Capital Multiplier 1.3 ▾   off peak Power Price ($/kWh) 0.32 ▾

Grid Sale Capacity (kW) 90 ▾   Max. Annual Capacity Shortage (%) 0 ▾

Double click on a system below for simulation results.   ⦿ Categorized ○ Overall   Export...   Details...

	PV (kW)	Hydro (kW)	Conv. (kW)	Grid (kW)	Initial Capital	Operating Cost ($/yr)	Total NPC	COE ($/kWh)	Ren. Frac.	Capacity Shortage
		227		120	$ 240,000	182,385	$ 2,331,940	0.080	0.81	0.00
	60	227	60	120	$ 570,444	180,693	$ 2,642,979	0.090	0.84	0.00

Figure 14. Results provided by Homer corresponding to the PV capital costs equal to, from top to bottom, USD$ 2000/kW, USD$ 2500/kW, USD$ 3000/kW, USD$ 3500/kW and USD$ 4000/kW.

and the cost of energy will be USD$ 0.059/kWh.

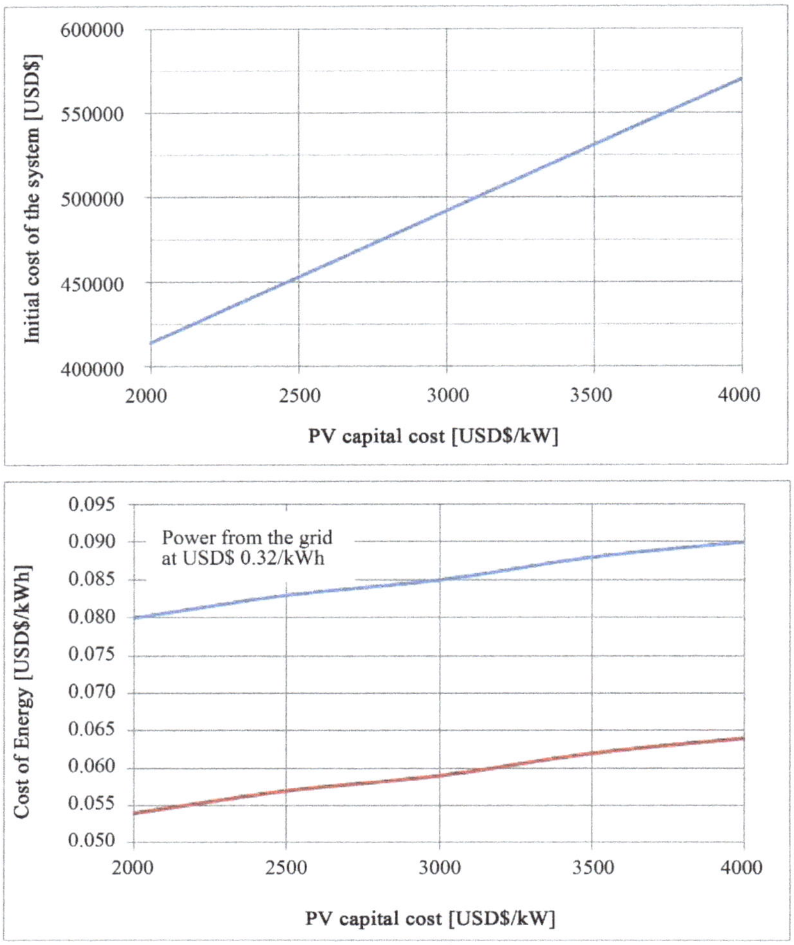

Figure 15. Initial costs of the system, above, and cost of energy, below, for a system with 227 kW hydropower and 60 kW photovoltaic as a function of PV capital cost. Below, there is curves for power from the grid at USD$ 0.32/kWh and USD$ 0.16/kWh.

8. Conclusions

This paper presented a pre feasibility study of a hydro PV hybrid system to be installed at the dam of Val de Serra, a dam for water supply in southern Brazil, aiming to identify conditions related to the components of PV generator for which the hybrid system becomes feasible. This paper considered the installation of PV modules on floating structures, but small, without influence on the evaporation of water from the reservoir or the amount of energy available.

The study was performed with the software Homer and results in a more conservative point of view indicated that it is not feasible to install the PV modules. However, an analysis of the "non optimal" solutions provided by Homer can identify a configuration that would present a cost slightly higher than the cost of only hydroelectric plant.

For this "non optimal" solution, for power purchased from the grid at USD$ 0.16/kWh and PV modules at USD$ 3000/kW, closer to actual values, the initial cost will be USD$ 492,444 (or USD$ 1715.83/kW) and the cost of energy will be USD$ 0.059/kWh. This solution was considered acceptable and it is the final outcome of this study.

Acknowledgements

The authors wish to thank the institutions involved for their support to research activities related to renewable energy, which resulted, among other things, in this article. Specifically, the third author would like to thank the partial financial support provided by CNPq through a support grant for research productivity.

Conflicts of Interest

The authors declare that there is no conflict of interests in relation to the issues addressed in this work, the data and information considered for preparation of this study and the publication of this paper.

References

[1] Muhida, R., Mostavan, A., Sujatmiko, W., Park, M. and Matsuura, K. (2001) The 10 Years Operation of a PV Micro Hydro Hybrid System in Taratak, Indonesia. *Solar Energy Materials & Solar Cells*, **67**, 621-627. http://dx.doi.org/10.1016/S0927-0248(00)00334-2

[2] Bekele, G. and Tadesse, G. (2012) Feasibility Study of Small Hydro PV Wind Hybrid System for Off Grid Rural Electrification in Ethiopia. *Applied Energy*, **97**, 5-15. http://dx.doi.org/10.1016/j.apenergy.2011.11.059

[3] Nfah, E.M. and Ngundam, J.M. (2009) Feasibility of Pico Hydro and Photovoltaic Hybrid Power Systems for Remote Villages in Cameroon. *Renewable Energy*, **34**, 1445-1450. http://dx.doi.org/10.1016/j.renene.2008.10.019

[4] Kenfack, J., Neirac, F.P., Tatietse, T.T., Mayer, D., Fogue, M. and Lejeune, E. (2009) Micro Hydro PV Hybrid System: Sizing a Small Hydro PV Hybrid System for Rural Electrification in Developing Countries. *Renewable Energy*, **34**, 2259-2263. http://dx.doi.org/10.1016/j.renene.2008.12.038

[5] Maher, P., Smith, N.P.A. and Williams, A.A. (2003) Assessment of Pico Hydro as an Option for Off Grid Electrification in Kenya. *Renewable Energy*, **28**, 1357-1369. http://dx.doi.org/10.1016/S0960-1481(02)00216-1

[6] Santarelli, M. and Macagno, S. (2004) Hydrogen as an Energy Carrier in Stand alone Applications Based on PV and PV Micro Hydro Systems. *Energy*, **29**, 1159-1182. http://dx.doi.org/10.1016/j.energy.2004.02.023

[7] Margeta, J. and Glasnovic, Z. (2011) Exploitation of Temporary Water Flow by Hybrid PV Hydroelectric Plant. *Renewable Energy*, **36**, 2268-2277. http://dx.doi.org/10.1016/j.renene.2011.01.001

[8] Beluco, A., Souza, P.K. and Krenzinger, A. (2008) PV Hydro Hybrid Systems. *IEEE Latin America Transactions*, **6**, 626-631. http://dx.doi.org/10.1109/TLA.2008.4917434

[9] Beluco, A., Souza, P.K. and Krenzinger, A. (2012) A Method to Evaluate the Effect of Complementarity in Time between Hydro and Solar Energy on the Performance of Hybrid Hydro PV Generating Plants. *Renewable Energy*, **45**, 24-30. http://dx.doi.org/10.1016/j.renene.2012.01.096

[10] Beluco, A., Souza, P.K. and Krenzinger, A. (2013) Influence of Different Degrees of Complementarity of Solar and Hydro Availability on the Performance of Hybrid Hydro PV Generating Plants. *Energy and Power Engineering*, **5**, 332-342. http://dx.doi.org/10.4236/epe.2013.54034

[11] Ehnberg, S.G.J. and Bollen, M.H.J. (2005) Reliability of a Small Power System Using Solar Power and Hydro. *Electric Power Systems Research*, **74**, 119-127. http://dx.doi.org/10.1016/j.epsr.2004.09.009

[12] Margeta, J. and Glasnovic, Z. (2010) Feasibility of the Green Energy Production by Hybrid Solar + Hydro Power System in Europe and Similar Climate Areas. *Renewable and Sustainable Energy Reviews*, **14**, 1580-1590. http://dx.doi.org/10.1016/j.rser.2010.01.019

[13] Denholm, P. and Margolis, R.M. (2007) Evaluating the Limits of Solar Photovoltaics in Traditional Electric Power Systems. *Energy Policy*, **35**, 2852-2861. http://dx.doi.org/10.1016/j.enpol.2006.10.014

[14] Hoffmann, W. (2006) PV Solar Electricity Industry: Market Growth and Perspective. *Solar Energy Materials & Solar Cells*, **90**, 3285-3311. http://dx.doi.org/10.1016/j.solmat.2005.09.022

[15] Glasnovic, Z., Rogosic, M. and Margeta, J. (2011) A Model for Optimal Sizing of Solar Thermal Hydroelectric Power Plant. *Solar Energy*, **85**, 794-807. http://dx.doi.org/10.1016/j.solener.2011.01.015

[16] Glasnovic, Z. and Margeta, J. (2011) Vision of Total Renewable Electricity Scenario. *Renewable and Sustainable Energy Reviews*, **15**, 1873-1884. http://dx.doi.org/10.1016/j.rser.2010.12.016

[17] https://goo.gl/maps/E9VJS

[18] https://goo.gl/maps/mj3bX

[19] https://goo.gl/maps/pOQC3

[20] https://goo.gl/maps/lNVlP

[21] https://goo.gl/maps/HdHoC

[22] https://goo.gl/maps/0yh8G

[23] AB Valves GmbH (2014) Howell Bunger Valves. http://abvalves.com/ab-valves/images/PDF/Automatic_Control_Valves/10._HJET_-_Text_new_cover.pdf

[24] Stelzer, R.S. and Walters, R.N. (1977) Estimating Reversible Pump Turbine Characteristics. U.S. Department of Interior, Bureau of Reclamation, Denver.

[25] Derakhshan, S. and Nourbakhsh, A. (2008) Experimental Study of Characteristic Curves of Centrifugal Pumps Working as Turbines in Different Specific Speeds. *Experimental Thermal and Fluid Science*, **32**, 800-807. http://dx.doi.org/10.1016/j.expthermflusci.2007.10.004

[26] Williams, A.A. (1996) Pumps as Turbines for Low Cost micro Hydro Power. *World Renewable Energy Congress*, **9**, 1227-1234. http://dx.doi.org/10.1016/0960-1481(96)88498-9

[27] Arriaga, M. (2010) Pump as Turbine, a Pico Hydro Alternative in Lao People's Democratic Republic. *Renewable Energy*, **35**, 1109-1115. http://dx.doi.org/10.1016/j.renene.2009.08.022

[28] Antwerpen, H.J. and Greyvenstein, G.P. (2005) Use of Turbines for Simultaneous Pressure Regulation and Recovery in Secondary Cooling Water Systems in Deep Mines. *Energy Conversion and Management*, **46**, 563-575. http://dx.doi.org/10.1016/j.enconman.2004.04.006

[29] Varun, N.H., Kumar, A. and Yadav, S. (2011) Experimental Investigation of Centrifugal Pump Working as Turbine for Small Hydropower Systems. *Energy Science and Technology*, **1**, 79-86.

[30] Ramos, H. and Borga, A. (1999) Pumps as Turbines: An Unconventional Solution to Energy Production. *Urban Water*, **1**, 261-263. http://dx.doi.org/10.1016/S1462-0758(00)00016-9

[31] Teixeira, L.E. (2014) Hydroelectric Power Generation in Water Supply dams. Master Dissertation, Programa de Pós Graduação em Recursos Hídricos e Saneamento Ambiental, Instituto de Pesquisas Hidráulicas, Universidade Federal do Rio Grande do Sul. (In Portuguese)

[32] Gisbert, C.M.F., Gonzálvez, J.J.F., Santafé, M.R., Gisbert, P.S.F., Romero, F.J.S. and Soler, J.B.T. (2013) A New Photovoltaic Floating Cover System for Water Reservoirs. *Renewable Energy*, **60**, 63-70. http://dx.doi.org/10.1016/j.renene.2013.04.007

[33] Santafé, M.R., Gsbert, P.S.F., Romero, F.J.S., Soler, J.B.T., Gonzálvez, J.J.F. and Gisbert, C.M.F. (2014) Implementation of a Photovoltaic Floating Cover for Irrigation Reservoirs. *Journal of Cleaner Production*, **66**, 568-570. http://dx.doi.org/10.1016/j.jclepro.2013.11.006

[34] HomerEnergy, L.L.C. (2014) Software HOMER, Version 2.68 Beta; the Micropower Opyimization Model. www.homerenergy.com

[35] Lilienthal, P.D., Lambert, T.W. and Gilman, P. (2004) Computer Modeling of Renewable Power Systems. In: Cleveland, C.J., Ed., *Encyclopedia of Energy*, Elsevier, Amsterdan, 633-647. http://dx.doi.org/10.1016/b0-12-176480-x/00522-2

[36] Lambert, T.W., Gilman, P. and Lilienthal, P.D. (2005) Micropower System Modeling with Homer. In: Farret, F.A. and Simões, M.G., Eds., *Integration of Alternative Sources of Energy*, John Wiley & Sons, Hoboken, 379-418. http://dx.doi.org/10.1002/0471755621.ch15

[37] Beluco, A., Souza, P.K. and Krenzinger, A. (2008) A Dimensionless Index Evaluating the Time Complementarity between Solar and Hydraulic Energies. *Renewable Energy*, **33**, 2157-2165. http://dx.doi.org/10.1016/j.renene.2008.01.019

[38] Brutsaert, W. (2005) Hydrology, an Introduction. Cambridge University Press, Cambridge, 618 p. http://dx.doi.org/10.1017/CBO9780511808470

[39] Agência Nacional de Águas (ANA) (2014) HidroWeb. www.hidroweb.ana.gov.br

9

Bidding Strategy in Deregulated Power Market Using Differential Evolution Algorithm

Veera Venkata Sudhakar Angatha[1], Karri Chandram[2], Askani Jaya Laxmi[3]

[1]EEE Department, SR Engineering College, Warangal, India
[2]Department of EEE & Instrumentation, BITS PilaniKK Birla Goa Campus, Goa, India
[3]EEE Department, JNTUCE, JNTUH, Hyderabad, India
Email: sudheavv@gmail.com, chandramk2006@gmail.com, ajl1994@yahoo.co.in

Abstract

The primary objective of this research article is to introduce Differential Evolution (DE) algorithm for solving bidding strategy in deregulated power market. Suppliers (GENCOs) and consumers (DISCOs) participate in the bidding process in order to maximize the profit of suppliers and benefits of the consumers. Each supplier bids strategically by choosing the bidding coefficients to counter the competitors bidding strategy. Electricity or electric power is traded through bidding in the power exchange. GENCOs sell energy to power exchange and in turn ancillary services to Independent System Operator (ISO). In this paper, Differential Evolution algorithm is proposed for solving bidding strategy problem in operation of power system under deregulated environment. An IEEE 30 bus system with six generators and two large consumers is employed to demonstrate the proposed technique. The results show the adaptability of the proposed method compared with Particle Swarm Optimization (PSO), Genetic Algorithm (GA) and Monte Carlo simulation in terms of Market Clearing Price (MCP).

Keywords

Bidding Strategy, Differential Evolution, Power market, Market Clearing Price

1. Introduction

The electric power industry worldwide is experiencing restructuring and deregulation of power market. Deregulation and restructuring of electric power industry around the world raises many challenging issues related to the operation of power system. The core issue of restructuring is deregulating the power industry and introducing the competition among the generating companies. Competition creates the opportunities for GENCOs to get more profit. In restructured power market, each GENCO reasonably builds strategic bid to maximize its own profit [1] [2]. The market efficiency is improved with the deregulated power market structure. The power exchange and ISO are independent non-profit organizations. Electricity is traded through bidding in the power exchange (PX). ISO controls and operates the transmission grid. GENCOs sell energy to PX and ancillary services to ISO. Energy is distributed to end users through distribution network and ancillary services are used to support the system organization. A framework for sub-optimal bidding strategy was presented in [3]. Genetic algorithm [4] [5] has been proposed for solving bidding strategy. A two level optimization algorithm was adopted in [6] for building market bidding strategy, where the market participants try to maximize their profit. A brief literature survey on bidding strategy is presented in [7]. Several classical techniques such as Markov decision process [8] [9], Lagrangian Relaxation [10], Monte Carlo based approach [11] have been proposed by various scholars in the past decade. Apart from these methods, few other techniques [12]-[21] are suggested to solve the bidding strategy. Recently, modern heuristic methods [22]-[32] like particle swarm optimization, gravitational search algorithm and hybrid algorithms have been proposed to solve the bidding strategy problem with wide variety.

It is noticed from the literature survey that several techniques are adopted to solve strategic bidding problem. However, there is a need to improve the quality of solution in terms of profit. The main objective of this paper is to suggest a new technique for solving bidding strategy problem. Therefore, DE is recommended in this paper. The remaining paper is formulated as follows: problem formulation for bidding in Section II; proposed methodology and implementation steps of DE in Section III; case studies in Section IV; conclusions of the paper in Section V.

2. Problem Formulation for Bidding

Consider a power system consists of "n" independent generators, an inter-connected transmission network controlled by ISO, a Power Exchange, an aggregated load, and "m" large consumers who join in demand side bidding. Further, assume that every generating company and large consumer is required to bid a linear non-decreasing supply and non-increasing demand functions respectively to power exchange.

The GENCO supply bid price is denoted by $G_i(P_i) = x_i + y_i P_i$ where $i = 1,2,3,\cdots,n$, P_i is the ith generator active power output, x_i is intercept and y_i is slope of the supply bid curve. The jth large consumer demand price is denoted by $W_j(L_j) = u_j - v_j L_j$ where $j = 1,2,3,\cdots,m$, L_j is the jth load active power demand. u_j is intercept and v_j is slope of the demand bid curve. x_i, y_i, u_j, and v_j are non-negative bidding coefficients.

Now, the main task of Power Exchange is to determine a set generation outputs, a set of consumer demands that minimizes the total purchasing cost, while meeting the security and reliability constraints with clear dispatch procedures. That is, Power Exchange determines a set of power outputs $P = (P_1, P_2, P_3, \cdots, P_n)^T$ and a set of demands $W = (W_1, W_2, W_3, \cdots, W_n)^T$ by solving the following Equations (1) to (5).

$$x_i + y_i P_i = MCP, i = 1,2,3,\cdots,n \tag{1}$$

$$u_j - v_j L_j = MCP, j = 1,2,3,\cdots,m \tag{2}$$

Power balance equation is as follows:

$$\sum_{i=1}^{n} P_i = Q(MCP) + \sum_{j=1}^{m} W_j \tag{3}$$

where, MCP is the electricity uniform market clearing price to be determined. $Q(MCP)$ is the forecasted aggregated pool load predicted by PX, made open to all the participants, and is dependent on the price elasticity.

Generation and load in equality constraints are:

$$P_{i_{\min}} \le P_i \le P_{i_{\max}}, i = 1,2,3,\cdots,n \tag{4}$$

$$W_{j_{\min}} \le W_j \le W_{j_{\max}}, j = 1,2,3,\cdots,m \tag{5}$$

where $P_{i_{\min}}$ and $P_{i_{\max}}$ are the lower and upper power limits of the ith generator respectively. $W_{j_{\min}}$ and $W_{j_{\max}}$ are the lower and upper demand limits of the jth consumer respectively. When the expression of Q(MCP)is available, Equations (1)-(3) can be solved directly. Assume that the aggregated forecasted pool load as follows in the linear form:

$$Q(MCP) = Q_0 - K * MCP \tag{6}$$

where Q_0: constant number and K: coefficient representing the price elasticity of the aggregated power demand. $K = 0$, if pool power demand is largely inelastic.

When the generation and load inequality constraints are neglected, the solution to equations are as follows:

$$MCP = \frac{Q_0 + \sum_{i=1}^{n} \frac{x_i}{y_i} + \sum_{j=1}^{m} \frac{u_j}{v_j}}{K + \sum_{i=1}^{n} \frac{1}{y_i} + \sum_{j=1}^{m} \frac{1}{v_j}} \tag{7}$$

Power awarded to generator and large consumer is calculated as follows:

$$P_i = \frac{MCP - x_i}{y_i}, i = 1,2,3,\cdots,n \tag{8}$$

$$L_j = \frac{u_j - MCP}{v_j}, j = 1,2,3,\cdots,m \tag{9}$$

When solution of Equations (8) violates the maximum power limit $P_{i_{max}}$, P_i is set to its maximum power limit $P_{i_{max}}$. If P_i is smaller than its lower power limit $P_{i_{min}}$, P_i should be set to zero instead of $P_{i_{min}}$ and the generator is removed from the bidding problem since the generator ceases to be competitive, similar treatment is applicable to L_j. Suppose that individual supplier reasonably aims at profit maximization, the ith supplier benefit maximization objective function for developing a bidding strategy may be defined as:

$$\text{Max. } F(x_i, y_j) = MCP * P_i - C_i(P_i) \tag{10}$$

Subject to Equations (1)-(5) where $C_i(P_i)$ is i^{th} supplier production cost function. The generation cost function is expressed as follows:

$$C_i(P_i) = f_i P_i^2 + e_i P_i + d_i \tag{11}$$

where f_i, e_i, and d_i are fuel cost coefficients of ith generator. The marginal cost of ith generator is calculated as follows:

$$\text{Marginal cost} = 2 f_i P_i + e_i \tag{12}$$

The objective is to determine x_i and y_i so as to maximize $F(x_i, y_j)$ while meeting the constraints (1)-(5). Similarly, for the jth large consumer, the benefit maximization objective function for developing a bidding strategy may be defined as follows:

$$\text{Max. } H(u_i, v_j) = B_j(L_j) - MCP * L_j \tag{13}$$

Subject to Equations (1)-(5) where $B_j(L_j)$ is j^{th} large consumer demand (benefit) function. The objective is to determine u_j and v_j so as to maximize $H(u_i, v_j)$ while meeting the constraints (1)-(5).

It is known that bidding participants can set MCP at the level that ensures maximum profit to them if they aware bidding strategy of rivals. Whereas in sealed bid auction based market, electricity data for the next bidding period is not openly available, hence GENCOs and large consumers cannot solve the problem due the information needed to solve the problem given in Equations (10) and (13) is not available. But, the bidding previous history is available, and hence the bidding coefficients of opponents can be roughly calculated. However, previous round bidding information will be made open to all participants, after market operator decides MCP, and market participants can make use of this information to develop their strategic bids for the next round of transaction. Whereas, each supplier can estimate their opponents bidding coefficients using probability distribution function (pdf).

3. Proposed Methodology

In this section, a brief description of DE [33] and the implementation steps of DE for bidding strategy are provided.

3.1. Differential Evolution

Differential Evolution algorithm [34] has been proposed by Storn and Price in 1995. It is a basic yet effective population-based stochastic search procedure for taking care of global optimization problems. The algorithm is named as Differential Evolution because of a unique kind of differential operator, which makes new off-springs from parent chromosomes rather than classical crossover and mutation.

A brief depiction about different operators of differential evolution is given beneath:

3.1.1. Initialization

First step of DE is initialization of population. The population should cover the whole search space. The upper and lower limits are taken as X_{max} and X_{min}:

$$X_{max} = \{X_{1max}, X_{2max}, X_{3max}, \cdots, X_{Dmax}\} \tag{14}$$

$$X_{min} = \{X_{1min}, X_{2min}, X_{3min}, \cdots, X_{Dmin}\} \tag{15}$$

where D represents number of variables. The jth component of the ith individual is initialized as follows:

$$X_{i,j} = X_{j,\min} + rand[0,1] * (X_{j,\max} - X_{j,\min}) \tag{16}$$

where $i = 1,2,3,\cdots,N$ p-individuals, $j = 1,2,\cdots,D$ and rand [0,1] represents a uniformly distributed random number in the interval [0,1], where 0 and 1 are the lower and upper limits.

3.1.2. Mutation

Off springs for the next generation are introduced into the population through transformation process. The transformation is performed by selecting three individuals from the populace in an arbitrary way.

Let X_{r1}, X_{r2}, X_{r3} and X_i represent three random individuals and target individual respectively such that $r1 \neq r2 \neq r3 \neq i$ upon which mutation is performed during the Gth generation as:

$$V_{i,G+1} = X_{r1,G} + F * (X_{r2}, G - X_{r3}, G) \tag{17}$$

where $V_{i,G+1}$ is the perturbed mutated individual. The difference of two random individuals is scaled by an element F, which controls the amplification of the contrast between two individuals in order to avoid search stagnation and to enhance convergence.

3.1.3. Crossover

New off-springs are reproduced through the crossover operation based on binomial distribution. The members of the current population (target individual) and the members of the mutated individuals are subjected to crossover operation in this manner delivering a trial vector as follows:

$$U_{i,j,G+1} = \begin{cases} V_{i,j,G+1} & \text{if } rand[0,1] \leq Cr \\ X_{i,j,G} & \text{otherwise} \end{cases} \tag{18}$$

where Cr is the crossover constant that controls the diversity of the population and prevents the algorithm from getting caught into the local optima. The crossover constant must be in the range of [0,1]. $Cr = 1$ infers the trial vector will be composed entirely of the mutant vector individuals and $Cr = 0$ infers that the trial vector members are composed of the individuals of parent vector. The mathematical statement can also be written as:

$$U_{i,j,\ G+1} = X_{i,j,G} * (1 - Cr) + V_{i,j,G} + 1 * Cr \tag{19}$$

3.1.4. Selection

In DE, selection technique is performed with the trial vector and the objective vector to get the best set of people for the next generation. In the proposed methodology, one and only population kept up and consequently the best individuals replace the object individual's in the present population. The objective values of the trial vector and the object vector obtained from the fitness function are evaluated and compared.

$$X_{i,G+1} = \begin{cases} U_{i,G} & \text{if } f(U_{i,G}) \le f(X_{i,G}) \\ X_{i,G} & \text{otherwise} \end{cases} \tag{20}$$

Mutation, crossover and selection continue until some stopping criterion is reached.

Flowchart of Differential Evolution is shown in **Figure 1**.

3.2. Differential Evolution for Bidding Strategy

Complete procedure of the differential evolution algorithm for solving bidding strategy is illustrated below (**Figure 2**):

1) Step I Input Data

(i) For Bidding problem: Number of suppliers, number of consumers, fuel cost data of suppliers and consumers. (ii) For Differential evolution: Population size, number of iterations, mutation and crossover ratio.

2) Step II Initialization

Initialization is one of the important steps in solving the bidding strategy problem. Thebidding coefficients are inter-dependent. Here, x and u are kept constant and y and v are selected randomly.

3) Step III Iterations Starts

(i) Calculate Market Clearing Price

(ii) Evaluate powers of suppliers and large consumers

(iii) Calculate profit of suppliers and large consumers.

4) Step IV Iterations Starts for DE

(i) Generate donors by mutation

(ii) Perform recombination

(iii) Evaluate MCP for updated values of b and d

(iv)Evaluate powers and check for limits of generators and consumers

(v) Check the error

(vi) Find the best solution

(vii) End of iterations for DE

(viii) End of iterations

5) Step V Final Results

4. Case Studies

The proposed DE algorithm is tested on a IEEE 30 bus system with six generators and two consumers. Coding of the algorithm is developed in MATLAB (version 2012 A) and executed in personal laptop (Dell vostro, i3,2310 M CPU 2.10 Ghz, 4GB RAM, 64 bit WINDOW operating system). Fuel cost coefficients of generators

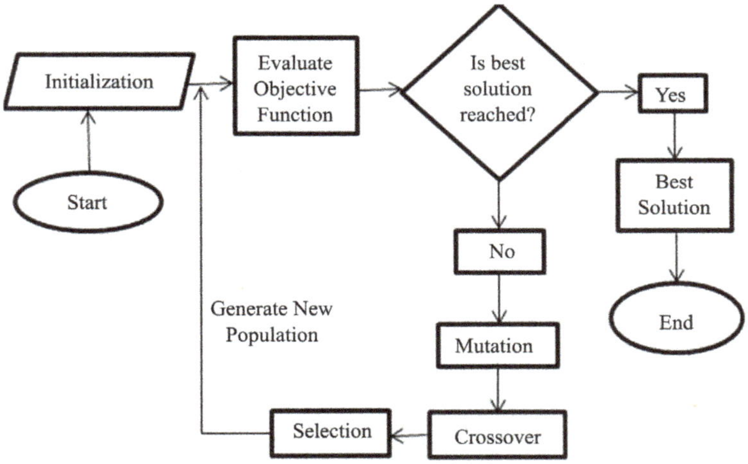

Figure 1. Flowchart of differential evolution.

(e in \$/MWh, and f in \$/MW^{2}h), demand cost coefficients (g and h), generator output power limits and large consumer load demands are taken from [31] and are provided in **Table 1**.

During execution of the proposed algorithm, the numerical values of various control parameters are used and are shown in **Table 2**.

Each rival supplier is assumed to have an estimated joint normal distribution for the bidding coefficients.

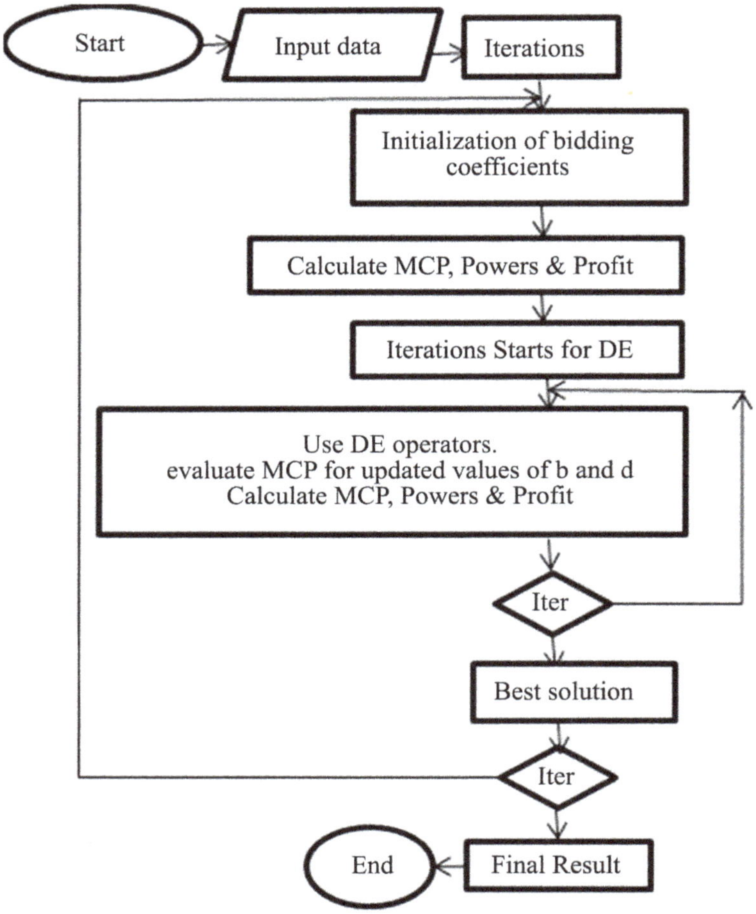

Figure 2. Flowchart of proposed method.

Table 1. Generator and large consumer data.

Generator	e	f	P_{min} (MW)	P_{max} (MW)
1	6	0.01125	40	160
2	5.25	0.0525	30	130
3	3	0.1375	20	90
4	9.75	0.02532	20	120
5	9	0.075	20	100
6	9	0.075	20	100
Consumer	g	h	L_{min} (MW)	L_{max} (MW)
1	30	0.04	0	200
2	25	0.03	0	150

After execution of the proposed algorithm, the end results of a and b are listed in **Table 3**. The market clearing price (MCP) by the proposed algorithm is 17.6405 $/MWh.

The powers of generators and consumers are given in **Table 4**. The estimated profit of six generators and benefit of two large consumers are shown in **Table 5**.

Market clearing price of different methods, which decides the profit of the bids, is given in **Figure 3**.

The estimated total profit of six generators and two large consumers with different methods are shown in **Figure 4**. The sum of profit obtained both for generators and consumers by the proposed method is $5076.70, which is higher than the profit obtained from PSO, GA and Monte Carlo methods.

Table 2. Numerical values of various control parameters.

S. No	Control Parameter	Value
1	Qo	300
2	K	5
3	Popoulation	1000
4	Iterations	500
5	F	0.005
6	Cr	0.85

Table 3. Bidding coefficients of generators and large consumers.

Generator	a	b
1	9.3326	0.0346
2	4.8758	0.1241
3	4.7321	0.3613
4	6.2295	0.0765
5	7.8143	0.176
6	7.1003	0.1600
Consumer	**c**	**d**
1	26.3122	0.0486
2	33.0692	0.0596

Table 4. Comparison of Generated Powers in MWs with different methods.

Generator	Monte Carlo	PSO	Proposed
1	160.0000	156.0	160.0000
2	105.8371	89.37	104.23
3	48.5923	45.67	36.66
4	120.0000	88.79	120.0000
5	49.0859	43.09	53.73
6	49.0859	43.09	68.49
Consumer	**Monte Carlo**	**PSO**	**Proposed**
1	170.4639	139.70	180.42
2	143.9578	112.06	150.00

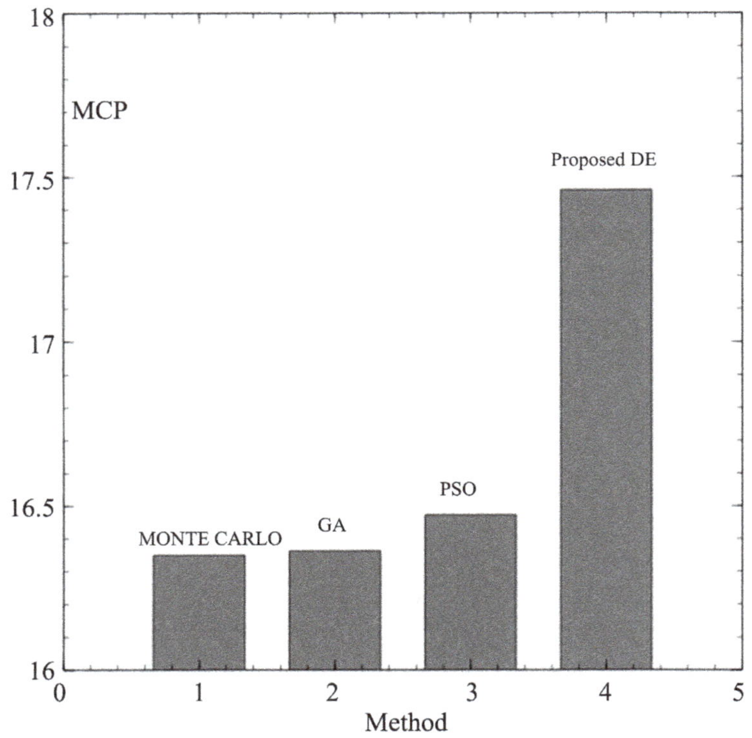

Figure 3. Comparison of Market Clearing Price (MCP) with different methods.

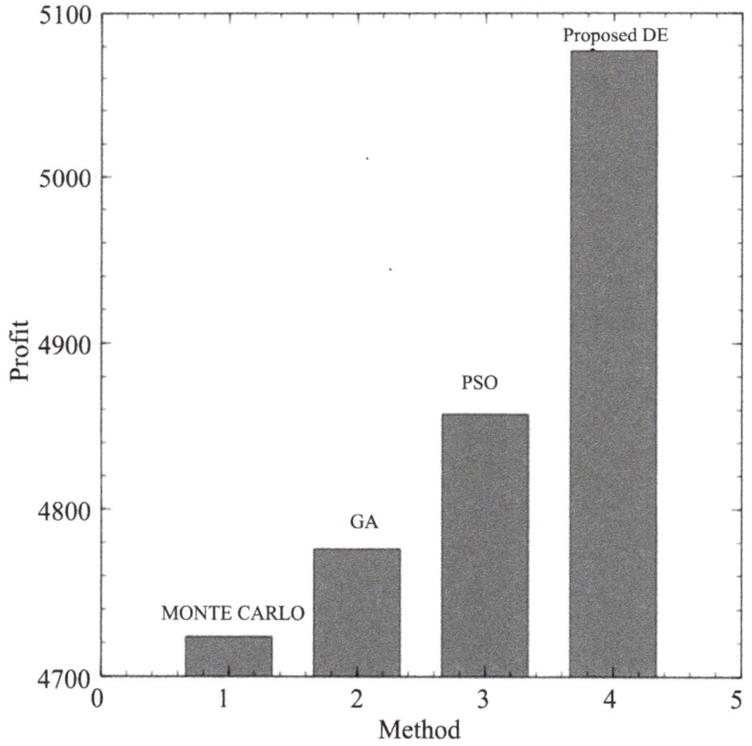

Figure 4. Comparison of Profits with different methods.

Table 5. Comparison of Benefits ($) with different methods.

Generator	Monte Carlo	PSO	Proposed
1	1370.12	1367.99	1056.2
2	588.12	572.69	702.7
3	324.8	322.90	345.5
4	428.90	386.40	561.1
5	180.71	177.45	238.3
6	180.71	177.45	227.9
Consumer	Monte Carlo	PSO	Proposed
1	1162.32	1126.26	959.69
2	621.73	592.60	455.41

5. Conclusion

Differential evolution for solving bidding strategy in deregulated power market is presented in this paper. The proposed algorithm is an effective method and easy to implement for solving bidding strategy problem. From the reported results it is observed that the proposed differential evolution is an effective method in terms of quality of solution. The simulation results show that the proposed differential evolution algorithm provides better solution in terms of profit compared to the existing methods available in the literature survey. More realistic strategic bidding problem can be developed for generating companies and consumers for better competition in the real time operation of power system under deregulation.

References

[1] Lai, L.L. (2001) Power System Restructuring and Deregulation-Trading, Performance and Information Technology. Wiley, New York. http://dx.doi.org/10.1002/0470846119

[2] Shahidepour, M., Yamin, H. and Li, Z. (2002) Market Operations in Electric Power Systems: Forecasting, Scheduling and Risk Management. Wiley, New York.

[3] Lamont, J.W. and Raman, S. (1997) Strategic Bidding in an Energy Brokerage. *IEEE Transactions on Power Systems*, **12**, 1729-1733. http://dx.doi.org/10.1109/59.627883

[4] Richter Jr., C.W. and Shible, G.B. (1998) Genetic Algorithm Evolution of Utility Bidding Strategies for the Competitive Marketplace. *IEEE Transactions on Power Systems*, **13**, 256-261. http://dx.doi.org/10.1109/59.651644

[5] Richter Jr., C.W., Shible, G.B. and Ashlock, D. (1998) Comprehensive Bidding Strategies with Genetic Programming: Finite State Automata. *IEEE Transactions on Power Systems*, **14**, 1207-1212. http://dx.doi.org/10.1109/59.801874

[6] Weber, J. and Overbye, T. (1999) A Two-Level Optimization Problem for Analysis of Market Bidding Strategies. *IEEE PES Summer Meeting*, **2**, 682-687.

[7] David, A.K. and Wen, F. (2000) Strategic Bidding in Competitive Electricity Markets: A Literature Survey. *IEEE Power Engineering Society Summer Meeting*, **4**, 2168-2173.

[8] Hao, S. (2000) A Study of Basic Bidding Strategy in Clearing Pricing Auctions. *IEEE Transactions on Power Systems*, **15**, 975-980. http://dx.doi.org/10.1109/59.871721

[9] Song, H., Liu, C.-C., Lawarree, J. and Dahlgren, R.W. (2000) Optimal Electricity Supply Bidding by Markov Decision Process. *IEEE Transactions on Power Systems*, **15**, 618-624. http://dx.doi.org/10.1109/59.867150

[10] Zhang D., Wang, Y. and Luh, P.B. (2000) Optimization Based Bidding Strategies in the Deregulated Market. *IEEE Transactions on Power Systems*, **15**, 981-986. http://dx.doi.org/10.1109/59.871722

[11] Wen, F.S. and David, A.K. (2001) Optimal Bidding Strategies and Modeling of Imperfect Information among Competitive Generators. *IEEE Transactions on Power Systems*, **16**, 15-21. http://dx.doi.org/10.1109/59.910776

[12] Wen, F.S. and David, A.K. (2001) Optimal Bidding Strategies for Competitive Generators and Large Consumers. *Electrical Power and Energy Systems*, **23**, 37-43. http://dx.doi.org/10.1016/S0142-0615(00)00032-6

[13] Wen, F.S. and David, A.K. (2001) Strategic Bidding for Electricity Supply in a Day-Ahead Energy Market. *Electric Power Systems Research*, **59**, 197-206. http://dx.doi.org/10.1016/S0378-7796(01)00154-7

[14] Guan, X., Ho, Y.-C. and Lai, F. (2001) An Ordinal Optimization Based Bidding Strategy for Electric Power Suppliers in the Daily Energy Market. *IEEE Transactions on Power Systems*, **16**, 788-797. http://dx.doi.org/10.1109/59.962428

[15] David, A.K. (2002) Competitive Bidding in Electricity Supply. *IEE Proceedings on Generation, Transmission and Distribution*, **140**, 421-426. http://dx.doi.org/10.1049/ip-c.1993.0061

[16] Gountis, V.P. and Bakirtzis, A.G. (2004) Bidding Strategies for Electricity Producers in a Competitive Electricity Market Place. *IEEE Transactions on Power Systems*, **19**, 356-365. http://dx.doi.org/10.1109/TPWRS.2003.821474

[17] Li, T. and Shahidehpour, M. (2005) Strategic Bidding of Transmission-Constrained GENCOs with Incomplete Information. *IEEE Transactions on Power Systems*, **20**, 437-447. http://dx.doi.org/10.1109/TPWRS.2004.840378

[18] Ma, X., Wen, F., Ni, Y. and Liu, J. (2005) Towards the Development of Risk- Constrained Optimal Bidding Strategies for Generation Companies in Electricity Markets. *Electric Power Systems Research*, **73**, 305-312. http://dx.doi.org/10.1016/j.epsr.2004.07.004

[19] Attaviriyanupap, P., Kita, H., Tanaka, E. and Hasegawa, J. (2005) New Bidding Strategy Formulation for Day-Ahead Energy and Reserve Markets Based on Evolutionary Programming. *Electrical Power and Energy Systems*, **27**, 157-167. http://dx.doi.org/10.1016/j.ijepes.2004.09.005

[20] Rahimiyan, M. and Mashhadi, H.R. (2008) Supplier's Optimal Bidding Strategy in Electricity Pay-as-Bid Auction: Comparison of the Q-Learning and a Model Based Approach. *Electric Power Systems Research*, **78**, 165-175. http://dx.doi.org/10.1016/j.epsr.2007.01.009

[21] Boonchuay, C., Ongsakul, W., Zhong, J. and Wu, F.F. (2010) Optimal Trading Strategy for GenCo in LMP-Based and Bilateral Markets Using Self-Organising Hierarchical PSO. *International Journal of Engineering, Science and Technology*, **2**, 82-93.

[22] Soleymani, S. (2011) Bidding Strategy of Generation Companies Using PSO Combined with SA Method in the Pay as Bid Markets. *Electrical Power and Energy Systems*, **33**, 1272-1278. http://dx.doi.org/10.1016/j.ijepes.2011.05.003

[23] Azadeh, A., Ghaderi, S.F., Nokhandan, B.P. and Sheikhalishahi, M. (2012) A New Genetic Algorithm Approach for Optimizing Bidding Strategy Viewpoint of Profit Maximization of a Generation Company. *Expert System with Applications*, **39**, 1565-1574. http://dx.doi.org/10.1016/j.eswa.2011.05.015

[24] Kumar, J.V., Kumar, D.M.V. and Edukondalu, K. (2013) Strategic Bidding Using Fuzzy Adaptive Gravitational Search Algorithm in a Pool Based Electricity Market. *Applied Soft Computing*, **13**, 2445-2455. http://dx.doi.org/10.1016/j.asoc.2012.12.003

[25] Qiu, Z., Gui, N. and Deconinck, G. (2013) Analysis of Equilibrium-Oriented Bidding Strategies with Inaccurate Electricity Market Models. *Electrical Power and Energy Systems*, **46**, 306-314. http://dx.doi.org/10.1016/j.ijepes.2012.10.036

[26] Wen, F.S. and David, A.K. (2000) Coordination of Bidding Strategies in Energy and Spinning Reserve Markets for Competitive Suppliers Using Genetic Algorithm. *Power Engineering Society Summer Meeting*, Vol. 4, Seattle, 16-20 July 2000, 2174-2179. http://dx.doi.org/10.1109/PESS.2000.866983

[27] Wen, F.S. and David, A.K. (2001) Strategic Bidding for Electricity Supply in a Day-Ahead Energy Market. *Electrical Power Systems Research*, **59**, 197-206. http://dx.doi.org/10.1016/S0378-7796(01)00154-7

[28] Wen, F.S. and David, A.K. (2002) Coordination of Bidding Strategies in Day-Ahead Energy and Spinning Reserve Markets. *Electrical Power and Energy Systems*, **24**, 251-261. http://dx.doi.org/10.1016/S0142-0615(01)00038-2

[29] Wen, F.S. and David, A.K. (2002) Optimally Co-Ordinated Bidding Strategies in Energy and Ancillary Service Markets. *IEE Proceedings on Generation, Transmission and Distribution*, **149**, 331-338. http://dx.doi.org/10.1049/ip-gtd:20020211

[30] Yang, L, Wen, F., Wu, F.F., Ni, Y. and Qiu, J. (2002) Development of Bidding Strategies in Electricity Markets Using Possibility Theory. *IEEE International Conference on Power System Technology*, **1**, 182-187. http://dx.doi.org/10.1109/ICPST.2002.1053529

[31] Kumar, J.V., Pasha, S.J. and Kumar, D.M.V. (2010) Strategic Bidding in Deregulated Market Using Particle Swarm Optimization. *Annual IEEE India Conference*, Kolkata, 17-19 December 2010, 1-6. http://dx.doi.org/10.1109/indcon.2010.5712648

[32] Zhang, G., Zhang, G.L., Gao, Y. and Lu, J. (2011) Competitive Strategic Bidding Optimization in Electricity Markets Using Bilevel Programming and Swarm Technique. *IEEE Transactions on Industrial Electronics*, **58**, 2138-2146. http://dx.doi.org/10.1109/TIE.2010.2055770

[33] Sudhakar, A.V.V., Chandran, K. and Laxmi, A.J. (2014) Differential Evolution for Solving Multi Area Economic Dispatch. *International Conference on Advances in Computing, Communications and Informatics* (*ICACCI*), New Delhi, 24-27 September 2014, 1146-1151. http://dx.doi.org/10.1109/ICACCI.2014.6968486

[34] Storn, R. and Price, K. (1997) Differential Evolution—A Simple and Efficient Heuristic for Global Optimization over Continuous Spaces. *Journal of Global Optimization*, **11**, 341-359. http://dx.doi.org/10.1023/A:1008202821328

10

Toward an Evolutionary Multi-Criteria Model for the Analysis and Estimation of Wind Potential

Fouad Amri, Omar Bouattane, Tajeddine Khalili, Abdelhadi Raihani, Abdelkader Bifadene

Lab SSDIA, ENSET Mohammedia, Hassan II University of Casablanca, Casablanca, Morocco
Email: amri.fouad@gmail.com

Abstract

The main objective of this paper is to model, analyze and estimate wind energy at East region of Mohammedia and other Moroccan sites. The basic data were taken from meteorological records of each region. In this context, this work is focused on a methodological approach of a decision support system for optimal choice of wind turbine using multi-criteria model that takes into consideration both the accurate Weibull distribution in the area (wind speed-ground roughness) and the technical parameters of the wind turbine. In this approach we realized an adapted modeling of each element of the turbine (rotor-multiplier-generator). This article also offers a way to forecast wind speed in a region where wind data are not accessible using an artificial neural network.

Keywords

Wind Potential, Meteorological Records, Energy Production, Weibull Statistical Distribution

1. Introduction

The wind is a promising sustainable energy source that can help reduce dependence on fossil fuels. Wind power has a very high world growth rate. In 2013 wind power worldwide grew by 35,572 MW, bringing the global installed capacity to over 318 GW [1] [2]. Although it can't replace completely the traditional energy sources, this energy can however offer an interesting renewable alternative that would help to reduce the exhaustion of fossil resources. It fits perfectly in the global effort to reduce CO_2 emissions and more generally in the context of sustainable development of the energy landscape. Choosing a wind farm in a wind energy production system is the most crucial part because the energy produced depends essentially on the wind resources available at the site. In order to achieve an accurate estimation of these resources, the implementation of a high performance computing tool is required to avoid wind intermittency misguidance [3]. In the first part of this paper we represented the wind speed probability distribution using Weibull distribution to estimate and analyze the energetic wind potential in many Moroccan sites. In the second part, we propose a decision support system model for an optimal choice of wind turbine technology using multi-criteria model that takes into account both the best Weibull distribution in a given location and the appropriate wind turbine. When wind speed data are not available in a given location we elaborated an artificial neural network that predicts wind speed records.

2. The Wind Analysis

In literature many probability density functions may be used to study the wind speed, although the most widespread is the Weibull distribution as it is more accurate in this field. According to Justus *et al.* [4], the Weibull distribution with two parameters is given by the following equation [4] [5]:

$$f(v)=\left(\frac{k}{c}\right)\cdot\left(\frac{v}{c}\right)^{k-1}\cdot\exp\left(-\left(\frac{v}{c}\right)^{k}\right) \quad (1)$$

where c (m/s) is the scale parameter that determinates the position of the curve, based on wind characteristics in the site, and k (unit less) is a shape parameter that indicates the shape of the curve, and v is the wind speed. There are several methods to estimate the c and k parameters of the Weibull distribution: the Standard deviation method (SD), the Method of maximum likelihood (MLH) and the least squares method (LSM). In this study, the two parameters k and c were determined using the SD, given by:

$$k=\left(\frac{\sigma}{V_m}\right)^{-1.086} \quad (2)$$

$$c=\frac{V_m}{\Gamma\left(1+\frac{1}{k}\right)} \quad (3)$$

V_m is the average speed, σ the standard deviation and Γ is the usual gamma function.

$$V_m=\int_0^{\infty} v\cdot f(v)\cdot \mathrm{d}v=c\cdot\Gamma\left(1+\frac{1}{k}\right) \quad (4)$$

In this study the wind speed data was collected at different heights H_0, it is then necessary to extrapolate the measures at a same height H.

$$v=v_m\cdot\left(\frac{H}{H_0}\right)^{\alpha} \quad (5)$$

where α the surface roughness and v is the wind speed at the desired height.

2.1. Wind Power Density

The wind power flowing at a speed v through the blade sweep area A [m^2] is given by the well known expression [6]:

$$P(v)=\frac{1}{2}\cdot\rho\cdot A\cdot v^3 \quad (6)$$

The wind power density, for a given theoretical probability distribution $f(v)$, can be calculated by the following integration:

$$P=\int_0^{\infty}P(v)f(v)\mathrm{d}v=\frac{1}{2}\cdot\rho\cdot A\cdot c^3\cdot\Gamma\left(\frac{k+3}{k}\right) \quad (7)$$

$\rho\left[\mathrm{kg/m^3}\right]$ is the air density.

2.2. Available Wind Energy Density Estimation

Likewise, the available energy per year E_{an} [W·h] is given by:

$$\frac{E_{an}}{A} = \frac{T}{2} \cdot \rho \cdot \int_{vi}^{vf} v^3 \cdot f(v) \cdot \mathrm{d}v = \frac{1}{2} \cdot \rho \cdot c^3 \cdot \Gamma\left(\frac{k+3}{k}\right) \cdot T \tag{8}$$

vi and vf are the cut-in and the cut-out wind speeds, T is the number of hours per year, and $f(v)$ is the Weibull function fitted for the specific wind site [7].

2.3. Wind Available Energy Estimation Results

The following results were obtained for the site of the ENSET Mohammedia and many other sites in the Moroccan territory, based on meteorological data. Owing to the fact that Sites in our study can be classified into three different types as explained further, we picked a site from every class to represent the Weibull distribution, as shown in **Figures 1-3**, while **Table 1** summarizes all the results.

The average wind speed, the Weibull parameters c and k, the wind power density and the available wind energy density, represented in **Figures 1-3** and **Table 1**, are the essential factors affecting site choice. As shown in **Table 1**, the most promising sites in terms of available wind energy density are: El Koudia Al Baida (Tetouan), Sendouk (Tanger), Cap Sim (Essaouira) Col Touahar (Taza), S. EL Garn (Tantan), Tarfaya, Laayoune,

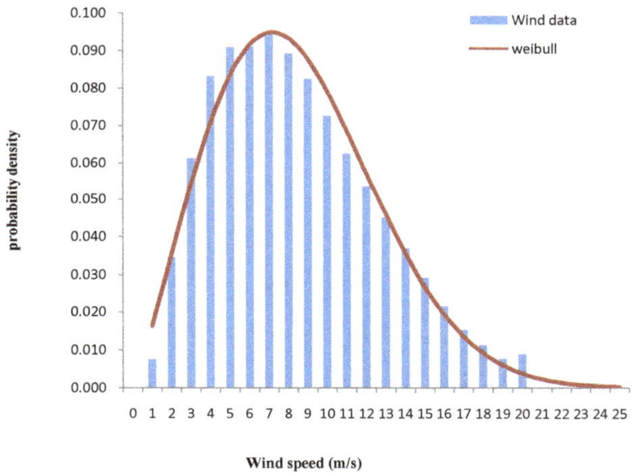

Figure 1. Histogram and Weibull distribution of wind speeds (Site: Col de Touahar Taza).

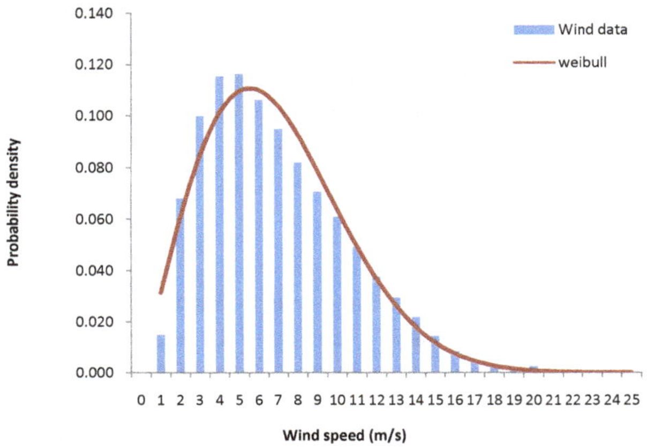

Figure 2. Histogram and Weibull distribution of wind speeds (Site: Had hrara Safi).

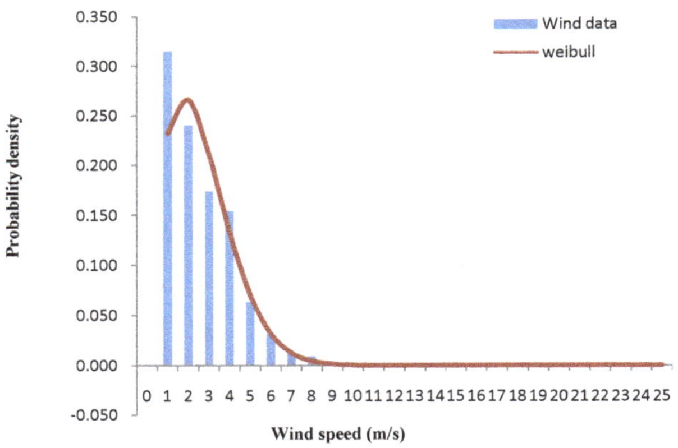

Figure 3. Histogram and Weibull distribution of wind speeds (station ENSET Mohammedia).

Table 1. Wind available energy estimation results.

Site	c (m/s)	k	P (W/m^2)	E (KWh/m^2)	H_0 m	Hm	V_{m0} (m/s)	V_m (m/s)
Station ENSET	2.93	1.72	48	420.3	8	78	2.61	5.17
El Koudia Al Baida (Tetouan)	12.45	2.58	5087.1	44563.1	9	78	11.06	15.23
Sendouk (Tanger)	10.28	2.36	2524.9	22118.8	9	78	9.11	13.01
Tiniguir (Dakhala)	9.96	3.10	3890.9	34084.8	9	78	8.91	12.80
MyBoussalham (Kenitra)	4.98	1.89	240.81	2109.5	9	78	4.42	7.24
Salouane (Nador)	4.41	1.96	169.5	1484.6	10	78	3.91	6.39
Had hrara (Safi)	7.79	2.02	950.6	8327.6	10	78	6.9	10.18
Col de Touaha (Taza)	9.54	2.16	1832.3	16051.4	20	78	8.44	10.66
Cap Sim (Essaouira)	10.91	2.18	2768.7	24253.4	40	78	9.66	10.75
S. ELGarn (Tantan)	7.38	2.74	1179.2	10329.4	40	78	6.57	7.48
Tarfaya	9.13	3.91	6841.3	59930.1	40	78	8.27	9.30
Laayoune	11.95	4.20	21650.5	189658.5	40	78	10.85	12.02

Tiniguir (Dakhala), therefore these locations are the most suited for wind farms.

Moulay Bousselham (Kenitra), Had hrara (Safi), Salouane (Nador) have a low average wind potential, which makes these location non suitable for massive wind farms and big wind turbines, however they can be very good spots for the installation of small wind turbines not exceeding 400 KW. The wind turbine installed in ENSET Mohammedia for example is destined to scientific research purpose, and is of domestic size because of the low potential in this region.

2.4. Selecting Best Fit Wind Turbines

The geographic location and wind turbine generator selection affect directly the wind farm economics. Appropriate wind turbine generators are chosen based on the annual energy production. As a first step we estimate the power production for many wind turbines, in order to choose the best matching one for each site in our study. **Figure 4** shows the curves of rated power of the first category of wind turbines: V52-850 KW, V80-2000 KW and V112-3000 KW. This type of wind turbines is destined to sites that are the most qualified for wind farms. For sites with moderate potential we decided to compare more adapted types of wind turbines: E33-335 KW, E40-600KW, and E44-910KW (**Figure 5**). In the ENSET Mohammedia region which has the lowest potential

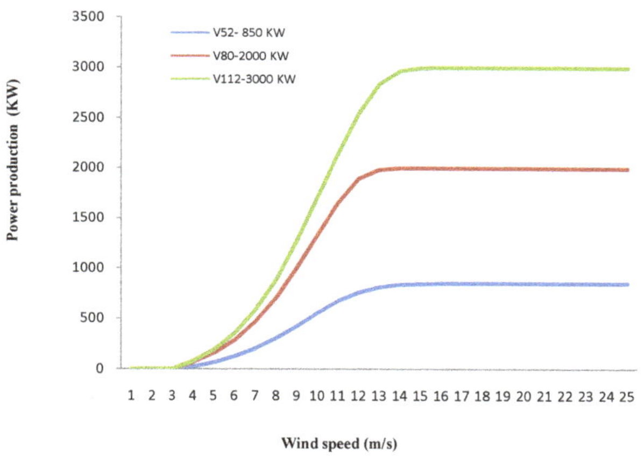

Figure 4. Power curves of Vestas wind turbines.

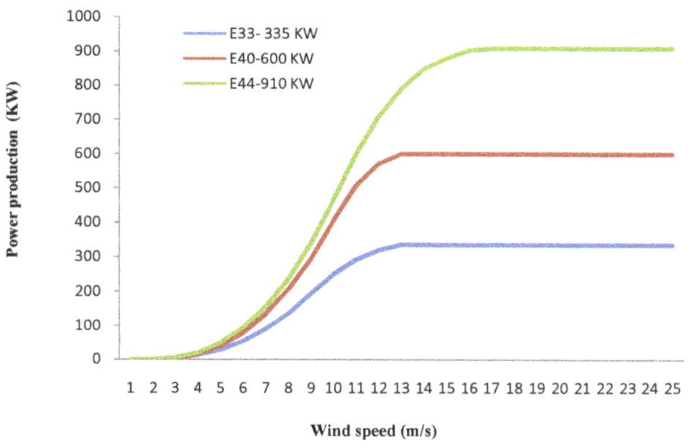

Figure 5. Power curves of Enercon wind turbines.

we used smaller wind turbines. All power curves were obtained from constructor data sheets.

In an attempt to identify the most efficient wind turbine in each site, we were forced to compute a yield factor (F_Y) obtained by dividing the annual energy production of the wind turbine (E_P) by the wind energy density (E_{an}):

$$F_Y = \frac{E_P}{E_{an}} \cdot 100 \tag{9}$$

In **Table 2**, we summarized computed values of the yield factor for each wind turbine. We noticed that efficient choice of the wind turbine is not established only by its annual production. Therefore in the 8 first sites the V80-2MW with an average annual energy production is the most effective investment, because it has the highest yield factor. On the same basis, the E33-335KW is the most adequate wind turbine for sites with average potential. Likewise in the ENSET station, the optimal wind turbine is the Ws-3,2KW.

3. Modeling of the Wind Turbine

In order to enhance the decision support system it is necessary to model the principal wind turbine components, using the V-80 as it appears to be the optimum choice in the windiest sites. In an aim to aggregate our decision system the following models (corresponding to the four main components of a horizontal axis wind turbine) were taken into account in our study for the modeling of the wind turbine.

3.1. Rotor Model

The power that can be produced by a wind turbine at every speed v is given by [3]:

Table 2. Annual energy and Yield factors.

	Wind Turbine energy (E_P)			Yield factor (E_Y)		
Site	V-52	V-80	V-112	V-52	V-80	V-112
El Koudia Al Baida (Tetouan)	4787.3	11464.5	16034.7	5.06	5.12	2.89
Sendouk (Tanger)	3725.1	8928.1	12178.7	7.93	8.03	4.42
Col de Touahar (Taza)	3289.9	7877.6	10682.1	9.64	9.76	5.34
Cap Sim (Essaouira)	3954.5	9466.7	13074.3	7.67	7.76	4.32
S. ELGarn (Tantan)	1951.5	4623.4	5915.9	8.89	8.90	4.59
Tarfaya	3201.9	7656.1	9908.4	2.51	2.54	1.33
Laayoune	5168.1	12479.8	17105.6	1.28	1.31	0.72
Tiniguir (Dakhala)	3703.6	8898.7	11875.9	5.11	5.19	2.79
	E-33	E-40	E-44	E-33	E-40	E-44
My Boussalham (Kenitra)	368.8	544.1	646.6	19.95	16.97	20.15
Had hrara (Safi)	1017.3	1649. 5	2033.7	13.94	13.02	16.06
Salouane (Nador)	249.2	351.8	420.5	19.16	15.58	18.62
	WS-3	AV-7	FL-30	WS-3	AV-7	FL-30
Station ENSET	1.38	7.816	6.532	20.79	14.30	11.95

$$Pot = \frac{1}{2} \cdot \rho \cdot Cp \cdot A \cdot v^3 \tag{10}$$

Cp is the power coefficient of the rotor which can be estimated by [8] [9]:

$$Cp = Cp_{\max} \cdot \exp\left[-\frac{(\ln v - \ln V_d)^2}{2 \cdot (\ln s_n)^2} \right] \tag{11}$$

$Cp_{\max}$ is the coefficient of maximum power for the wind turbine.
V_d, s_n are the operation optimal speed, and operating range parameter of for the wind turbine.

$$V_d \text{ is given by: } V_d = \frac{v_n}{\exp\left[3(\ln s_n)^2\right]} \tag{12}$$

v_n is the nominal wind speed. $Cp_{\max}$ is given by [3] [6]:

$$Cp_{\max} = 0.593\left[\frac{\lambda_{\max} \cdot B^{0.67}}{1.48 + (B^{0.67} - 0.04) \cdot \lambda_{\max} + 0.0025\lambda_{\max}^2} - \frac{1.92\lambda_{\max}^2 \cdot B}{1 + 2\lambda_{\max} \cdot B} \cdot \frac{C_D}{C_L} \right] \tag{13}$$

where

B is the number of pales.

$\lambda_{\max} = \frac{\pi \cdot N}{V_d} \cdot \frac{D_n}{60}$, N is the rotor angular speed.

$\frac{C_D}{C_L}$ is the accuracy given by the ratio between the coefficients of drag and lift, set to 120 [3] [10].

3.2. Gearbox Model

Total Gearbox efficiency ηm depends on the rotor output power $Pot\,[\mathrm{W}]$, the nominal power $Pn\,[\mathrm{W}]$, and the Gearbox's efficiency factor Fm [11] [12].

$$\eta m = 1 + \left[(1 - Fm) \left(\frac{Pn}{4Pot} + \frac{3}{4} \right) \right] \tag{14}$$

$Fm = 0.89 Pn^{0.012}$ and Pn is in kW.

3.3. Generator Model

The total power $Potm$ generated is:

$$Potm = \eta m \cdot Pot \tag{15}$$

The total efficiencyηgof the generator is given by [11] [12]:

$$\eta g = 1 + \left[(1 - Fg) \left(5 \left(\frac{Potm}{Png} \right)^2 + 1 \right) \left(\frac{Png}{6Potm} \right) \right] \tag{16}$$

With:

$Fg = 0.87 Pn^{0.014}$ (Pn kW)

$Png = Pn \cdot Fm \cdot Fg \cdot F_s$,

F_s is service factor of the multiplier [11].

$$Png = PnF_s = \begin{cases} 2 & \text{if } SCS \\ 1.75 & \text{if } PCS \\ 1.25 & \text{if } PVS \end{cases}$$

SCS is the stall-constant-speed, *PCS* is the pitch-constant-speed and *PVS* is the pitch-variable-speed.

By combining the three component models presented above we get the overall power coefficient:

$$Cptot = Cp \cdot \eta m \cdot \eta g \tag{17}$$

Once the total power coefficient determined, it is now possible to calculate the annual electric energy production.

3.4. Energy Model

The annual production of electrical energy E_{pa} W·h/year by a wind turbine rotor working at a range of wind speed $v \in [v_i, v_f]$ having a surface A is given by [3]:

$$E_{pa}(\mathrm{W \cdot h}) = \frac{T}{2} \cdot \rho \cdot A \cdot \int_{v_i}^{v_f} v^3 f_{hub,i} \cdot Cptot \cdot \mathrm{d}v \tag{18}$$

where v_i [m/s] is the cut-in wind speed, and v_f [m/s] is the cut-off wind speed. $f_{hub,i}$ is The Weibull function adapted to the height of the hub.

The boundary layer is the energy exchange area between the atmosphere and the land surface. The Weibull parameters c and k depend of the site characteristics. With these parameters and the height of the hub H_{hub}, it is possible to define the new Weibull parameters c_{hub} and k_{hub} [3]:

$$c_{hub} = c \cdot \left(\frac{H_{hub}}{H_0} \right)^{\alpha} \tag{19}$$

$$k_{hub} = k + 0.03 \cdot H_{hub} + 0.02 \tag{20}$$

with:

$$\alpha = 0.37 - 0.088 \ln c \tag{21}$$

H_0 (m) is the measurement height of the wind, H_{hub} is the height of the hub. Thus, the Weibull function defined at H_{hub} is given by:

$$f_{hub,i} = \left(\frac{k_{hub}}{c_{hub}}\right) \cdot \left(\frac{v}{c_{hub}}\right)^{k_{hub}-1} \cdot \exp\left(-\left(\frac{v}{c_{hub}}\right)^{k_{hub}}\right) \tag{22}$$

4. Results and Discussion

Our decision support model aim to select the optimal wind turbine technology at a specific location, in order to minimize cost and maximize benefits by selecting the most homogenous correlation between technical gear and geographic location. To achieve this objective we developed a method including input variables that represent the technical characteristics of the wind turbines, and other input variables representative of the wind energy available in the site computed from the wind model.

The evolution of the power coefficient Cp is a specific data in every wind turbine. From records made in a wind turbine, we use the power curve given by the manufacturer of the V80- 2MW as a reference. This curve is obtained by the use of procedures and recommendations in the international standards. **Table 3** represents the technical characteristics of the V80-2MW used in this study.

Figure 6 represents a Comparison between the power curve given by the manufacturer and the one computed by the model. The wind power curve shows that between the starting speed and nominal wind speed, the machine extracts the maximum power of wind on the rotor disc, as the two curves are matching. We notice that the two power curves are different for high speeds; this is due to the mechanical constraints that are applied on pales, it is necessary to block the wind turbine running in case of high wind speed.

In **Figure 7**, we plotted the power coefficient Cptot provided by (17).

Knowing the total power coefficient, it is possible to calculate the annual electric energy produced. Thus, we took the site Col de Touahar as an example for detailed study with the wind turbine V80-2MW, and we summarized the rest in the **Table 5**.

- Site: Col de Touahar

First we carefully realized an extrapolation of Weibull parameters at the Hub height given by the **Table 4**. The Weibull distribution at the Hub height is plotted in the **Figure 8**.

According to Equation (18), annual energy produced can be calculated per year for this site:

$$E_{pa} = 10063.95\ \mathrm{MW} \cdot \mathrm{h}$$

E_{pa} Is the annual energy produced in the Col Touahar site using the V80-2MW wind turbine. The **Table 5** summarizes the results of other sites taken into account in our study.

5. The Artificial Neural Network (ANN) to Forecast Wind Speed in Other Regions in Morocco

Our study took into consideration many sites in Morocco; however our model and the data acquired are not sufficient to provide results in sites where the wind speed data is not available. In an attempt to predict the wind speed in regions where data is missing or not complete, we realized an artificial neural network to forecast monthly wind speed in the region of Boujdour as an example of location where metrological instrumentation is not available [13]. Our ANN **Figure 9** establishes a correspondence between longitude, latitude, temperature, elevation and wind speed, based on measures gathered from 21 stations inside the Moroccan territory.

The ANN used is a multilayer Perceptron (MLP) type that provides FeedForward architecture [14] [15]; the input and output layer are respectively constituted of 4 and 12 neurons respectively, the hidden layer contains 20

Table 3. Wind turbine design variables.

Wind turbine V-80						
D (m)	Pn (kW)	Vn (m/s)	H (m)	N (tr/mn)	Type of control	p
80	2000	16	78	16.7	PVS	3

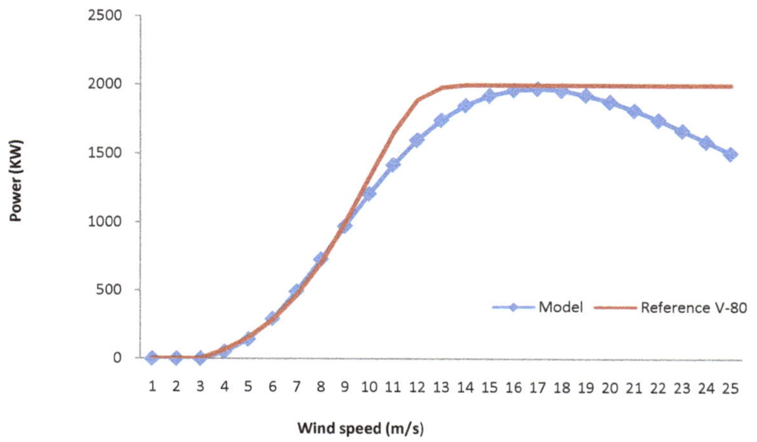

Figure 6. Power, comparison of model results with measured data for wind turbine V-80.

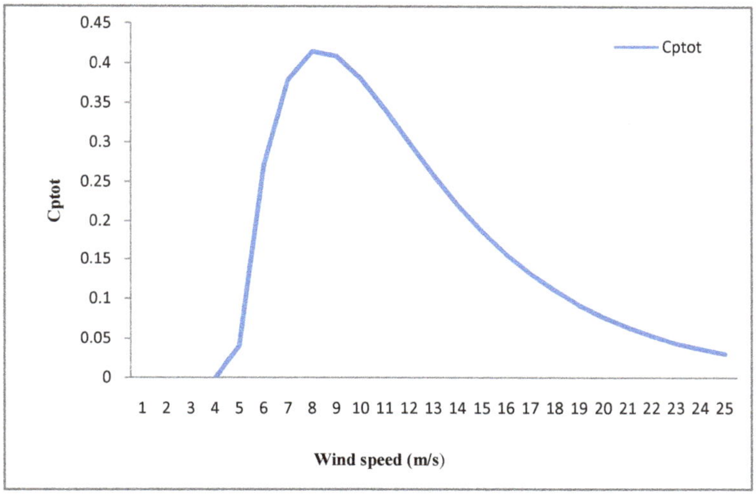

Figure 7. Coefficient of maximum power for wind turbine as obtained by Equation (17).

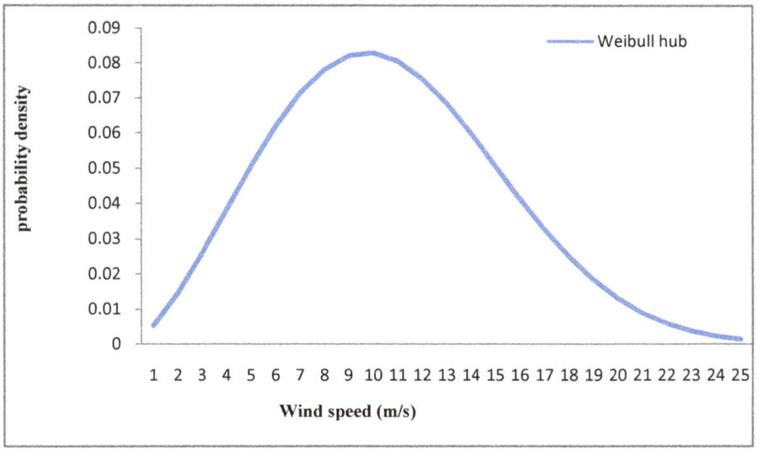

Figure 8. Weibull distribution of wind speeds at the hub height Site "Col de Touahar".

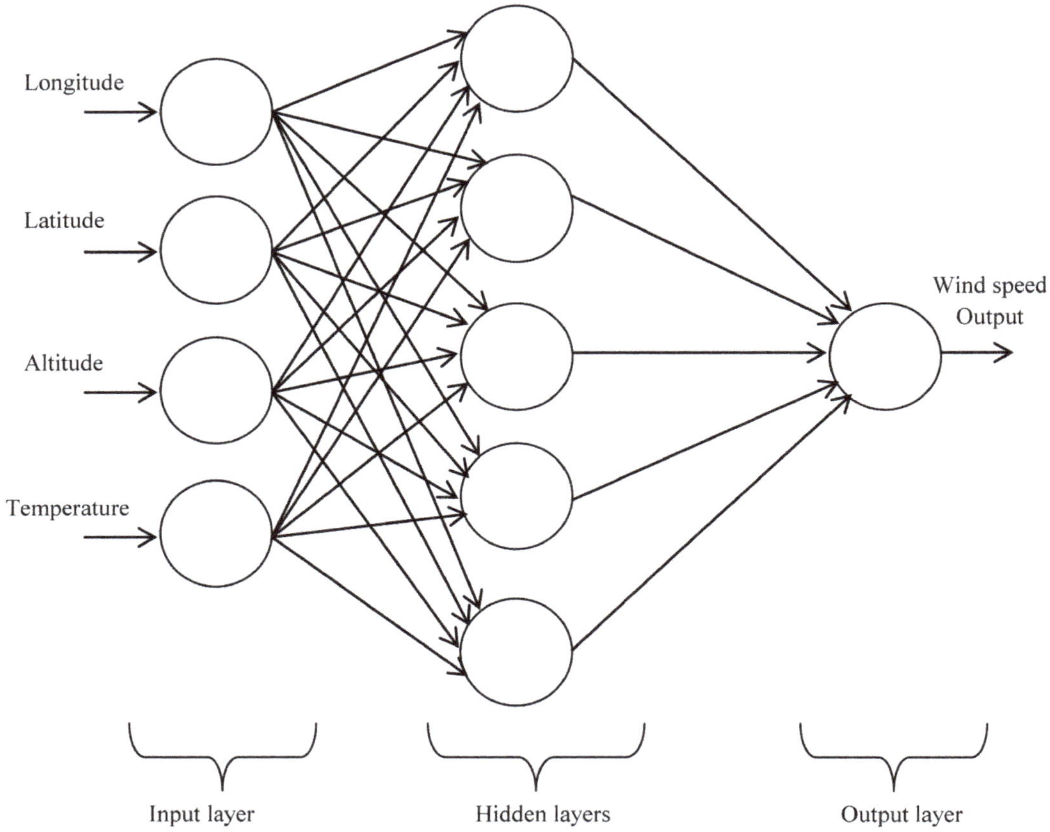

Figure 9. Conceptual idea of the built artificial neural network.

Table 4. The site characteristics at the hub height: Col de Touahar.

The site characteristics at the hub height: Col de Touahar		
k_{hub}	c_{hub} (m/s)	H_{hub} (m)
2.46	12.03	78

Table 5. The results of the wind energy available at hub height.

Site	H_0 (m)	H_{hub} (m)	c_{hub} (m/s)	k_{hub}	E_{pa} (MWh)
El Koudia Al Baida (Tétouan)	9	78	17.16	3.19	13354.7
Sendouk (Tanger)	9	78	14.67	2.92	12472.84
Cap Sim (Essaouira)	40	78	12.12	2.33	9935.36
Col de Touahar (Taza)	20	78	12.03	2.46	10063.95

neurons. This configuration is optimal because we can get a minimal error $\varepsilon(n)$ in the entire output given by the following relation [16]:

$$\varepsilon(n)=\frac{1}{2}\sum_{J}e_j^2(n) \tag{23}$$

Where, e_j is the error for the output node j in the nth data point, given by the following relation:

$$e_j(n)=d_j(n)-y_j(n) \tag{24}$$

d and *y* are successively the target value and the value produced by the perceptron.

Geographical coordinates **Table 6**: longitude, latitude, altitude, and average temperature monthly observed represent the input for the ANN used in this study.

Temperatures used for this study, displayed in **Table 7**, are the average monthly recorded degrees token from the NASA: Atmospheric Science Data Center [17].

The average wind speed recorded monthly from several years of data **Table 8** in the 21 sites used for this ANN approach will be in the output layer.

The data collected in the previous table was also collected at different heights, it is then also necessary to extrapolate the measures at a same height too, using the relation (5). We will use the same relation (21) to get the surface roughness. **Table 9** summarizes the results obtained after extrapolation to 10 meters.

Normalization is a very important procedure in the training phase and the testing of the artificial neural network, because it guarantees that outputs and inputs are on the same weight scale [18]. Data was normalized linearly in order to values ranging from 0 to 1 using the following relation:

$$X_{Norm} = \frac{X - X_{\min}}{X_{\max} - X_{\min}} \quad (25)$$

Predicted wind speed results in the unknown site of Boujdour, using the ANN for a height of 10 m like the data used before are displayed in **Table 10**.

As performed previously we can extrapolate the obtained results to a desired height, we give the following

Table 6. Geographical coordinates of the different sites used for the study [2].

	Sites	Longitude	Latitude	Altitude (m)
1	Akhfennir	−12.05	28.09	35
2	Bouznika	−7.17	33.79	0
3	Cap cantin (Safi)	−9.23	32.31	150
4	Cap sim (Essaouira)	−9.39	31.81	100
5	Dakhla	−15.83	23.62	100
6	El gaada (Tiznit)	−6.58	33.56	150
7	Fnidek	−5.35	35.84	80
8	El Koudia al baida (Tetouan)	−5.51	35.82	400
9	Lamdint (Taroudant)	−7.38	30.25	1700
10	Myboussalham	−6.4	34.52	133
11	Rabat	−6.74	33.98	132
12	Sadane	−5.7	35.78	400
13	Sahb el harcha (Tan-tan)	−11.38	28.42	30
14	Salouane (Nador)	−2.93	35.16	150
15	Sidi garn (Tan-tan)	−11.3	28.52	48
16	Tagant (Essaouira)	−9.83	31.12	265
17	Tamagrout (Zagoura)	−7.48	31.38	800
18	Tan-tan (Port)	−11.34	28.48	60
19	Tarfaya	−12.93	27.91	100
20	Torreta (Tetouan)	−5.4	35.54	208
21	Touahar (Taza)	−4	34.22	510

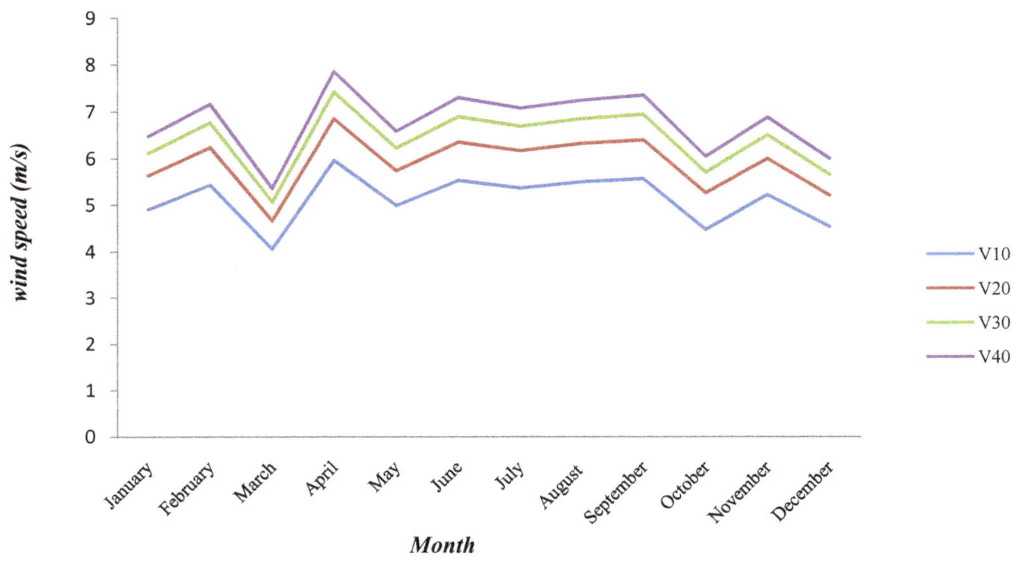

Figure 10. Plotted wind speed data obtained by the ANN for Boujdour.

Table 7. Temperaturesused for the study [17].

	January	February	March	April	May	June	July	August	September	October	November	December
1	16.7	17.3	18.3	18.5	19.5	21	22.8	23.5	22.9	21.9	20.1	18
2	13.5	14.2	15.8	16.6	18.7	21.2	23.5	23.6	22.4	20.1	17.3	15.1
3	14.7	15.2	16.4	17.2	18.9	21	23.2	23.5	22.5	20.5	18.2	16.2
4	13.4	14.7	16.6	17.6	19.6	22.3	25.3	25.3	23.4	20.7	17.5	14.8
5	17.8	19.1	20.4	21.3	22.8	25.3	26.9	27.6	27.1	25	22.1	19.2
6	10.4	12.1	14.9	16.3	19.4	23.2	26.4	25.9	23.1	19.5	15.2	12
7	12.5	13.2	15.1	16.5	19	22.4	25.1	25	22.9	19.8	16.3	13.8
8	12.5	13.2	15.1	16.5	19	22.4	25.1	25	22.9	19.8	16.3	13.8
9	7.13	9.83	13.3	16.3	20	24.8	28.6	28.2	24.1	18.6	12.6	8.42
10	12.2	13.5	16	17.4	20.2	23.8	26.6	26	23.9	20.6	16.7	13.8
11	10.4	12.1	14.9	16.3	19.4	23.2	26.4	25.9	23.1	19.5	15.2	12
12	12.5	13.2	15.1	16.5	19	22.4	25.1	25	22.9	19.8	16.3	13.8
13	16.3	17.3	18.4	18.6	19.3	20.8	23	23.7	22.9	21.9	20	17.8
14	12.1	13	14.8	16.6	19.3	22.8	25.5	25.8	23.4	20	16.1	13.4
15	16.3	17.3	18.4	18.6	19.3	20.8	23	23.7	22.9	21.9	20	17.8
16	13.4	14.7	16.6	17.6	19.6	22.3	25.3	25.3	23.4	20.7	17.5	14.8
17	7.74	10.1	13.3	15.4	18.7	22.9	26.7	26.3	22.8	18.2	12.9	9.13
18	16.3	17.3	18.4	18.6	19.3	20.8	23	23.7	22.9	21.9	20	17.8
19	15.8	17.5	19.4	20	21.4	23.4	25.8	26.5	25.3	23.4	20.3	17.3
20	12.5	13.2	15.1	16.5	19	22.4	25.1	25	22.9	19.8	16.3	13.8
21	8.3	10.3	13.4	15.8	19.3	23.8	27.1	26.5	22.6	18.4	13.3	9.75

Table 8. Average wind speed recorded monthly used for the study [2].

	H (m)	January	February	March	April	May	June	July	August	September	October	November	December
1	10	5.1	5.8	6.1	6.5	5.2	4.8	4.5	4.3	4.2	4.18	4.9	5.1
2	9	3.8	4.3	3.9	4.5	4.2	3.9	3	3.5	3.4	4.3	3.3	3.8
3	10	5.64	5.64	6	5.6	5.4	4.4	5.2	5.6	4.4	5.64	5.64	5.64
4	10	6	6.9	5.8	7.6	7.9	8.9	9.6	9.4	7	6.2	6.7	5.7
5	9	7.4	7.49	7.54	8.54	9.14	10.1	10.5	9.3	8.49	7.16	6.75	6.23
6	9	4.7	4	5.7	4.5	5	4.9	4.3	4.6	4.6	3.5	5.3	4.3
7	9	5.81	5.24	5.53	6.54	5.1	6.13	5.96	6.02	5.99	5.81	5.81	5.81
8	9	10.4	11.7	9.71	10.2	8.4	11	11	10.4	10.5	8.2	10.9	9.6
9	9	4.7	4.5	6.1	6.3	6.6	5.9	5.3	5.3	6.8	4.4	4.5	8.1
10	9	2.5	3.71	4.89	4.02	4.32	4.16	4.35	3.97	3.74	3.96	2.96	2.72
11	9	4.7	3.58	3.72	4.1	4	3.72	3.63	3.62	3.55	3.85	3.85	3.85
12	10	9.2	8.5	8.8	8	8	6.5	7.9	6.8	6	7.9	8.9	9.2
13	10	5.73	5.7	4.82	5.5	5.28	4.87	4.26	4.4	3.76	4.24	5.29	5.17
14	10	4.2	4.67	4.17	5.03	3.2	3.17	2.93	3.37	3.43	3	3.27	3.34
15	10	7.62	7.98	8.23	7.73	5.79	5.3	5.1	5.12	4.3	4.9	5.32	4.41
16	10	6	5.2	6	5.9	7.3	7.6	8.5	7.1	6	5.5	6.2	5.6
17	9	2.5	3.6	4.6	4.3	5.3	5.5	4.8	4.2	4.4	3.5	2.7	2.3
18	10	6	5.6	6.05	5.8	5.7	5.3	4.58	4.5	4.22	4.32	5.35	5.27
19	10	5.39	7.08	6.34	7.39	6.5	7.23	6.73	7.18	5.84	6.23	6.16	5.86
20	9	5.81	5.24	5.53	6.54	5.1	6.13	5.96	6.02	5.99	5.81	5.81	5.81
21	10	6.87	8.4	7.5	8.57	7.07	8.67	8.33	8.23	8.3	7.27	7.8	7.6

Table 9. Results obtained after extrapolation to 10 meters.

	January	February	March	April	May	June	July	August	September	October	November	December
1	5.1	5.8	6.1	6.5	5.2	4.8	4.5	4.3	4.2	4.18	4.9	5.1
2	3.88	4.39	3.98	4.6	4.29	3.98	3.06	3.57	3.47	4.39	3.37	3.88
3	5.64	5.64	6	5.6	5.4	4.4	5.2	5.6	4.4	5.64	5.64	5.64
4	6	6.9	5.8	7.6	7.9	8.9	9.6	9.4	7	6.2	6.7	5.7
5	7.56	7.65	7.7	8.72	9.33	10.3	10.8	9.5	8.67	7.31	6.89	6.36
6	4.8	4.09	5.82	4.6	5.11	5	4.39	4.7	4.7	3.57	5.41	4.39
7	5.93	5.35	5.65	6.68	5.21	6.26	6.09	6.15	6.12	5.93	5.93	5.93
8	10.6	11.9	9.92	10.4	8.58	11.2	11.2	10.6	10.7	8.37	11.1	9.8
9	4.8	4.6	6.23	6.43	6.74	6.03	5.41	5.41	6.94	4.49	4.6	8.27
10	2.55	3.79	4.99	4.11	4.41	4.25	4.44	4.05	3.82	4.04	3.02	2.78
11	4.8	3.66	3.8	4.19	4.09	3.8	3.71	3.7	3.63	3.93	3.93	3.93

Continued

12	9.2	8.5	8.8	8	8	6.5	7.9	6.8	6	7.9	8.9	9.2
13	5.73	5.7	4.82	5.5	5.28	4.87	4.26	4.4	3.76	4.24	5.29	5.17
14	4.2	4.67	4.17	5.03	3.2	3.17	2.93	3.37	3.43	3	3.27	3.34
15	7.62	7.98	8.23	7.73	5.79	5.3	5.1	5.12	4.3	4.9	5.32	4.41
16	6	5.2	6	5.9	7.3	7.6	8.5	7.1	6	5.5	6.2	5.6
17	2.55	3.68	4.7	4.39	5.41	5.62	4.9	4.29	4.49	3.57	2.76	2.35
18	6	5.6	6.05	5.8	5.7	5.3	4.58	4.5	4.22	4.32	5.35	5.27
19	5.39	7.08	6.34	7.39	6.5	7.23	6.73	7.18	5.84	6.23	6.16	5.86
20	6.3	6.54	6.46	7.05	6.1	6.67	7.38	6.57	6.51	6.4	7.18	7.73
21	6.87	8.4	7.5	8.57	7.07	8.67	8.33	8.23	8.3	7.27	7.8	7.6

Table 10. Wind speed Results obtained by the ANN at a height of 10 m.

	January	February	March	April	May	June	July	August	September	October	November	December
Wind speed (V_{20}) (m/s)	4.91	5.43	4.06	5.96	4.99	5.53	5.36	5.49	5.56	4.57	5.21	4.53

Table 11. Extrapolated wind speed Results obtained by the ANN at a height 20, 30, 40 m.

	January	February	March	April	May	June	July	August	September	October	November	December
V_{20}	5.64	6.24	4.66	6.85	5.73	6.35	6.16	6.31	6.39	5.25	5.98	5.2
V_{30}	6.12	6.76	5.06	7.42	6.22	6.89	6.68	6.84	6.93	5.69	6.49	5.64
V_{40}	6.48	7.16	5.36	7.86	6.58	7.3	7.07	7.24	7.34	6.03	6.87	5.98

extrapolated data at different heights (20, 30, 40 m), **Table 11** summarizes the results, **Figure 10** displays the wind velocity prediction for the four different heights.

In this part we obtained the wind speed data in a region where it was impossible to get wind speed records. The artificial neural network used has proven its reliability in predicting wind speed for the site of Boujdour. This approach allowed us to get the missing wind data, which is the most important parameter in our decision support system for optimal choice of wind turbine, when obtained; the process of assessing wind potential in order to install the right technology can be continued using the same procedure.

6. Conclusions

In this study, we have modeled and evaluated the available wind energy per year in many Moroccan sites. To achieve this objective, we have modeled the wind speed data using the Weibull probability density function and determined values for both the shape and scale parameters, in order to estimate wind energy potentials for the each studied location. Those parameters were extrapolated at a same height to assess the available wind potential in these sites.

Thus, we can classify the studied areas into three categories: the greatest wind potential regions, the medium potential regions, and the lowest potential regions, like the ENSET Mohammedia station, where the average wind speed and the available energy do not exceed respectively 2.61 (m/s) and 420.3 (KWh/m^2).

Through this study, we could choose the appropriate wind turbine for each site. Medium wind potential regions like Moulay Bousselham (Kenitra), and salouane (Nador) are not suited for the installation of big turbines but can accommodate wind turbine not exceeding 400 KW.

Sites of Al Koudia Al Baida (Tetouan) Sendouk (Tanger), Cap Sim (Essaouira) and Col de Touahar (Taza) have very substantial wind resources with an average annual wind speeds between 8 and 11 (m/s). The energy that can be produced in these sites using the V80-2MW wind turbine model is between 9935.36 and 13354.7 MWh, which makes them more suited for the installation of wind farms for electricity production connected to the national grid. This study has also shown that when wind speed data are not available in regions like Boujdour, the use of an artificial neural network designed to predict wind speed records can be a very reliable forecasting tool.

In terms of perspectives our prime objective is to propose effective methods to forecast wind speed in real time. This very important wind-related engineering topic is crucial in the right management of a wind farm.

References

[1] Wind Energy Barometer. EUROBSERV'ER, February 2014. http://www.energies-renouvelables.org/observ-er/stat_baro/observ/baro-jde14-gb.pdf

[2] Enzili, M., Affani, F. and Nayssa, A. (2007) Les ressources éoliennes du Maroc.

[3] Ouammi, A., Ghigliotti, V., Robba, M., *et al.* (2012) A Decision Support System for the Optimal Exploitation of Wind Energy on Regional Scale. *Renewable Energy*, **37**, 299-309. http://dx.doi.org/10.1016/j.renene.2011.06.027

[4] Justus, C.G., Hargraves, W.R. and Yalcin, A. (1976) Nationwide Assessment of Potential Output from Wind-Powered Generators. *Journal of Applied Meteorology*, **15**, 673-678. http://dx.doi.org/10.1175/1520-0450(1976)015%3C0673:naopof%3E2.0.co;2

[5] Gualtieri, G. and Secci, S. (2012) Methods to Extrapolate Wind Resource to the Turbine Hub Height Based on Power Law: A 1-h Wind Speed vs. Weibull Distribution Extrapolation Comparison. *Renewable Energy*, **43**, 183-200. http://dx.doi.org/10.1016/j.renene.2011.12.022

[6] Diveux, T., Sebastian, P., Bernard, D., *et al.* (2001) Horizontal Axis Wind Turbine Systems: Optimization Using Genetic Algorithms. *Wind Energy*, **4**, 151-171. http://dx.doi.org/10.1002/we.51

[7] Ouammi, A., Dagdougui, H., Sacile, R., *et al.* (2010) Monthly and Seasonal Assessment of Wind Energy Characteristics at Four Monitored Locations in Liguria Region (Italy). *Renewable and Sustainable Energy Reviews*, **14**, 1959-1968. http://dx.doi.org/10.1016/j.rser.2010.04.015

[8] Kiranoudis, C.T. and Maroulis, Z.B. (1997) Effective Short-Cut Modelling of Wind Park Efficiency. *Renewable Energy*, **11**, 439-457.http://dx.doi.org/10.1016/s0960-1481 (97)00011-6

[9] Kiranoudis, C.T., Voros, N.G. and Maroulis, Z.B. (2001) Short-Cut Design of Wind Farms. *Energy Policy*, **29**, 567-578. http://dx.doi.org/10.1016/s0301-4215 (00)00150-6

[10] Arbaoui, A. (2006) Aide à la décision pour la définition d'un système éolien, adéquation au site et à un réseau faible. Thèse de Doctorat, ENSAM, Paris. https://tel.archives-ouvertes.fr/pastel-00002722/document

[11] Harrison, R. and Jenkins, G. (1994) Cost Modelling of Horizontal Axis Wind Turbines (Phase 2). ETSU W/34/00170/ REP, University of Sunderland, Sunderland. http://www.opengrey.eu/item/display/10068/633220

[12] Harrison, R., Jenkins, G. and Taylor, R.J. (1989) Cost Modelling of Horizontal Axis Wind Turbines—Results and Conclusions. *Wind Engineering*, **13**, 315-323.http://www.opengrey.eu/item/display/10068/633220

[13] Ouammi, A., Zejli, D., Dagdougui, H. and Benchrifa, R. (2012) Artificial Neural Network Analysis of Moroccan Solar Potential. *Renewable and Sustainable Energy Reviews*, **16**, 4876-4889. http://dx.doi.org/10.1016/j.rser.2012.03.071

[14] Ata, R. (2015) Artificial Neural Networks Applications in Wind Energy Systems: A Review. *Renewable and Sustainable Energy Reviews*, **49**, 534-562. http://dx.doi.org/10.1016/j.rser.2015.04.166

[15] Gardner, M.W. and Dorling, S.R. (1998) Artificial Neural Networks (the Multilayer Perceptron)—A Review of Applications in the Atmospheric Sciences. *Atmospheric Environment*, **32**, 2627-2636. http://dx.doi.org/10.1016/s1352-2310(97)00447-0

[16] Velo, R., López, P. and Maseda, F. (2014) Wind Speed Estimation Using Multilayer Perceptron. *Energy Conversion and Management*, **81**, 1-9. http://dx.doi.org/10.1016/j.enconman.2014.02.017

[17] NASA Atmospheric Science Data Center. http://eosweb.larc.nasa.gov

[18] Chaturvedi, D.K. (2008) Soft Computing. *Studies in Computational Intelligence*, **103**. http://dx.doi.org/10.1007/978-3-540-77481-5

11

Efficient Choice of a Multilevel Inverter for Integration on a Hybrid Wind-Solar Power Station

Tajeddine Khalili, Abdelhadi Raihani, Hassan Ouajji, Omar Bouattane, Fouad Amri

Lab. SSDIA, ENSET Mohammédia, Hassan II University of Casablanca, Morocco
Email: khalili.tajeddine@gmail.com

Abstract

DC/AC converters are very important components that have to be chosen efficiently for each type of power station. In this article, we present in details, a comparison between three different architectures of multilevel inverters, the flying capacitor multilevel inverter (FCMLI), the diode clamped multilevel inverter (DCMLI), and the cascaded H-bridge multilevel inverter (CHMLI). Thus the comparison is focused on the output voltage quality, the complexity of the power circuits, the cost of implementation, and the influence on a power bank inside the renewable power station. We also investigate trough simulation the efficient number of levels and suitable characteristics for the CHMLI that showed the most promising performance. The study uses Matlab Simulink platform as a tool of simulation, and aim to choose the most qualified inverter, for a potential insertion on a hybrid renewable energy platform (wind-solar). In all the simulations we use the same PWM control type (SPWM).

Keywords

Multilevel Inverters, FCMLI, DCMLI, CHMLI, Hybrid Power Station, SPWM, Power Bank

1. Introduction

Multilevel inverters have emerged, as a new option for DC/AC conversion [1] for high power and renewable energy applications [2] [3]. The multilevel inverter synthesizes basically a stair wave voltages staged in several DC levels [4]. There are many topologies of multilevel inverters, some are basic, others are a combination of different types, and however they can be classified into three basic structures: the FCMLI, the DCMLI, and the

CHMLI. All multilevel inverter topologies produce almost a similar output, which consists of voltage steps, thus providing a higher quality voltage [5], compared to classic 2 level inverters, and as close as possible to the waveform intended to be produced. Multilevel inverters are an efficient alternative to solve problems posed by conventional inverters; their main characteristics are having lower switching losses, lower solicitation on switching devices, higher power applications, and they filter easier since the output consists of DC voltage levels. In literature, there are many research activities dealing with a special topology trying to raise the number of levels [6], or developing new methods for harmonic minimization [7] [8]. In this paper we decided to focus our study on the three 5 level basic multilevel inverters, as they are the most widespread [9], and investigate the most efficient parameters for the CHMLI, as it was the best candidate in our comparative study. An analysis using Matlab Simulink tool was made, to compare between the performances of the different 5 levels architectures (voltage quality, the complexity of the power circuits, the cost of implementation [10], and the impact on a power bank inside the renewable power station) using the same PWM techniques [11] in order to choose the best inverter suited for a hybrid wind-solar renewable power station. An evaluation of the effect of levels configuration on the CHMLI performance was made to determinate the most effective parameters of this inverter. Thus a simulative testing of the CHMLI inverter was made from level 5 to 15.

2. The Hybrid Renewable Power Station with a Classic Inverter

The hybrid renewable power station that the inverter is intended to, is composed basically of two sources; photovoltaic panels and a wind turbine. In this part we simulate the work of the power station, with a normal full bridge inverter, in order to compare its performance to the rest of multilevel inverters. The **Figure 1** presents the diagram used on Matlab Simulink to simulate the running of the inverter.

Figure 2 represents the wave form of the output voltage, and current of the full bridge used in the simulation. The output current of the full bridge inverter was simulated with a (R = 50 Ω, L = 100 mH) load.

The total harmonic distortion is given by the well-known expression:

$$THD = \frac{\sqrt{\sum_{k=2}^{n} V_k}}{V_1} \tag{1}$$

V_k is the k harmonic and V_1 is the fundamental. The **Figure 3** shows the output voltage spectrum; we found that the classic full bridge inverter that we used has an output voltage THD of 76.51%, which is very high compared to the signal purity wanted in the hybrid power station.

In the power bank where we used a model of 20 lead acid batteries of 12 Vdc, and a rated capacity of 6.5 Ah each, we noticed a general power drop of 57.3 mv during a 1 sec time simulation.

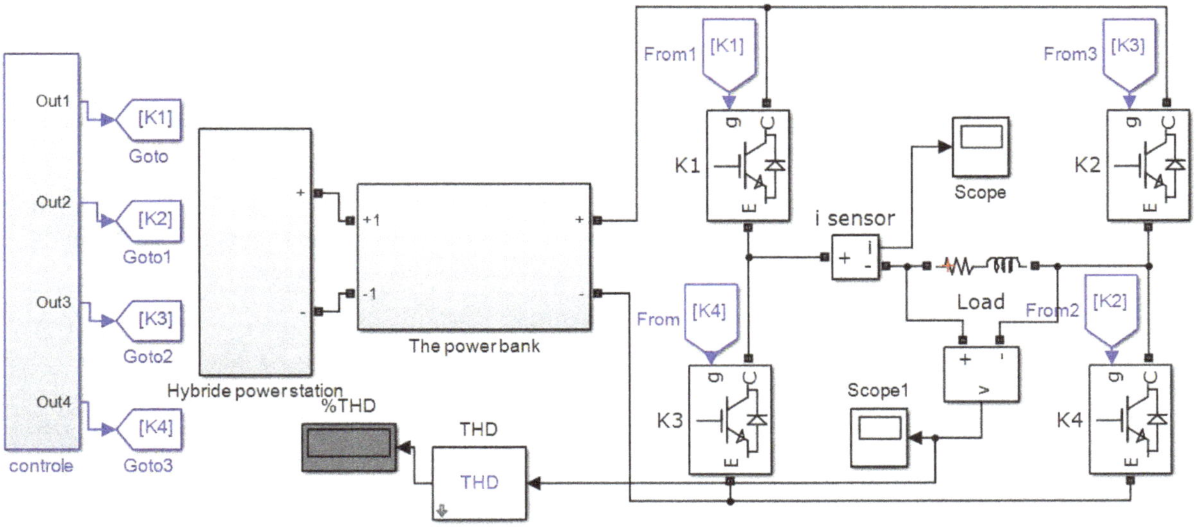

Figure 1. Diagram of the full bridge inverter used on Matlab Simulink.

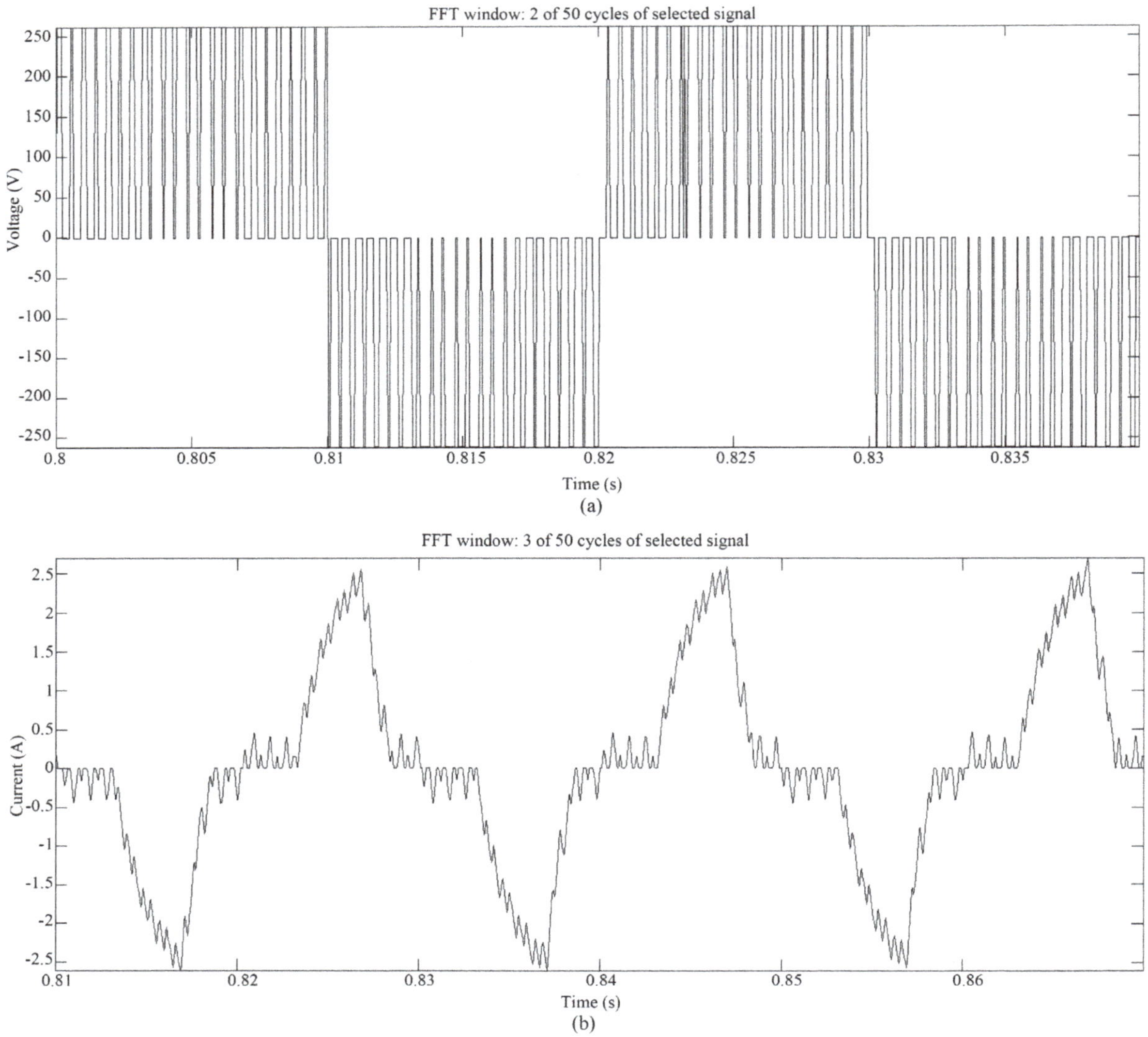

Figure 2. Output simulation results of the full bridge: (a) voltage; (b) current.

3. The Hybrid Renewable Power Station with the Studied Multilevel Inverters

3.1. The PWM Law Command and Power Bank Used for All Multilevel Inverters in the Simulation

In this entire article we used the same SPWM law command, which is the most widespread command in multilevel inverters field. Its principle is to compare a reference signal V_{ref} to four saw tooth carriers **Figure 4**.

The signal references for the three phases used in the simulation are expressed as follows:

$$V_{1ref} = A \cdot m \cdot \sin(wt) \tag{2}$$

$$V_{2ref} = A \cdot m \cdot \sin\left(wt - \frac{2\pi}{3}\right) \tag{3}$$

$$V_{3ref} = A \cdot m \cdot \sin\left(wt - \frac{4\pi}{3}\right) \tag{4}$$

$A = 10$: The magnitude.

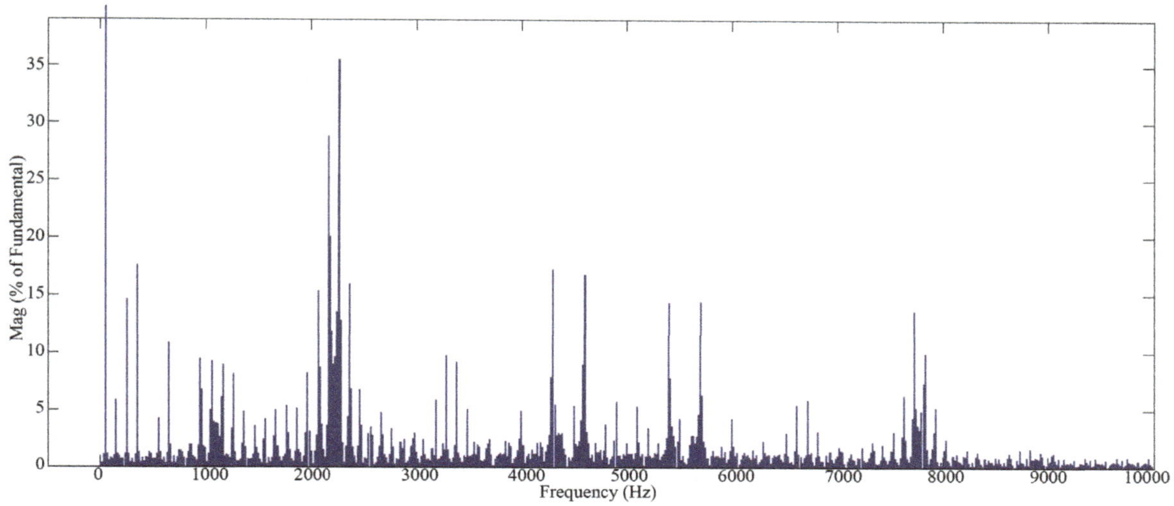

Figure 3. Output voltage spectrum analyze of the full bridge inverter.

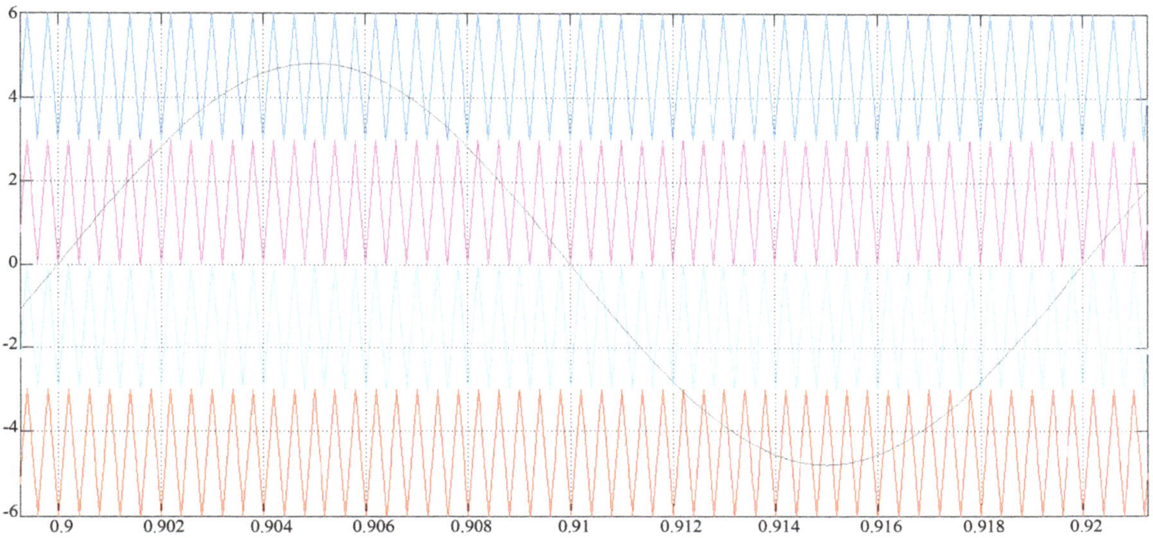

Figure 4. The reference signal with the four saw tooth carriers used for the command.

$m = 0.6$: The modulation index.

$f = w/2\pi$: The frequency.

In order to see the influence of each multilevel inverter on storage units, we modeled a power bank out of 20 lead-acid batteries on Matlab Simulink having the following simulation characteristics:

- Nominal voltage: 12 V
- Rated capacity: 6.5 A·h
- Initial state of charge: 100%
- Maximum capacity: 6.77 A·h
- Fully charged voltage: 13.06 V
- Nominal discharge current: 1.3 A
- Internal resistance: 0.02 Ω

3.2. The Flying Capacitor Multilevel Inverter (FCMLI)

In this type of multilevel inverters, the output can be expressed as the different possible combinations of connection of the capacitors [12]. This structure is similar to the DCMLI but uses capacitors instead of diodes to set the voltage levels.

If we examine the power circuit of the FCMLI **Figure 5(a)** we can find some useful relations; if m is the number of levels, k the number of capacitors at the DC side, l the number of switches per phase, and N_c the number of clamping capacitors then:

$$m = k + 1 \tag{5}$$

$$l = 2(m-1) \tag{6}$$

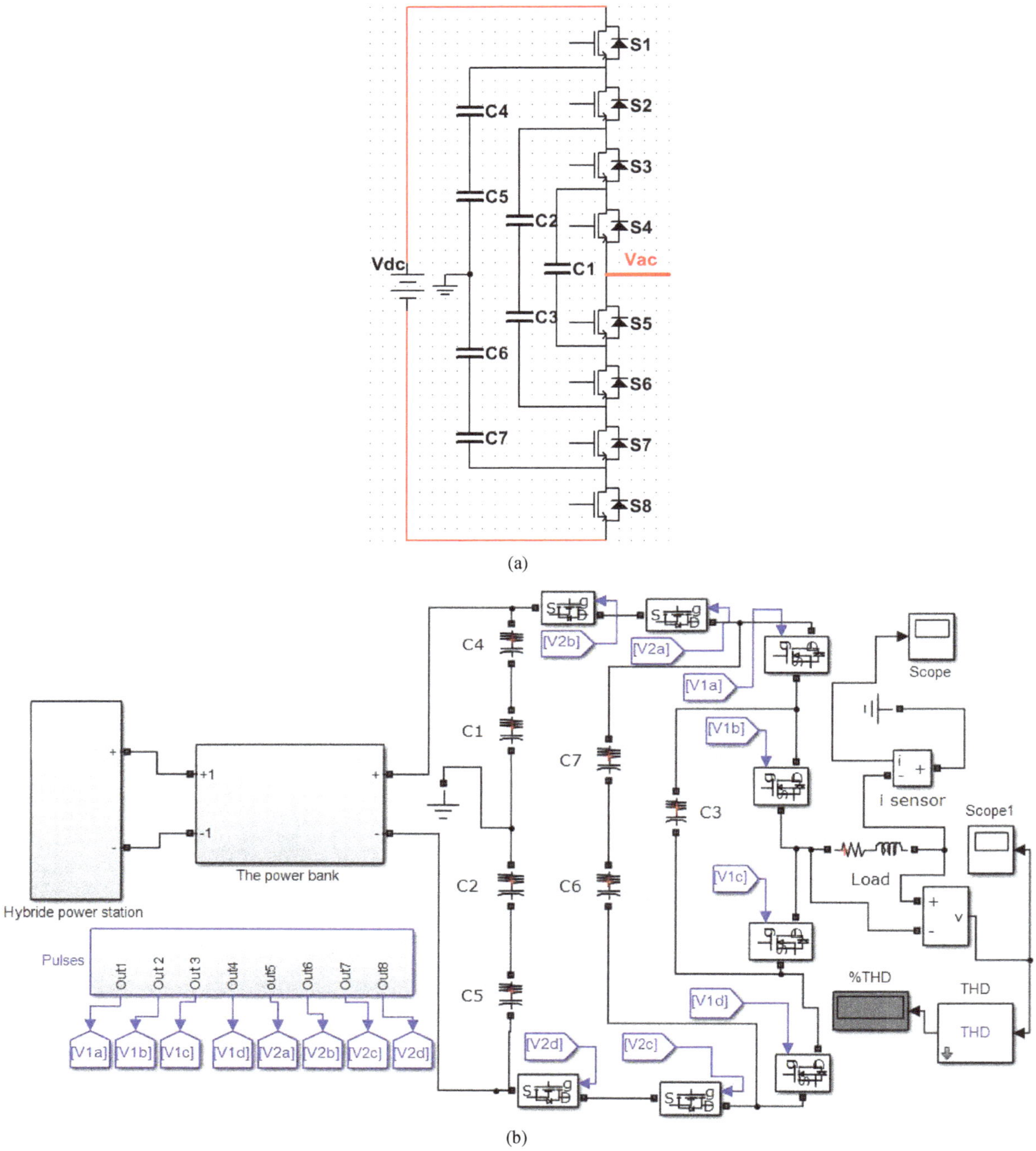

Figure 5. The architecture of the FCMLI: (a) basic diagram; (b) simulation diagram used in Matlab Simulink.

$$N_c = \frac{(m-1)(m-2)}{2} \tag{7}$$

In this structure, the boot is more complex, because this topology has the disadvantage of requiring the pre-charge of the capacitors before beginning to operate as an inverter. However the stress on voltage switches is balanced with the number of levels, and it provides different combinations of switching states for the same output level.

To analyze this topology, only one phase will be taken in consideration. We used the diagram in **Figure 5(b)** on Matlab Simulink to study the 5 level FCMLI.

Figure 6 represents the output simulation results of the voltage, current, and voltage spectrum analyze successively. The simulation showed that the flying capacitor multilevel inverter that we used has an output voltage level THD of 35.34%. In the same power bank, and using the same load (R = 50 Ω, L = 100 mH), we have noticed a general power drop of 50.4 mv during a 1sec time simulation.

3.3. The Diode Clamped Multilevel Inverter (DCMLI)

This topology **Figure 7(a)** was first presented by [13] [14]; its main function is to use locking diodes to synthesize a sine wave from several voltage levels, typically obtained from capacitors that work as DC sources. The capacitors used are connected in series to divide the voltage.

If j is the number of clamping diodes per phase; we can find the following relations:

$$m = k + 1 \tag{8}$$

$$l = 2(m-1) \tag{9}$$

$$j = (m-1)(m-2) \tag{10}$$

In this structure the effort that each device must handle decreases as the number of levels increases. And yet interlocking diodes may handle the stress of more than one level. We used a single phase diagram to analyze this structure; the diagram **Figure 7(b)** was implemented on Matlab Simulink, Using the same command pattern as previously.

Figure 8 represents the output simulation results of the voltage, current, and voltage spectrum analyze successively. The simulation showed that the diode clamped multilevel inverter that we used has an output voltage level THD of 38.21%. In the same power bank, and using the same load (R = 50 Ω, L = 100 mH), we have noticed a general power drop of 12.3 mv during a 1sec time simulation.

3.4. The Cascaded H-Bridge Multilevel Inverter (CHMLI)

This topology [15] [16] performs the same function as the above, a sinusoidal voltage is generated from different DC sources; it is based on cascading full bridge inverters **Figure 9(a)**. This configuration is widely used in applications where the AC sources are destined to variable speed drives, and high voltage.

Each module may generate for itself different output voltage levels, 0, +V, or –V. By using different combinations we obtain the stepped AC voltage output. Generally if we consider Ns the number of independent DC sources per phase, then:

$$m = 2N_s + 1 \tag{11}$$

$$l = 2(m-1) \tag{12}$$

This converter can avoid the use of interlocking diodes or balancing capacitors. We can also get a low harmonic distortion by controlling the firing angles of different voltage levels. Furthermore this significant topology offers great flexibility to increase the number of levels. We used a single phase CHMLI to simulate the inverter; the diagram **Figure 9(b)** was implemented on Matlab Simulink.

Figure 10 represents the output simulation results of the voltage, current, and voltage spectrum analyze successsively. The simulation showed that the diode clamped multilevel inverter that we used has an output voltage level THD of 35.12%. To simulate the impact of this inverter on the power bank, we divided it into two parts of 10 batteries each in a way that each section gives a 120 Vdc. We noticed that in the first section there is a general power drop of 58 mV while there is in the second section a power drop of 67 mV, during 1 sec time simulation.

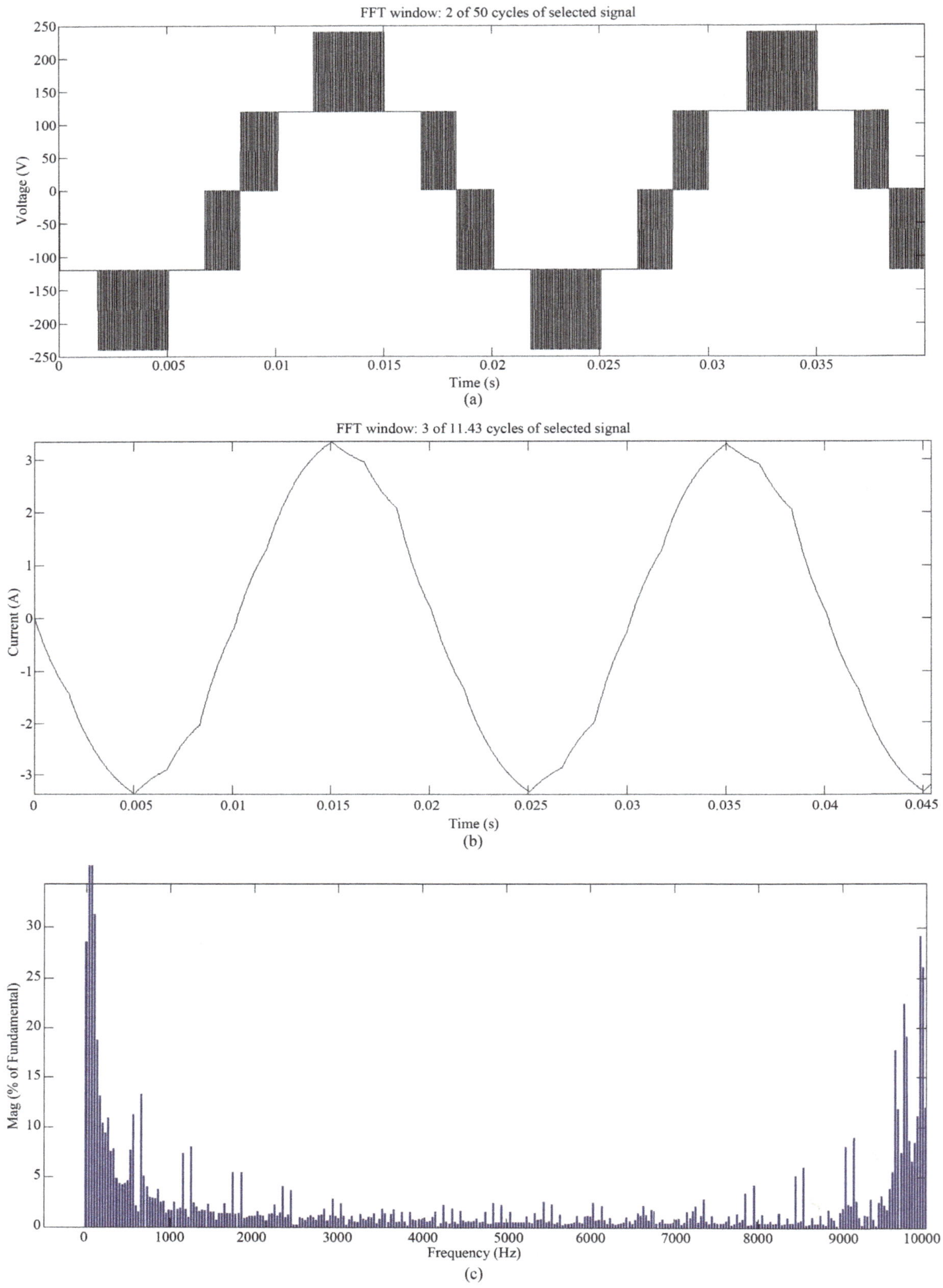

Figure 6. Output simulation results of the FCMLI: (a) voltage; (b) current; (c) output voltage spectrum analyze.

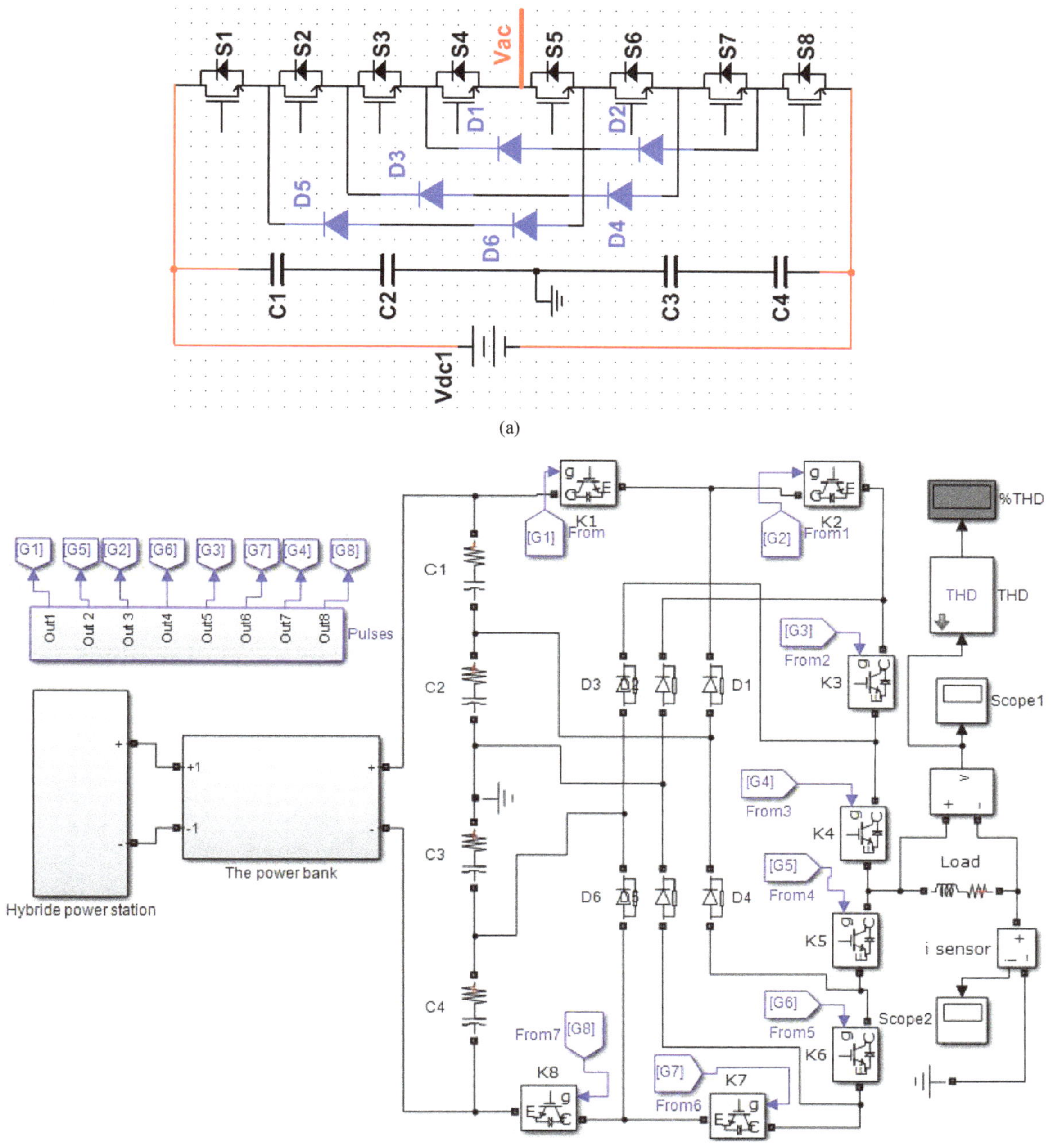

Figure 7. The architecture of the DCMLI: (a) basic diagram; (b) simulation diagram used in Matlab Simulink.

4. Effect of Level State on the CHMLI Performance

The CHMLI has shown the most promising features and performances, because it has the lowest THD in term of output voltage. Compared to the rest of multilevel inverters the cascaded H-bridge multilevel inverter requires the fewest number of components. CHMLI possesses a very importer characteristic; if we need to change the number of levels, this means increasing or decreasing the number of levels, there is no need to change the command laws, and there is no major change on the circuits, we only add or remove cells with a slight time shift on carriers. Thus we picked this inverter for more detailed testing. In attempt to establish the influence of the number of levels on the CHMLI performance, we simulated this multilevel inverter in different level states ranging from 5 to 15. **Table 1** summarizes all the simulation results obtained for the CHMLI from 5 to 15.

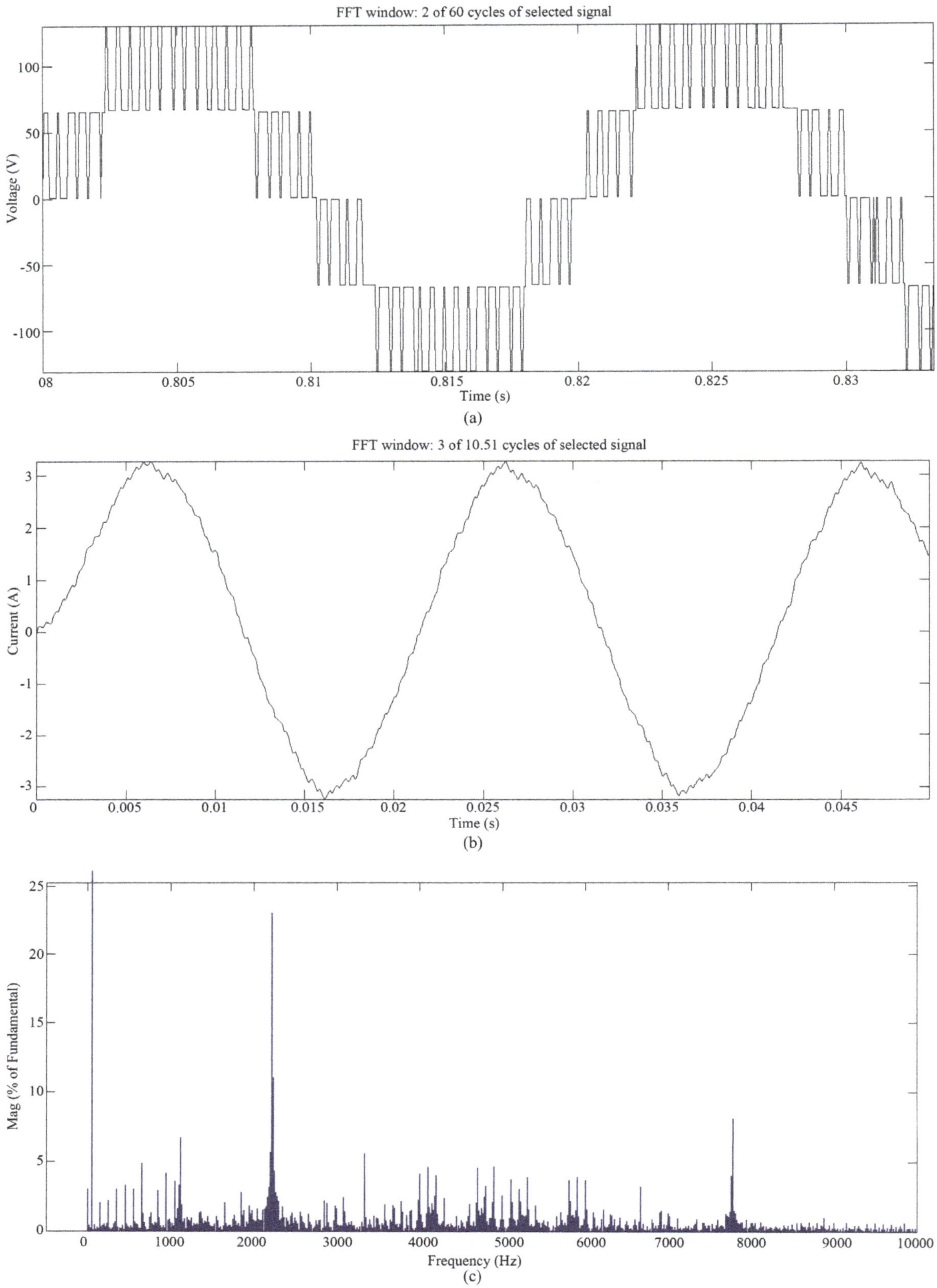

Figure 8. Output simulation results of the DCMLI: (a) voltage; (b) current; (c) output voltage spectrum analyze.

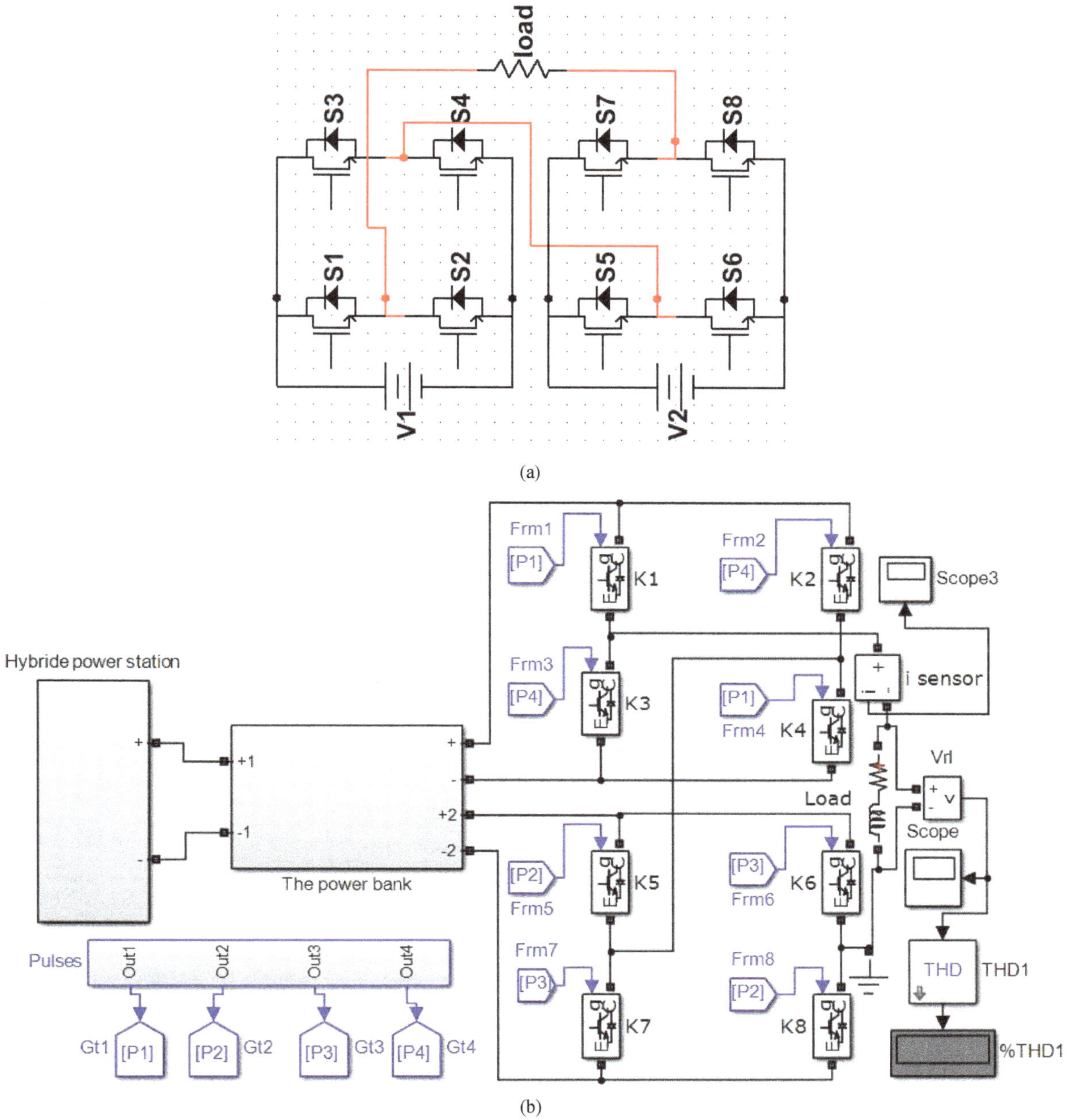

Figure 9. The architecture of the CHMLI: (a) basic diagram; (b) simulation diagram used in Matlab Simulink.

As it can be clearly observed in **Table 1**, starting from level 11, the THD decrease is slightly noticeable. The power drop is also more affecting. **Figure 11** represents the evolution of THD and power drop with the rising of levels.

As it can be observed in **Figure 11**, level 11 is the limit from where the THD's decrease is note very noticeable and the power drop is more important. Our primary objective is to choose the most efficient configuration of the CHMLI, which means getting a reasonable output current and voltage, yet at the same time the inverter has to be energy efficient. Therefore a CHMLI of more than 11 would not be a preferable investment for our hybrid renewable power station. Due to these conclusions we decided to aim our study toward the CHMLI with 11 levels in order to draw the most accurate results, **Figure 12** represents the diagram of the 11 levels CHMLI used in the study. We used the same simulative conditions as before in all the multilevel inverters concerned by this study.

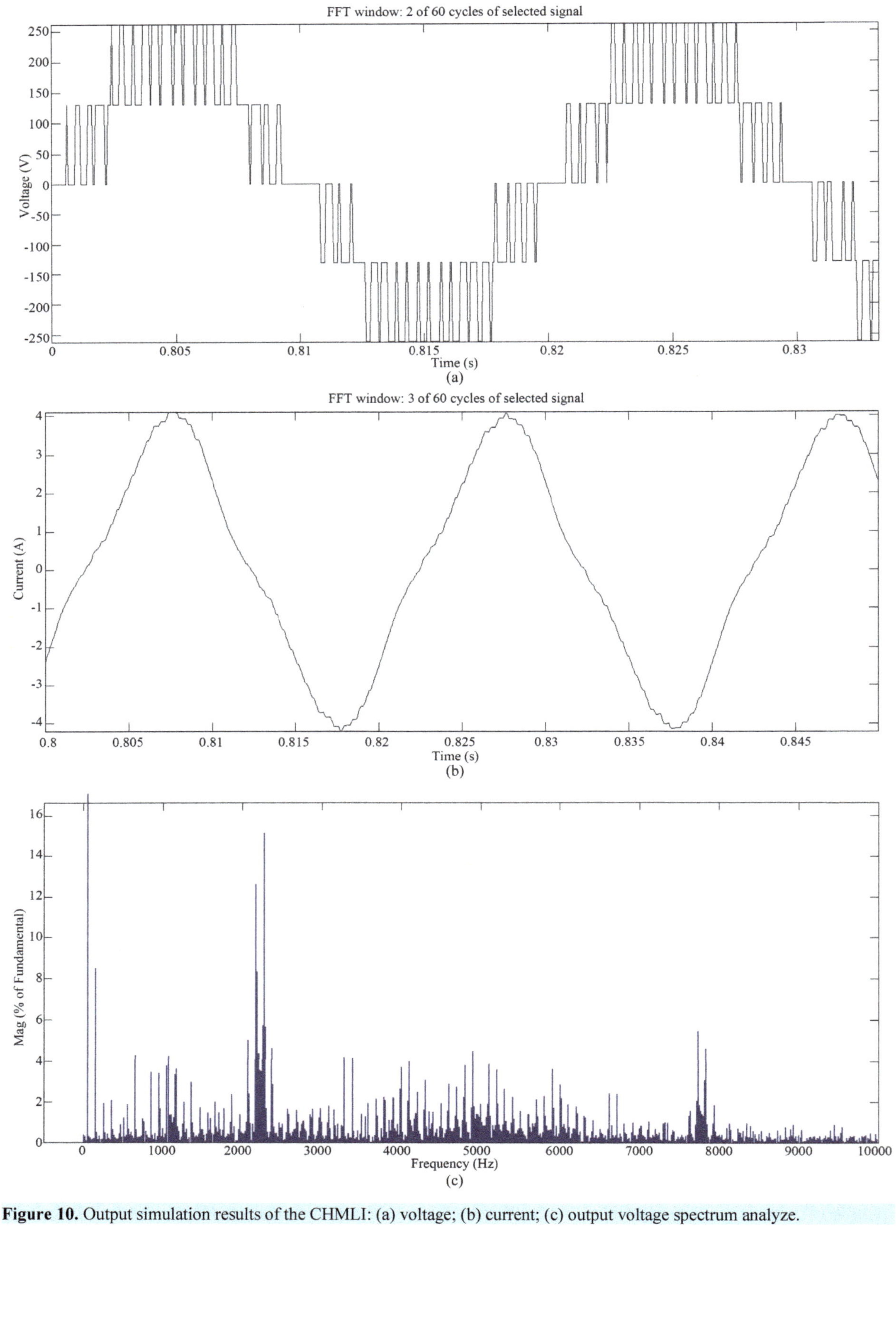

Figure 10. Output simulation results of the CHMLI: (a) voltage; (b) current; (c) output voltage spectrum analyze.

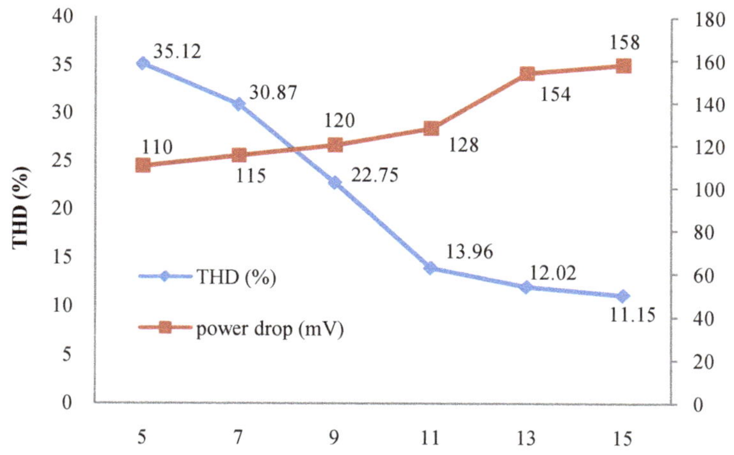

Figure 11. Simulation results of the CHMLI in many levels ranging from 5 to 15.

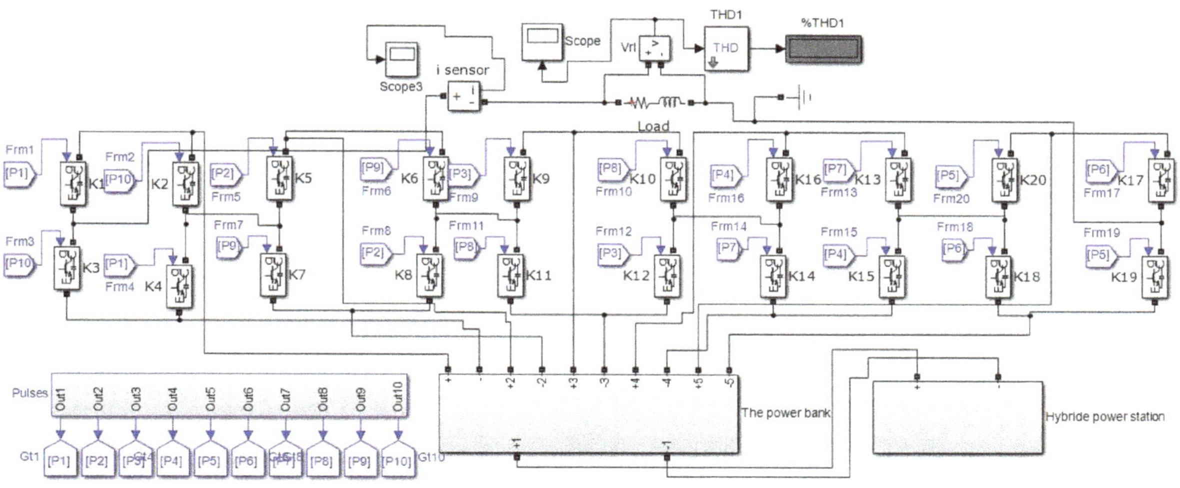

Figure 12. Diagram of the 11 levels CHMLI used on Matlab simulink.

Table 1. Simulation results of the CHMLI in many levels ranging from 5 to 15.

Level	Number of switching devices	THD (%)	Power drop (mV)
5	8	35.12	110
7	12	30.78	115
9	16	22.75	120
11	20	13.96	128
13	24	12.02	154
15	28	11.15	158

Figure 13 represents the output simulation results of the voltage, current, and voltage spectrum analyze successively. During the simulation of the CHMLI from level 5 to 15 we ensured that the previous simulation parameters were tightly respected, in order to have uniform results.

5. Discussion of the Simulation Results

In our study we found that the classic full bridge inverter has a THD of 76.51%. Despite its few components we

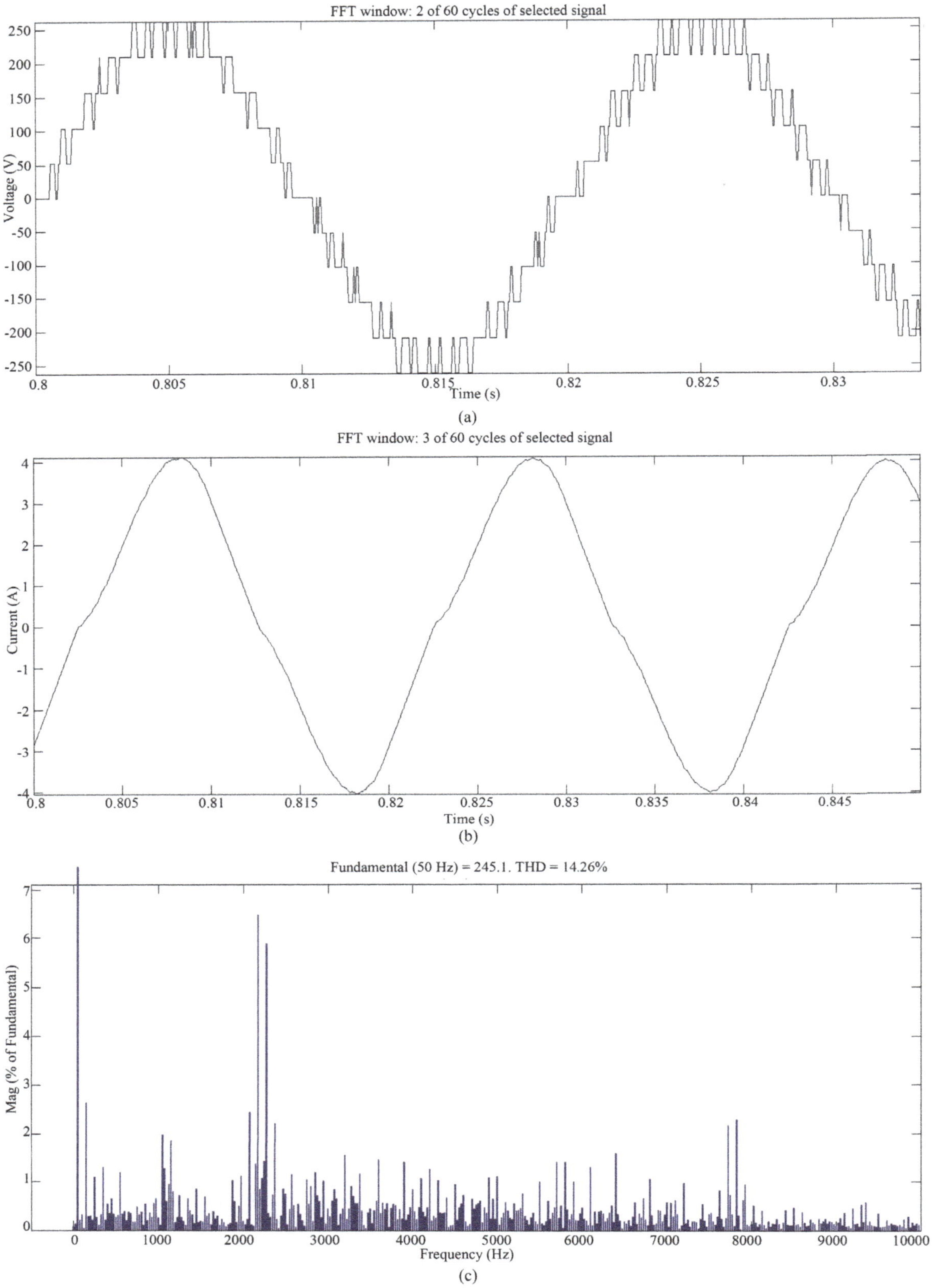

Figure 13. Output simulation results of the 11 levels CHMLI: (a) voltage; (b) current; (c) output voltage spectrum analyze.

noticed a 57.3 voltage decrease inside the power bank for the use of a (R = 50 Ω, L = 100 mH) load, during 1sec, which makes this inverter non effective for the use in the hybrid PowerStation.

The FCMLI circuitry, offers the advantage of providing extra switching combinations to balance the voltage levels, which may be used also to balance the switching losses, it also has a better signal quality (THD = 35.34%) than the DCMLI. But still the main default of the FCMLI is requiring excessive number of capacitors when the number of levels is high, which makes it difficult and expensive to implement. We noticed despite the components, the use of the FCMLI with a (R = 50 Ω, L = 100 mH) load causes a 50.4 mV general power decrease in the power bank.

We noticed in the DCMLI that it is possible to achieve a high efficiency. The use of filters can also be avoided when the number of levels is high enough. The simulation showed that the use of the DCMLI during a 1sec time, on a (R = 50 Ω, L = 100 mH) load causes a voltage drop in the power bank of approximately 12.3 mV; this allows us to say that the DCMLI has the lightest impact on the power bank. Despite these conclusions the DCMLI used still has a high THD (38.21%) compared to the rest of inverters. The DCMLI requires excessive locking diodes when the number of levels is high.

As it can be seen from the study, the CHMLI requires the fewest number of power components. thus to increase the number of levels, no diodes or capacitors are needed, each semiconductor switch manages a uniform stress, whatever is the level of the inverter, which makes it the easiest to implement. The 5 level CHMLI has also the lowest THD (35.12%); on the one hand this inverter needs separated DC sources, which can be used to divide the power bank in multiple sections, on the other hand the use of such inverter causes bigger voltage decrease in the power bank (58 mV in the first section, and 67 mV in the other one for a (R = 50 Ω, L = 100 mH) during 1 sec time simulation); this is normally due to the fact that the CHMLI is basically a cascaded set of full bridge inverters. Therefore we decided to raise the number of levels in an attempt to study the impact of level increasing upon the performance of the CHMLI. Thus the optimal level found was 11. In this level state the CHMLI in our study had a 13.96% THD, and caused a 128 mv power decrease inside the power bank during a 1 sec time simulation. On the basis of these conclusions we can affirm that the optimal choice for our hybrid power station is a CHMLI with 11 levels.

6. Conclusion

In our study a comparison was made between the three basic 5 levels inverters topologies: the FCMLI, the DCMLI, and the CHMLI using the a PWM law command, in order to draw accurate conclusions to develop the most adequate multilevel inverter for a hybrid wind-solar power station. In addition to the THD and the number of components this study reveals a unique comparison result on the impact of the multilevel inverter type on storage units. The CHMLI was identified as the most suited multilevel inverter for the hybride renwable power station, even if it causes a more important power drop. The FCMLI and the the DCMLI require both more components and have higher THD on the output voltage. This study has also shown unique findings related to the effect of level state on the efficiency of the CHMLI, in this countext we found that the CHMLI with 11 levels is the most efficient configuration for this type of multi level inverters. Furthermore we have found that impact on power storage units with level increasing for the CHMLI is less consistent compared to the signal purity obtained. These results pushes us to focus our effort on realizing an apropriate CHMLI for the renewable power station experimentaly speaking. In terms of perspectives a real time obtimisation command would allow us to get more sutisfying results witout altering the power consumption. Eliminating spesific harmonics is also a fast way to get a better power quality, but is still hard to implement in real time mode. Our prime goal is to get best performance by accomodation between real time optimization and harmonic elimination without compromising power storage eficiency.

References

[1] Fuchs, E.F. and Masoum, M.A.S. (2011) Power Conversion of Renewable Energy Systems. Springer, New York, 187-197. http://dx.doi.org/10.1007/978-1-4419-7979-7

[2] Tolbert, L.M., Peng, F.Z. and Habetler, T.G. (1999) Multilevel Converters for Large Electric Drives. *IEEE Transactions on Industry Applications*, **35**, 36-44. http://dx.doi.org/10.1109/28.740843

[3] Khomfoi, S., Praisuwanna, N. and Tolbert, L.M. (2010) A Hybrid Cascaded Multilevel Inverter Application for Renewable Energy Resources Including a Reconfiguration Technique. *Energy Conversion Congress and Exposition*

(*ECCE*), 2010 *IEEE*, Atlanta, 12-16 September 2010, 3998-4005. http://dx.doi.org/10.1109/ecce.2010.5617803

[4] Kang, F.S., Park, S.J., Cho, S.E., Kim, C.U. and Ise, T. (2005) Multilevel PWM Inverters Suitable for the Use of Stand-Alone Photovoltaic Power Systems. *IEEE Transactions on Energy Conversion*, **20**, 906-915. http://dx.doi.org/10.1109/tec.2005.847956

[5] Hochgraf, C., Lasseter, R., Divan, D. and Lipo, T.A. (1994) Comparison of Multilevel Inverters for Static VAr Compensation. *Industry Applications Society Annual Meeting*, 1994, *Conference Record of the* 1994 *IEEE*, Denver, 2-6 October 1994, 921-928. http://dx.doi.org/10.1109/ias.1994.377528

[6] Kang, F.S., Park, S.J., Lee, M.H. and Kim, C.U. (2005) An Efficient Multilevel-Synthesis Approach and Its Application to a 27-Level Inverter. *IEEE Transactions on Industrial Electronics*, **52**, 1600-1606. http://dx.doi.org/10.1109/tie.2005.858715

[7] Mythili, M. and Kayalvizhi, N. (2013) Harmonic Minimization in Multilevel Inverters Using Selective Harmonic Elimination PWM Technique. 2013 *International Conference on Renewable Energy and Sustainable Energy* (*ICRESE*), Coimbatore, 5-6 December 2013, 70-74. http://dx.doi.org/10.1109/icrese.2013.6927790

[8] Al-Othman, A.K. and Abdelhamid, T.H. (2009) Elimination of Harmonics in Multilevel Inverters with Non-Equal dc Sources Using PSO. *Energy Conversion and Management*, **50**, 756-764. http://dx.doi.org/10.1016/j.enconman.2008.09.047

[9] Rodriguez, J., Lai, J.S. and Peng, F.Z. (2002) Multilevel Inverters: A Survey of Topologies, Controls, and Applications. *IEEE Transactions on Industrial Electronics*, **49**, 724-738. http://dx.doi.org/10.1109/tie.2002.801052

[10] Fujii, K., Schwarzer, U. and De Doncker, R.W. (2005) Comparison of Hard-Switched Multi-Level Inverter Topologies for STATCOM by Loss-Implemented Simulation and cost Estimation. *Power Electronics Specialists Conference*, 2005. *PESC'*05. *IEEE* 36*th*, Recife, 16-16 June 2005, 340-346. http://dx.doi.org/10.1109/pesc.2005.1581646

[11] Dixit, A., Mishra, N., Sinha, S.K. and Singh, P. (2012) A Review on Different PWM Techniques for Five Leg Voltage Source Inverter. 2012 *International Conference on Advances in Engineering, Science and Management* (*ICAESM*), Nagapattinam, 30-31 March 2012, 421-428.

[12] Lee, W.K., Kim, S.Y., Yoon, J.S. and Baek, D.H. (2006) A Comparison of the Carrier-Based PWM Techniques for Voltage Balance of Flying Capacitor in the Flying Capacitor Multilevel Inverter. *Applied Power Electronics Conference and Exposition*, 2006. *APEC'*06. *Twenty-First Annual IEEE*, 19-23 March 2006, 6 p. http://dx.doi.org/10.1109/apec.2006.1620763

[13] Nabae, A., Takahashi, I. and Akagi, H. (1981) A New Neutral-Point-Clamped PWM Inverter. *IEEE Transactions on Industry Applications*, **IA-17**, 518-523. http://dx.doi.org/10.1109/tia.1981.4503992

[14] Arulkumar, K., Vijayakumar, D. and Palanisamy, K. (2015) Modeling and Control Strategy of Three Phase Neutral Point Clamped Multilevel PV Inverter Connected to the Grid. *Journal of Building Engineering*, **3**, 195-202. http://dx.doi.org/10.1016/j.jobe.2015.06.001

[15] Peng, F.Z., McKeever, J.W. and Adams, D.J. (1998) A Power Line Conditioner Using Cascade Multilevel Inverters for Distribution Systems. *IEEE Transactions on Industry Applications*, **34**, 1293-1298. http://dx.doi.org/10.1109/28.739012

[16] Alonso, O.S., Gubia, P. and E Marroyo, L. (2003) Cascaded H-Bridge Multilevel Converter for Grid Connected Photovoltaic Generators with Independent Maximum Power Point Tracking of Each Solar Array. *Power Electronics Specialist Conference*, 2003. *PESC* '03. 2003 *IEEE* 34*th Annual*, **2**, 731-735. http://dx.doi.org/10.1109/pesc.2003.1218146

12

Sensitivity Analysis of a 50+ Coal-Fired Power Unit Efficiency

Katarzyna Stępczyńska-Drygas, Sławomir Dykas, Krystian Smołka

Institute of Power Engineering and Turbomachinery, Silesian University of Technology, Gliwice, Poland
Email: slawomir.dykas@polsl.pl

Abstract

The coal-fired power unit integration with a CO_2 capture and compression installation involves a considerable rise in the costs of electricity generation. Therefore, there is a need for a continuous search for methods of improving the electricity generation efficiency in steam power plants. One technology which is especially promising is the advanced ultra-supercritical (A-USC) power unit. Apart from steam parameters upstream the turbine, the overall efficiency also depends on the efficiency values of individual elements of the plant and the size of energy consumption of the process of CO_2 sequestration from the boiler flue gases. These problems are considered herein to emphasize that without specifying the efficiency values of the power plant main elements the information concerning its electricity generation efficiency is incomplete. This paper presents the influence of the efficiency of individual elements of the power plant on its electricity generation efficiency. The lack of information of the efficiencies of the power plant individual elements, by presenting its overall efficiency, may lead to the false conclusions.

Keywords

Sensitivity Analysis, A-USC, Power Unit Efficiency, CO_2 Capture

1. Introduction

The electricity generation efficiency in coal-fired power plants has an enormous impact on the consumption of fossil fuels and on emissions into the environment. The European reference standard is now set by the conceptual coal-fired Reference Power Plant North Rhine-Westphalia (RPP NRW). Steam parameters upstream the turbine are 28.5 MPa/600˚C/620˚C. The power unit gross and net capacity is 600 MW and 556 MW, respectively. The plant achieves the net electricity generation efficiency of 45.9% at the condenser pressure at the level of 4.5 kPa [1].

The power industry worldwide has set itself a goal of shifting the net electricity generation efficiency limit for the reference conditions from the current 46% to 50% and higher. Reaching this level will require substantial technological changes, especially in the area of the design materials used to make the plant basic elements [2]. Perfecting the power unit steam-water cycle, together with an improvement in the steam turbine internal efficiency and the boiler power efficiency, may contribute essentially to a further rise in overall power efficiency.

The most promising method of improving the efficiency of state-of-the-art coal-fired power units is to raise both live and reheated steam parameters. For every rise in the temperature of live and reheated steam by 20°C, there is an increase in efficiency by 1 percentage point, and for each increment in the live steam pressure by 1 MPa—an increase in efficiency by 0.2% [3]. A rise in the efficiency of the Rankine steam-water cycle can be achieved by: lowering the condenser pressure, raising the final temperature of feed water, increasing the number of stages in the feed water regenerative heaters and using steam superheating and reheating systems correctly. It should also be noted that overall electricity generation efficiency of a steam power unit is affected by internal efficiencies of individual elements of the power engineering machinery and equipment used in the plant. The impact of the internal efficiency of the steam turbine on the overall efficiency of the plant is bigger than that of other elements, and the key factors are as follows: advanced three-dimensional blade design, state-of-the-art manufacture technology and the use of large-size last stage blades to limit stack losses. On the other hand, the boiler power efficiency may be improved by lowering the flue gas temperature, reducing the incomplete and imperfect combustion losses (unburnt carbon loss—UBC), minimizing the pressure and temperature losses and using coal drying. A further rise in the power unit net efficiency can be achieved by a reduction in own-needs electricity consumption (e.g. by using pump and fan drives with rotational speed adjustment). In the case of a coal-fired power unit integrated with an installation of carbon dioxide capture by means of chemical absorption the electricity generation efficiency is hugely affected by the energy consumption of the process of CO_2 sequestration from flue gases, which may vary during the power unit service life due to the use of new and better amines [4] [5].

The increment in the electricity generation efficiency from 35% (for the subcritical power units currently operating in Poland) to 50% will cause a reduction in unit CO_2 emissions from about 984 to 689 kgCO_2 per 1 MWh net generated electricity, *i.e.* by 30% (**Figure 1**). It should be emphasized that the values of emissions and fuel consumption presented in **Figure 1** are characteristic of a specific hard coal type (here: hard coal with the calorific value of 23 MJ/kg and the elemental carbon (C) content of 60%). For a power unit with the net electric power of 832.5 MW (gross: 900 MW) this means a reduction in fuel consumption per year (assuming the annual operation time of 7000 h) by 782 thousand tons and CO_2 emissions smaller by 1720 thousand tons (**Figure 2**).

Globally, the average efficiency of coal-fired power plants is about 30%. The average efficiency of power stations in the European Union is close to 38%. By contrast, the average efficiency of plants in the USA is only 33% and in China—37% [6]. However, the issues related to the power plant efficiency have to be considered with care because the actual efficiency of a given unit depends on its location, fuel quality and operating conditions. The basis factor is the cooling conditions and the pressure in the steam turbine condenser resulting from that.

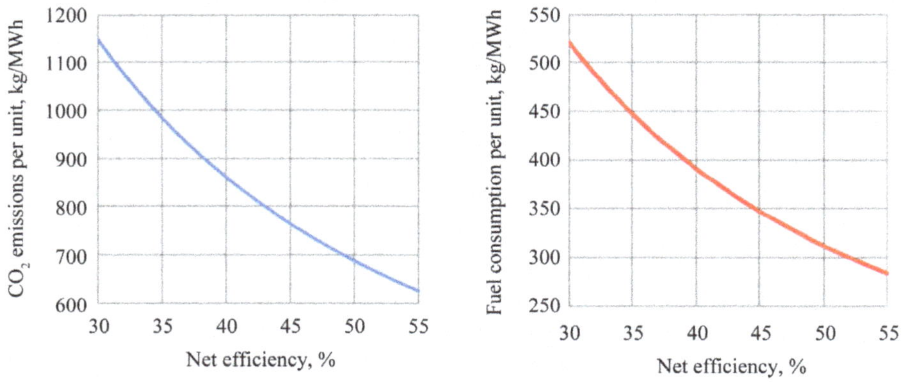

Figure 1. Unit CO_2 emissions (left) and unit fuel consumption (right) per unit of net generated electricity depending on net electricity generation efficiency (hard coal with calorific value of 23 MJ/kg and a 60% content of C).

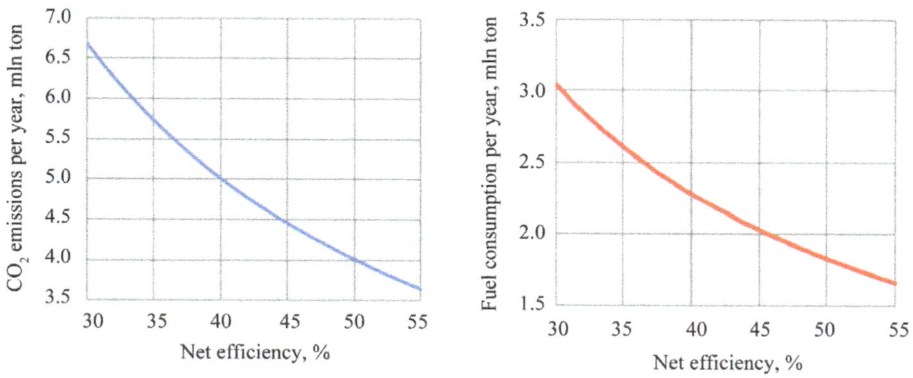

Figure 2. Annual CO_2 emissions (left) and annual fuel consumption (right) for a coal-fired power unit with net electric power of 832.5 MW (gross: 900 MW) assuming the annual operation time of 7000 h depending on net electricity generation efficiency (hard coal with calorific value of 23 MJ/kg and a 60% content of C).

In inland locations in the USA the steam pressure in the condenser is 7 - 9 kPa, whereas in European conditions —especially if sea water is used as coolant—the condenser steam pressure may reach the value of up to 3 kPa. For this reason, European power stations are characterized by efficiency values by about 2% higher compared to their American counterparts. The coal quality and the once-through boiler structure, which is common in Europe, may cause a reduction in own-needs energy consumption. Attention should also be paid to the type of the coal calorific value assumed while determining the power unit efficiency—whether it is the lower or higher calorific value (LCV or HCV). The cumulative effect of all these factors, which condition the levels of achieved efficiency, may lead to differences in the obtained values as high as 4 percentage points for seemingly identical plants [7]. Thus a typical subcritical power unit in the USA may be characterized by a 37% efficiency, whereas a modern supercritical unit may have an efficiency of 42%. For the European conditions, the same values of efficiency may be 41% and 46%, respectively.

In the case of thermodynamic calculations of cycles of coal-fired power units, the obtained values of the gross and net efficiency of electricity generation depend on assumed values of the input data, *i.e.* on the boiler power efficiency, the turbine internal efficiency, the condenser pressure, the power unit own-needs index, the pressure loss in the boiler and steam pipelines, the heat loss in regenerative heaters. Therefore, it is essential that all assumptions made for the calculations should be determined precisely.

In order to assess achievable values of the electricity generation gross and net efficiency, a comparison was made of the basic indices of the operation of an ultra-supercritical coal-fired power unit with gross electric power of 900 MW [5] [8] [9]. The impact of the values of efficiency of the power unit individual elements of machinery and equipment and of heat and flow losses in pipelines and heat exchangers on the overall plant electricity generation efficiency was determined. An analysis was also conducted of the sensitivity of the coal-fired power unit efficiency to the energy consumption of the process of CO_2 separation from flue gases in the case of a coal-fired power unit integrated with an installation of carbon dioxide capture by means of chemical absorption.

2. Definitions of Efficiency of Individual Elements of Machinery and Equipment

The basic parameters and indices of the power unit operation are defined as follows:

- The boiler power efficiency:

$$\eta_K = \frac{\dot{Q}_{uz}}{\dot{P}W_d} \quad (1)$$

where:

$\dot{Q}_{uz}$: flux of useful heat supplied to the cycle medium in the boiler,
$\dot{P}$: fuel mass flow,
W_d: fuel calorific value in the as-received state.

- The turbine isentropic efficiency:

$$\eta_{iT} = \frac{h_1 - h_2}{h_1 - h_{2s}} \tag{2}$$

where:
h_1: steam specific enthalpy at the turbine inlet,
h_2: steam specific enthalpy at the turbine outlet,
h_{2s}: steam specific enthalpy at the turbine outlet in the isentropic process.

- The efficiency of regenerative heaters:

$$\eta_{PR} = \frac{\dot{m}_1 \left(h_2 - h_1\right)}{\dot{m}_2 \left(h_3 - h_4\right)} \tag{3}$$

where:
$\dot{m}_1$: mass flow of heated water,
h_1: water specific enthalpy at the heater inlet,
h_2: water specific enthalpy at the heater outlet,
$\dot{m}_2$: mass flow of heating steam,
h_3: steam specific enthalpy at the heater inlet,
h_4: condensate specific enthalpy at the heater outlet.

- The efficiency of pipelines:

$$\eta_R = 1 - \frac{h_1 - h_2}{h_1} \tag{4}$$

where:
h_1: medium specific enthalpy at the pipeline inlet,
h_2: medium specific enthalpy at the pipeline outlet.

- Flow losses:

$$\xi = \frac{p_1 - p_2}{p_1} \tag{5}$$

where:
p_1: medium pressure at the pipeline/heat exchanger inlet,
p_2: medium pressure at the pipeline/heat exchanger outlet.

- The cycle efficiency:

$$\eta_o = \frac{\dot{Q}_d - \dot{Q}_w}{\dot{Q}_d} \tag{6}$$

where:
$\dot{Q}_d$: heat flux supplied to the cycle,
$\dot{Q}_w$: heat flux extracted from the cycle.

- Gross electricity generation efficiency:

$$\eta_{elB} = \frac{N_{elB}}{\dot{P}W_d} \tag{7}$$

where:
N_{elB}: gross electric power of the power unit.

- Net electric power:

$$N_{elN} = N_{elB} - N_{PW} = N_{elB}\left(1 - \varepsilon\right) \tag{8}$$

where:
N_{PW}: power unit own-needs electric power,
ε: power unit own-needs index.

- Net electricity generation efficiency:

$$\eta_{elN} = \frac{N_{elN}}{\dot{P}W_d} \tag{9}$$

3. The Reference 900 MW Power Unit for Advanced Ultra-Supercritical Steam Parameters

3.1. Basic Parameters of the Conceptual 900 MW Power Unit

The flowchart of the conceptual reference advanced ultra-supercritical power unit is presented in **Figure 3**. The conceptual power unit with gross electric power of 900 MW is fired with hard coal with a calorific value of 23 MJ/kg. The composition of coal in the as-received state is presented in **Table 1**. Complete and perfect combustion is assumed. It is further assumed that the excess air factor in the boiler is $\lambda = 1.2$. The composition of wet flue gases is presented in **Table 2**. The live and reheated steam parameters before the turbine are 35 MPa/700°C and 7.43 MPa(a)/720°C, respectively. The power unit basic parameters are listed in **Table 3**. **Table 4** presents parameters of the feed water regenerative heaters. The basic indices of the power unit operation are listed in **Table 5**. For the presented system with the gross electric power of 900 MW, the values of the achieved gross and net electricity generation efficiency are 52.61% and 49.04%, respectively.

3.2. Reference Structure of a Power Unit Integrated with a CO_2 Capture Installation

The basic diagram of a cycle integrated with a CO_2 capture and compression installation is presented in **Figure 4**. The steam needed for the sorbent regeneration is extracted from the main turbine IP/LP crossover pipe. Due to the fact that more than half of the mass flow from the IP/LP crossover pipe is directed to the CO_2 capture installation, the low-pressure turbine is reduced from two to one double-flow part. The reboiler feed steam condensate is returned and introduced into the cycle in the low-pressure regeneration region. Taking account of the limitations concerning the maximum size of the absorption columns, the capture installation is composed of four parallel absorber-desorber-compressor lines. The steam needed for the sorbent regeneration is extracted from the

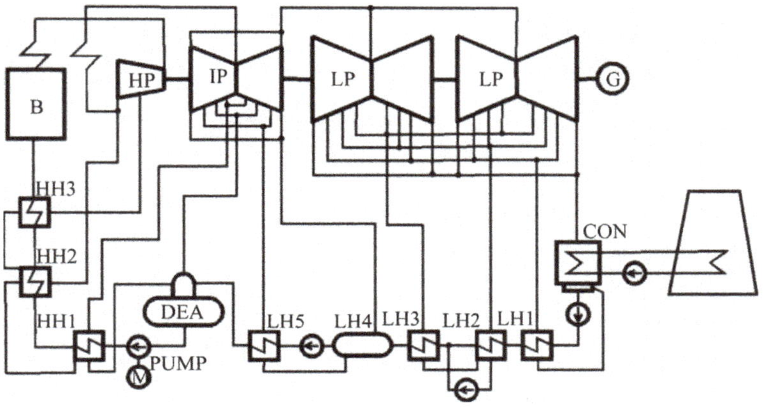

Figure 3. Flowchart of the reference advanced ultra-supercritical power unit.

Table 1. Fuel composition—coal 23.

w	p	c	h	o	n	s
0.09	0.2	0.6	0.038	0.054	0.013	0.01

Table 2. Wet flue gas composition.

CO_2	SO_2	O_2	N_2	H_2O	Ar
0.1416	0.0009	0.0329	0.7378	0.078	0.0088

Table 3. Basic figures for the reference 900 MW advanced ultra-supercritical power unit developed within the project.

Parameter		Unit
Live steam temperature at the boiler outlet	702	°C
Live steam temperature at the turbine inlet	700	°C
Live steam pressure at the boiler outlet	35.8	MPa
Live steam pressure at the turbine inlet	35	MPa
Reheated steam temperature at the boiler outlet	721	°C
Reheated steam temperature at the turbine inlet	720	°C
Reheated steam pressure at the boiler outlet	7.5	MPa
Steam pressure in the IP/LP crossover pipe	0.5	MPa
Feed water temperature	330	°C
Internal efficiency of the turbine HP part stage groups	90	%
Internal efficiency of the turbine IP part stage groups	92	%
Internal efficiency of the turbine LP part stage groups (the efficiency value is corrected due to the stack loss)	92	%
Stack loss	20	kJ/kg
Flue gas temperature at the boiler outlet	110	°C
The boiler power efficiency	95	%
Excess air factor	1.2	-
Hard coal calorific value	23	MJ/kg
Generator efficiency	98.8	%
The turbine mechanical losses	0.9	MW
Efficiency of feed water pumps	85	%
Efficiency of regenerative exchangers	99.5	%
Flow losses in steam pipelines to regenerative exchangers	2	%
Losses of the feed water flow through regenerative exchangers	1	%
The cycle medium pressure drop in the boiler	4.3	MPa
Pressure drop of interstage steam in the boiler reheater	0.2	MPa
Flow losses in reheated steam pipelines	1	%
Flow losses between the turbine IP and LP part	1	%
The power unit gross electric power (at the generator terminals)	900	MW
Internal efficiency of the feed water and condensate pumps	85	%
Internal efficiency of the cooling water pumps	82	%
Efficiency of air and flue gas fans	85	%
Energy consumed by coal mills per kg of coal	90	kJ/kg
Efficiency of electric motors driving auxiliary equipment	97	%
Efficiency of the rotational speed adjustment	96	%
Efficiency of the unit transformer	99.5	%
Ambient temperature	14	°C
Ambient pressure	98	hPa
The condenser cooling water temperature	19.1	°C
Increment in the cooling water temperature	9	K
Temperature difference in the condenser	2.8	K
Pressure in the condenser	4.5	kPa
Heat from the machinery and equipment cooling as a percentage of heat extracted in the condenser	4	%

Table 4. Parameters of the regenerative water heaters.

Heater	PN1	PN2	PN3	PN4	PN5	ODG	PW1	PW2	PW3
Bleed steam pressure, kPa	22.2	78.5	212	490	1228	2364	4542	7622	13,192
Temperature difference, K	3	3	3	-	3	-	2	2	2
Condensate supercooling, K	-	-	10	-	10	-	10	10	10

Table 5. Basic indices of the 900 MW power unit operation.

Live steam mass flow	578.42 kg/s
Heat flux given up in the condenser	743.1 MW
Auxiliary equipment cooling heat (4% of heat in the condenser)	29.5 MW
Cooling water mass flow	20,532 kg/s
Fuel mass flow	74.4 kg/s
Flue gas mass flow at the boiler outlet	776.4 kg/s
CO_2 mass flow at the boiler outlet	163.3 kg/s
CO_2 emissions per unit of net generated electricity	701 gCO_2/kWh
Gross electric power	900 MW
Gross electricity generation efficiency	52.61%
Net electric power	838.8 MW
Net electricity generation efficiency	49.04%
Own-needs index	6.79%

turbine IP/PL crossover pipe. The temperature difference in the evaporator (REB) between condensing steam and the sorbent regeneration temperature (124˚C) is 10 K. Therefore, the required parameters of the heating steam are 134˚C and 0.33 MPa (considering flow losses at the level of 8%). Such parameters have to be kept constant in the entire range of the power unit load. It is assumed in the basic calculations that the pressure in the IP/LP crossover pipe of the integrated and the reference power units is identical and totals 0.5 MPa.

Table 6 presents the basic operating indices of the reference power unit (with no CO_2 capture installation) and of the integrated one. Assuming an identical mass flow of live steam as in the reference power unit (578.42 kg/s) and an identical pressure in the turbine IP/LP crossover pipe, the integrated power unit achieves the net electric power of 636.9 MW and the net efficiency of 37.23%. The drop in the net efficiency totals 11.82% and in the net electric power—202 MW, 57 MW of which is the power needed to drive CO_2 compressors and 6 MW—the driving power of the capture installation auxiliary equipment (fans, pumps). The other 139 MW is the effect of the reduction in the steam turbine power due to the considerable mass flow of steam extracted from the IP/LP crossover pipe for the sorbent regeneration and of the rise in power needed to drive the cooling water pumps.

3.3. Assessment of the Power Unit Efficiency for Different Assumptions

In order to present changes in the steam power unit efficiency resulting from changes in the efficiency of its individual elements, the typical range of changes in efficiency of the basic components of the steam turbine installation was applied. The minimum and maximum efficiency values of the power unit elements are listed in **Table 7** and **Table 8**. The calculations were performed for the analyzed structure of the 50+ power unit using the following variants:

1) The efficiencies of all machinery and equipment elements total 100% and the heat and flow losses in pipelines and exchangers are 0.

2) The efficiencies of all machinery and equipment elements are assumed at the maximum currently achievable level (**Table 7**) and the heat and flow losses in bleed steam pipelines and in exchangers are assumed

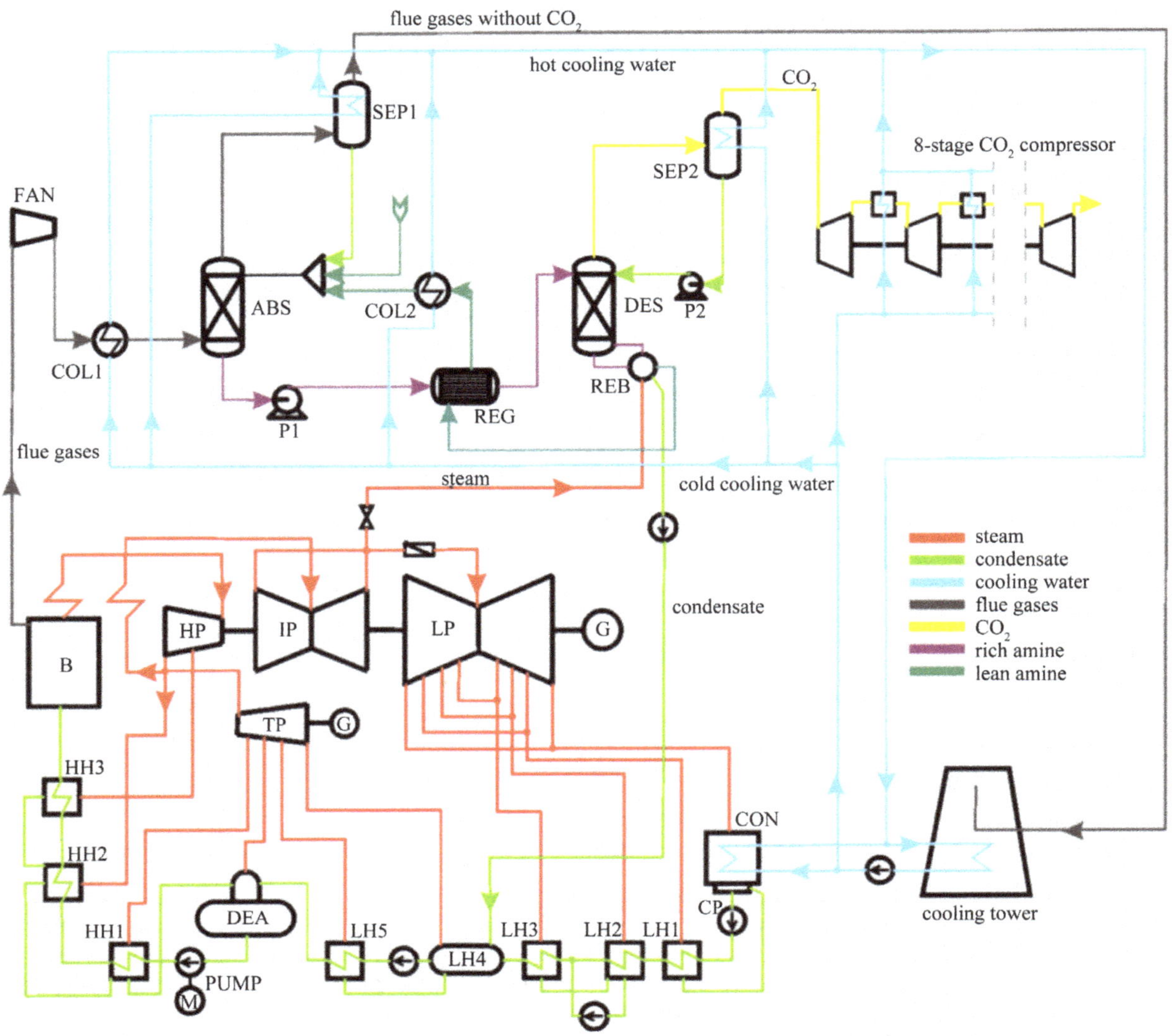

B - steam boiler; HP, IP, LP - the turbine HP, IP and LP part, CON - condenser; HH - high-pressure regeneration heater; LH - low-pressure regeneration heater; PUMP - feed pump; CP - condensate pump; G - generator; ABS - absorber; DES - desorber; REB - reboiler; REG - cross-flow heat exchanger; FAN - flue gas fan; COL1 - flue gas precooler; COL2 - lean amine cooler; SEP1 - flue gas moisture separator, flue gas cooler; SEP2 - CO_2 moisture separator and CO_2 cooler; P1 - rich amine pump; P2 - lean amine pump

Figure 4. Diagram of the 900 MW power unit with the CO_2 capture and compression installation.

according to **Table 3**.

3) The efficiencies of all machinery and equipment elements and the heat and flow losses in bleed steam pipelines and in exchangers are assumed according to **Table 3**.

4) The efficiencies of all machinery and equipment elements are assumed at the minimum level (**Table 8**) and the heat and flow losses in bleed steam pipelines and in exchangers are assumed according to **Table 3**.

The calculation results are listed in a table. **Figure 5** presents the gross and net electricity generation efficiency for three calculation variants: MAX (maximum), REF (reference) and MIN (minimum). **Figure 6** presents the share of the power unit individual elements in the difference between the cycle efficiency determined assuming 100% efficiency of all the machinery and equipment and zero heat and pressure losses (calculation variant 1, **Table 9**) and the gross and net electricity generation efficiency for the REF variant. **Figure 7** presents the share of the power unit individual elements in the increase in the net electricity generation efficiency from the calculation variant MIN to variant MAX.

Table 6. Basic operating indices of the reference power unit (with no CO_2 capture installation) and of the integrated one (hard coal 23).

Parameter	Unit	Reference power unit	Integrated power unit
Nominal pressure in the IP/LP crossover pipe	MPa	0.5	0.5
Live steam mass flow	kg/s	578.42	578.42
Steam mass flow directed to the CO_2 capture installation	kg/s	0	205.5
Heat flux given up in the turbine condenser	MW	741.9	363.2
Waste heat flux from the CO_2 capture and compression installation	MW	0	552
Heat flux given up in the cooling tower	MW	771.4	944.7
Gross electric power	MW	900	765.6
Gross electricity generation efficiency	%	52.58	44.66
Net electric power	MW	838.8	636.9
Net electricity generation efficiency	%	49.04	37.23

Table 7. Maximum achievable efficiencies of individual elements of machinery and equipment.

Steam boiler	95%
HP turbine	94%
IP turbine	97%
LP turbine	95%
Generator	99%
Live and reheated steam pipelines	99.9%

Table 8. Minimum assumed efficiencies.

Steam boiler	92%
HP turbine	88%
IP turbine	90%
LP turbine	88%
Generator	98.5%
Live and reheated steam pipelines	99.5%

Table 9. Gross electricity generation efficiency (η_{elB}) and drop in gross efficiency ($\Delta\eta_{elB}$) compared to variant 1.

Calculation variant	Gross electricity generation efficiency, [%]	Drop in efficiency compared to variant 1, percentage points
1	**59.54**	-
(MAX) 2	**53.84**	**5.7**
(REF) 3	**52.61**	**6.93**
(MIN) 4	**49.18**	**10.36**

3.4. The Impact of Energy Consumption of the CO_2 Capture Process on the Power Unit Operating Parameters

The energy consumption of the CO_2 capture process (or of the sorbent regeneration process to be exact) is a very

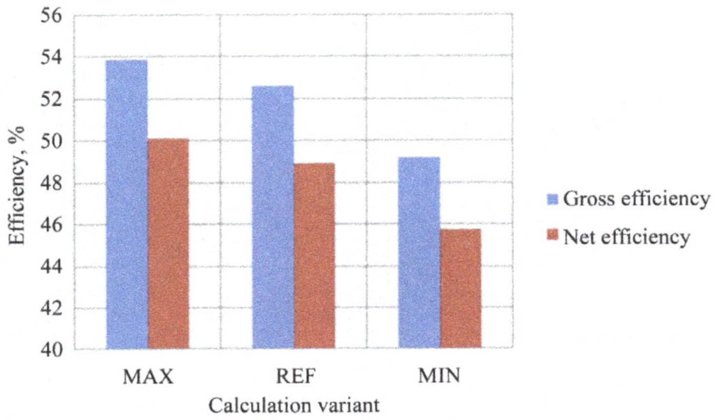

Figure 5. Gross and net electricity generation efficiency (own-needs index: 7%) for three calculation variants: MAX, REF and MIN.

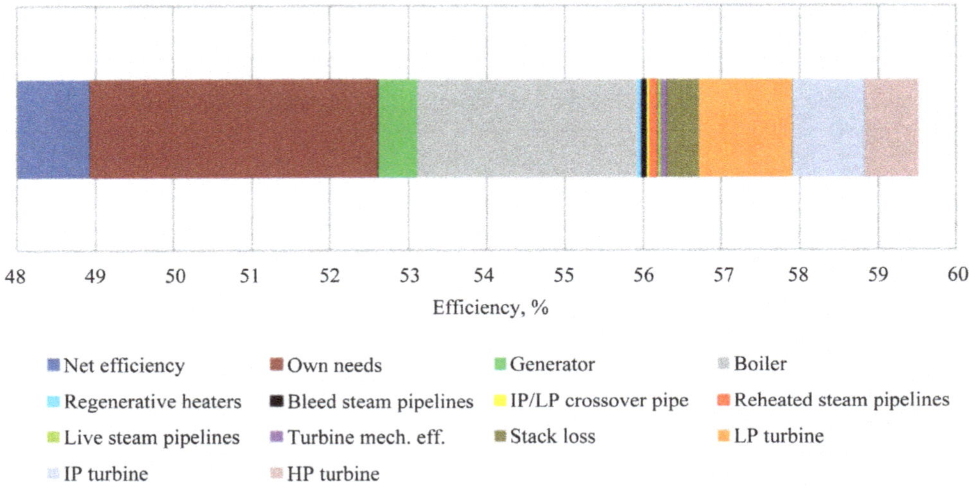

Figure 6. Share of individual elements in the difference between the reference Clausius-Rankine cycle efficiency (calculation variant 1 from **Table 9**) and net electricity generation efficiency for the REF calculation variant.

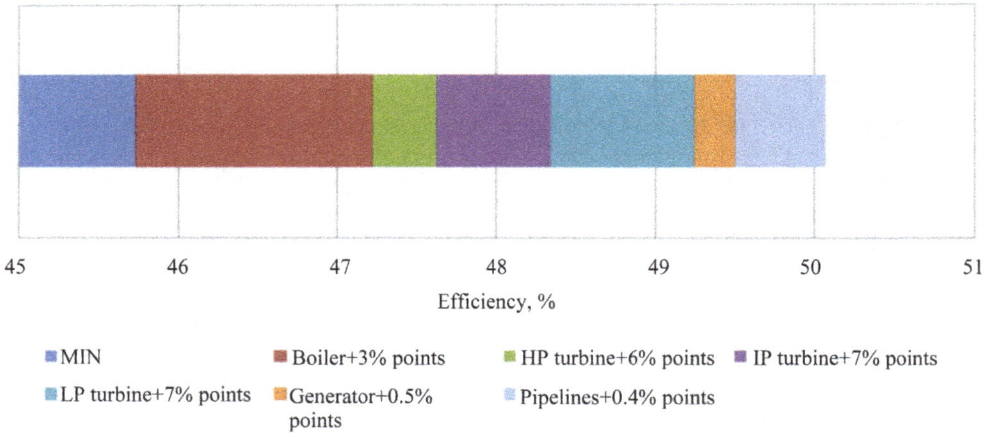

Figure 7. Share of individual elements in the increase in net electricity generation efficiency from the MIN calculation variant to variant MAX (in the legend the rise in efficiencies of the power unit individual elements is marked in respect of variant MIN).

important factor that affects both the integrated power unit electricity generation efficiency and the economic indices. **Figure 8** presents the impact of the capture process energy consumption on the decrease in the net electric power and on the drop in the net electricity generation efficiency after the power unit integration with the CO_2 capture installation. **Figure 9** shows costs of electricity generation in a power unit integrated with a CO_2 capture installation for four values of the capture process energy consumption assuming that the price of CO_2 emissions allowances is at the level of 40 €. For energy consumption of 3.5 MJ/kgCO_2 the costs are 90.10 €/MWh, and for the consumption of 2 MJ/kgCO_2 – they drop to 82.99 €/MWh. The chart in **Figure 10** illustrates changes in marginal costs of electricity generation depending on the price of CO_2 emissions allowances for a power unit with no capture installation and for a power unit integrated with such an installation with different values of the process energy consumption. For the energy consumption index of 3.5 MJ/kg, the limit price of the allowances is 60 €/t, and for the index of 2 MJ/kg it is lower and totals 48 €/t.

The charts presented above indicate clearly that the CO_2 capture process energy consumption has a huge impact on the economic aspect of operating a coal-fired power unit integrated with a CCS system. In view of the

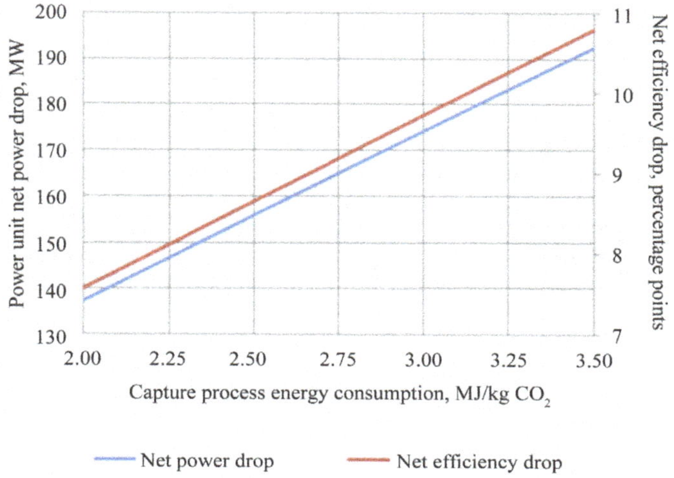

Figure 8. Drop in net electric power and in net electricity generation efficiency after the power unit integration with a CO_2 capture installation depending on the capture process energy consumption.

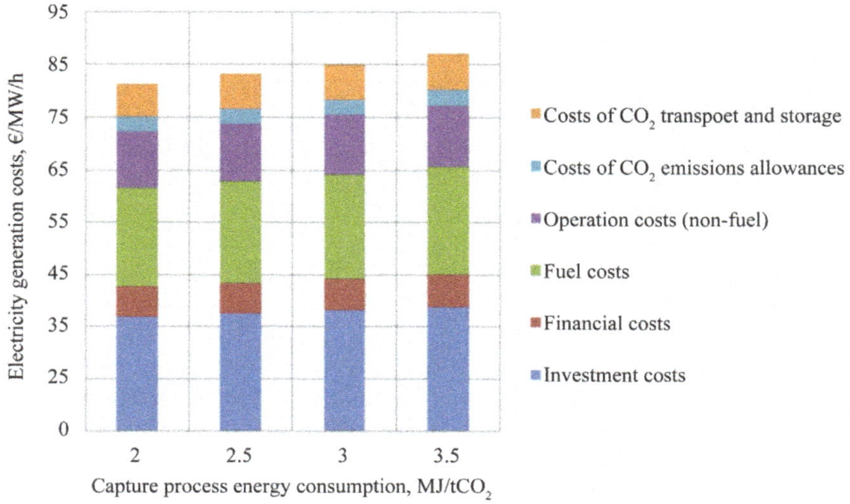

Figure 9. Costs of electricity generation—power unit with a CO_2 capture installation (price of CO_2 emissions allowances: 40 €)—for different values of the CO_2 capture process energy consumption.

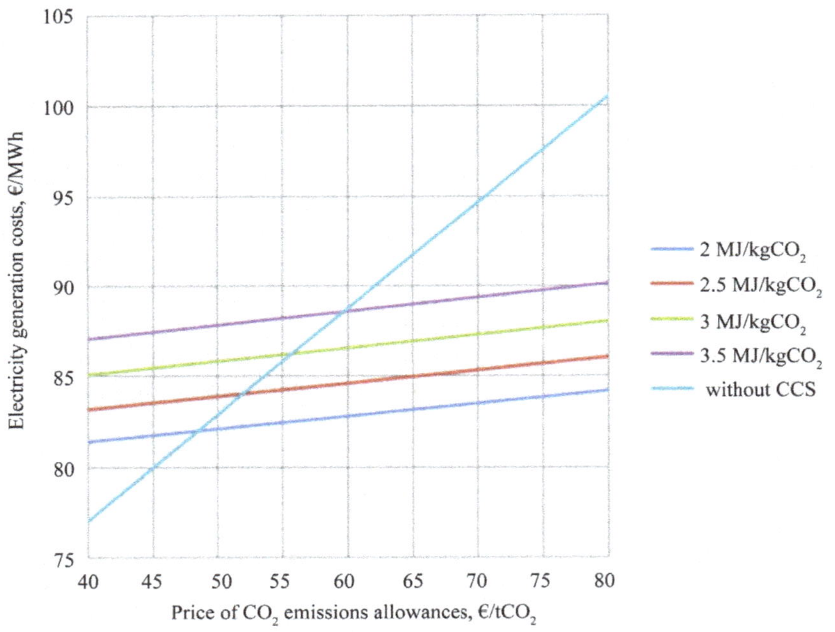

Figure 10. Marginal costs of electricity generation for a power unit with and without a CO_2 capture installation for different values of the CO_2 capture process energy consumption depending on the price of CO_2 emissions allowances.

continuous development and improvement in the technologies of carbon dioxide capture from flue gases by means of chemical absorption, e.g. by using better sorbents, it is very unlikely that the CO_2 capture process energy consumption will remain constant during the power plant entire service life. On the contrary, it is rather bound to decrease.

4. Conclusions

In this paper it shows an influence of efficiency of turbine, boiler and other elements of power cycle on its overall efficiency. Presenting the electricity generation efficiency without any information about the efficiency of the individual elements, main elements like turbine or boiler of the power plant are unfounded and may give wrong information about the effectiveness of considered technology of electricity generation.

The need to reduce emissions of greenhouse gases and improve the economy of electricity generation resulted in substantial and on-going progress in the field of condensing coal-fired power units. The development of coal-based technologies is now oriented towards achieving higher and higher powers and electricity generation efficiencies. The great step forward in materials engineering has made implementation of the advanced ultra-supercritical (A-USC) power unit technology more and more common. Due to the current state of knowledge concerning the turbine design and the strength properties of available operating materials, the maximum achievable steam parameters in steam power units are 30 MPa and 600°C for live steam and 620°C for reheated steam. It is estimated that using such steam parameters, the net electricity generation efficiency may reach 48%, depending on the power plant location (cooling conditions) and on the solutions applied to bring about a further improvement in efficiency.

A significant increase in the live and reheated steam parameters (from current 30 MPa/600°C/620°C to e.g. 35 MPa/700°C/720°C) will result in a rise in the net electricity generation efficiency by 2.5 ~ 3 percentage points, which is essential in terms of the reduction in the fossil fuel consumption, emissions of greenhouse gases and profitability of investments related the power unit integrated with a CO_2 capture installation. A further rise in efficiency will be impossible without optimization of the steam cycle structure, improvement in the design of turbines, boilers and regeneration systems, optimization of the cold end, minimization of the consumption of energy needed to satisfy the power unit own needs and the use of low-temperature waste heat.

Ecologically, respecting the limits of discharge and emission of pollutants from power engineering installations and—economically—satisfying the requirements related to the investment profitability should make it possible for the net electricity generation efficiency of coal-fired "capture ready" plants with a power capacity higher than 300 MW_{el} to exceed 48 percentage points before the plant is actually integrated with the CO_2 capture and compression installation so that the integration carried out later should bring economic profits. Such efficiency values can only be achieved by optimum technological structures if the best available technologies (BATs) are applied in the field of power engineering machinery and equipment. The paper presents the impact of the efficiency of individual elements of the power unit machinery and equipment on the plant overall efficiency. It can be seen that giving values of the power unit electricity generation efficiency without specifying the efficiencies of the unit main components is burdened with a considerable margin of uncertainty which may even be as high as several percentage points.

Acknowledgements

The results presented in this paper were obtained from research work co-financed by the Polish National Centre of Research and Development in the framework of Contract SP/E/1/67484/10—Strategic Research Programme —Advanced technologies for energy generation: Development of a technology for highly efficient zero-emission coal-fired power units integrated with CO_2 capture.

References

[1] Rosenkranz, J. and Wichtmann, A. (2005) Balancing Economics and Environmental Friendliness—The Challenge for Supercritical Coal-Fired Power Plants with Highest Steam Parameters in the Future. http://www.energy.siemens.com/mx/pool/hq/energy-topics/pdfs/en/steam-turbines-power-plants/2_Balancing_economics.pdf

[2] Smołka, K., Stępczyńska-Drygas, K., Dykas, S. and Wróblewski W. (2014) Machinery and Equipment for an Advanced Ultra-Supercritical Coal-Fired Power Unit. *Proceedings of the 8th International Science and Technology Conference "Energetyka* 2014", Wrocław, 5-7 November 2014, 107-117 (in Polish).

[3] Breeze, P. (2012) Raising Steam Plant Efficiency—Pushing the Steam Cycle Boundaries. *PEI Magazine*, **20**, 16-20.

[4] Pfaff, I., Oexmann, J. and Kather, A. (2010) Optimised Integration of Post-Combustion CO_2 Capture Process in Greenfield Power Plants. *Energy*, **35**, 4030-4041. http://dx.doi.org/10.1016/j.energy.2010.06.004

[5] Stępczyńska-Drygas, K., Dykas, S., Łukowicz, H. and Czaja, D. (2014) Assessment of the Impact of a Coal-Fired Power Unit Integration with a CO_2 Capture Installation on Operation under Varied Load Conditions. *Proceedings of the International Science and Technology Conference GRE* 2014, Szczyrk, 16-18 June 2014, 98 (in Polish).

[6] Electric Power Research Institute (2012) Outlook 2011. Electric Power Research Institute, Palo Alto.

[7] Leyzerovich, A. (2005) Steam Turbines for Modern Fossil-Fuel Power Plants. The Fairmont Press Inc., Liburn.

[8] Stępczyńska-Drygas, K., Łukowicz, H. and Dykas, S. (2013) Calculation of an Advanced Ultra-Supercritical Power Unit with a CO_2 Capture Installation. *Energy Conversion and Management*, **74**, 201-208. http://dx.doi.org/10.1016/j.enconman.2013.04.045

[9] Stępczyńska-Drygas, K., Bochon, K., Dykas, S., Łukowicz, H. and Wróblewski, W. (2013) Operation of a Conceptual A-USC Power Unit Integrated with a CO_2 Capture Installation at Part Load. *Journal of Power Technologies*, **93**, 383-394.

13

Test Standards for Direct Steam Generating Solar Concentrators

Mahesh M. Rathore[1], Ravi M. Warkhedkar[2]

[1]Research Scholar, Department of Mechanical Engineering, Govt. Engineering College, Aurangabad, India
[2]Department of Mechanical Engineering, Govt. College of Engineering, Karad, India
Email: mmrathore@gmail.com

Abstract

There are a few standards reported in the literature for testing and evaluation of thermal performance of solar concentrators based on sensible heating of working fluid. The preceding standard measures only the cooking efficiency and cooking capacity. Apart from thermal efficiency, there is an imperative need for other important parameters of the solar concentrators such as its stagnation temperature, cooking capacity, cost per watts delivered, weight of the cooker, ease of handling and aesthetics. The characterization of a concentrator at its operating temperature settles appropriate size and type of concentrator for any thermal application. The performance test is conducted at Chandwad (20.3292°N, 74.2444°E), Maharashtra and the proposed protocol aims for evaluation of thermal performance of solar cooking system and standardization of reporting the test results so that anyone can easily recognize and use it.

Keywords

Solar Concentrator Tests, Efficiency, Cooking Power, Standardization in Tests and Reporting

1. Introduction

The direct focusing solar cookers are called solar concentrators. The solar concentrators include a reflecting surface (collector), a receiver and a tracking mechanism. The reflecting surface may be constructed with the help of low iron glass mirror pieces or specially treated metallic surface like anodized aluminium sheet. The incident sun-radiation on the collector surface is reflected towards the small receiver located at the focus [1]. The reflected solar radiation is concentrated on the focus thus increasing energy flux. The working fluid in the receiver absorbs this concentrated solar energy thus subjected to heat gain. The increased energy flux makes the solar energy suitable for thermal applications and power generation.

These concentrators are broadly classified as line concentration type and point concentration type. The line-focus concentrators are usually two-dimensional parabolic reflector where the focal point becomes a line. On the other hand, a point-focus concentrator is a paraboloid dish, which is formed by rotating the parabola about its axis; the focus remains a point as shown in **Figure 1**. It attains higher stagnation temperature at the receiver, thus point-focus solar concentrating cookers are gaining popularity because of their capability to deliver operations like frying, roasting, stewing steaming and baking along with boiling. Also they offer faster cooking speed competing with conventional cooking systems.

2. Discrepancies in Existing Test Standards for Solar Concentrating Cookers

Different test standards are developed and followed worldwide for testing of the "Solar Concentrators". The test standards normally deliver technical information like thermal efficiency, cooking power, heating/cooling rates etc. under standard or normalized conditions. This information is not useful for the field conditions.

In case of solar concentrators, as shown in **Figure 1**, the cooking vessel is open to atmosphere without a greenhouse. The operating conditions are totally different from those considered during testing of box solar cookers. The solar concentration ratio of about 75 gives an operating temperature of 400˚C [2]. Usually the solar concentrators operate on mostly latent heating principle that is completely dissimilar from sensible heating behavior. Testing of the solar concentrators in sensible heating range gives misleading results. In sensible heat regimes, some quantity of water always evaporates to steam during testing period that gets released to atmosphere. This steam takes away latent heat with it. Further, the sensible heat gain is recorded in transient state. Temperature measurements are tricky and recording of temperature depends on the location and position of thermocouple in the pot. The convective current inside the pot causes time delay in actual heat gain and reported temperature rise. The temperature recording in such transient state leads to error proneness.

Further, the receiver of these concentrators has major radiation losses in addition to convective losses [3]. The radiation heat losses are proportional to fourth power of temperature. For this reason the loss characterization in case of concentrating solar cookers can't be treated as linear.

With these reasons, authors believe that the task of developing a test protocol for standardization and certification is challenging because the performance of concentrating cookers is very sensitive to design parameters and operating conditions. Presently, no specific testing procedure is available. All these developments indicate that there is an urgent need of improvement in test standards for solar concentrating cookers.

3. Review of Existing Test Standards for Solar Concentrating Cookers

A test procedure for testing of concentrating type solar concentrators had been laid down by center for energy studies of Indian Institute of Technology Delhi and Ministry of New and Renewable Energy, Government of India in 2006 [4]. This test draft was based on testing methods suggested by Mullick *et al.* [5]. This work was fur-

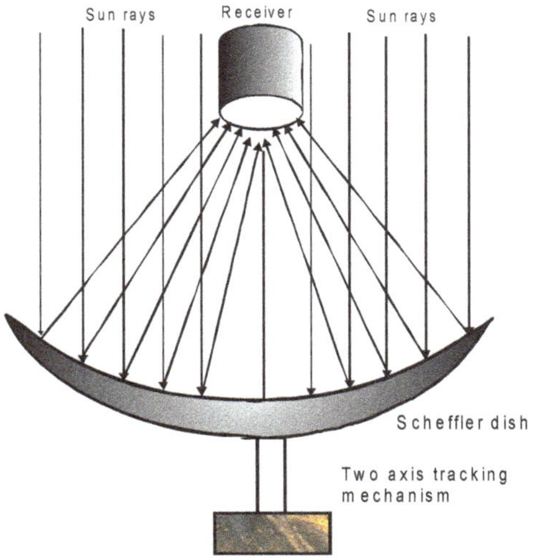

Figure 1. Principle of point focus concentration.

ther carried by Subodh Kumar *et al.* [6] [7] in the area of heat losses due to reflector orientation and effect of wind from the receiver. This proposed test protocol was designed with the help of heating and cooling tests. The parameters considered are, heat loss factor ($F'U_L$), optical efficiency factor ($F'\eta_o$), and standardized cooking power (P_s). Further, this test technique uses calculations in the sensible heating of working fluid that is transient heating mode for performance parameters. Such calculations lead to high error proneness.

American Society for Testing of Materials (ASTM) has published a Standard Test Method for Determining Thermal Performance of Tracking Concentrating Solar Collectors [8]. This test standard appears more universal and appropriate for line and point focus concentrators. Further, this standard is suitable for outdoor conditions and is valid only in sensible heat regimes and steam generation does not fall under purview of this standard.

Shaw [9] worked extensively to analyze the outline for evaluating the performance of solar concentrators and compared test procedures proposed by various researchers. He reported that no test standard fulfill all the criteria that a user expects and for this reason he proposed a new standard that accounts for technical parameters like efficiency along with other parameters like reproducibility, understandability and objectivity.

Kundapur and Sudhir [10] have also proposed a new standard for testing solar concentrators which has consideration of nine parameters including ergonomics, cooking test, user interaction and cost.

Sardeshpande *et al.* [11] have developed a procedure to examine the performance of a 25 sqm solar concentrator. Their results appeared to be rational, consistent and satisfactory. Pillai *et al.* [12] have also used above procedure for evaluating the performance of a Scheffler concentrator of 16 m^2 and got reasonable results. These both trials were conducted with latent heat exchange only; not with sensible heat exchange.

4. Proposed Test Standard for Concentrating Solar Cookers

Proposed test method deviates from conservative idea of recording heat gain in sensible heat regimes only. Instead authors propose a test protocol setup as shown in **Figure 2**, which can take care of all limitations as well as to serve a very reliable method for testing, performance prediction, monitoring and verifications programs.

Further, it is recommended that the experimentation for evaluation of thermal performance must be carried out when sky is clear and solar radiation intensity I_{bn} is above 550 W/m^2 and average wind speed during test duration should be less than 3 m/sec [13].

4.1. Principle of Operation

The heat energy supplied to working fluid is used to change the phase of water at constant pressure. The operating pressure of working fluid regulates boiling temperature. The enthalpy of vaporization of water can be obtained from steam tables at operating pressure. The product of dryness fraction of steam and enthalpy of vaporization is the amount of heat supplied for phase change of one kilogram of water.

4.2. Test Setup

The proposed experimental setup is shown in **Figure 2**. Its specifications are given in **Table 1**. Setup consists of

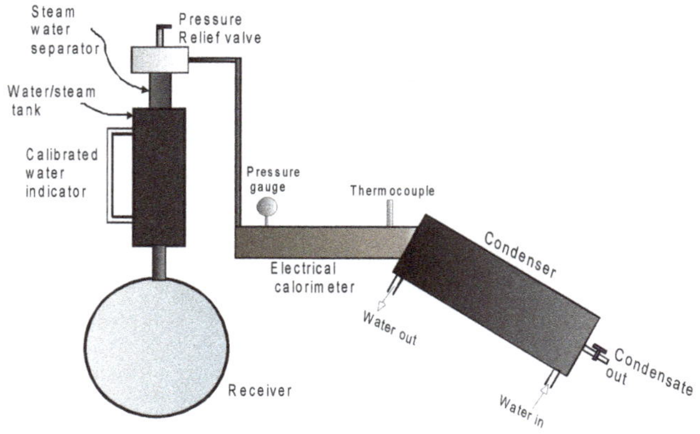

Figure 2. Proposed setup for evaluation of thermal performance of direct steam generating solar concentrators.

Table 1. Solar concentration system configuration.

Sr. No.	Parameter	Make	Specification
1	Solar reflector	Essential Equipment Dhule	16 sq m area
2	Receiver dish Plane	Essential Equipment Dhule	0.5 m dia
3	Reflective surface	Low iron glass mirrors	100 mm × 10 0mm
4	Selective black paints	0.85% reflective	

a 16 m^2 parabolide Scheffler reflector dish fitted with low iron glass mirror. A mild steel structure supports the reflector dish and sun tracking system. The tracking system swivels the reflector throughout the day to ensure maximum solar radiation on to the reflector. A receiver is installed at the focus of reflector dish to receive the concentrated solar heat flux, which in turn transferred to water present in the receiver. The water circulating system is equipped with pressure relief valve, an air vent and moisture separator. A steam/water tank supplies water to receiver and stores generated steam. The operating pressure can be set with the help of relief valve. If pressure of generated steam exceeds the operating pressure, some quantity of steam escapes through pressure relief valve to bring the steam pressure to pre-set value. The air vent removes the air and dissolved gases during initial heating of water. A moisture separator is mounted between receiver and pressure relief valve to avoid moisture droplets carry over with steam.

The instrumentation requires for measurement of pressure and temperature of steam, ambient temperature, solar radiation intensity, wind velocity, receiver temperature, mass of steam generation, etc. The instruments used during testing are given in **Table 2**.

4.3. Test Procedure

The following measuring steps should be followed during trial period.

1. Fill the measured quantity of water m_1 in the receiver system.
2. Record water temperature, T_w, ambient temperature, T_∞, pressure of water and normal beam radiation, I_{bn}, with a small interval of time till the water reaches boiling point.
3. Take one hour test after water gets boiling temperature and it starts vaporizing at preset pressure value.
4. Measure quantity of residual water m_2 in the receiver system.
5. Measure the pressure and temperature of superheated steam coming electrical calorimeter.
6. Measure electrical energy input to electrical heaters.
7. Tracking of the solar concentrator should also be made continuously during trial period to ensure normal sunrays on the collector.

4.4. Calculation

The energy balance on Scheffler concentrator and receiver is shown graphically in **Figure 3**.

Energy incident on Scheffler dish

$$Q_s = \frac{A_a \times I_{bn}}{1000} \times 3600 \ \left(\text{KJ/h}\right) \tag{1}$$

where I_{bn} is average of solar beam normal radiation over one hour test period.

Under steady state conditions, the useful energy delivered by solar collector is equal to energy absorbed by working fluid.

Actual mass of water evaporated during test period, $m_s = m_1 - m_2$ (kg/h)

The total heat energy of steam coming out electrical calorimeter

$$Q_{\text{sup}} = m_s \times h_{\text{sup}} \ \left(\text{KJ/h}\right) \tag{2}$$

Electrical work input,

$$Q_{\text{electrical}} = \frac{W_{\text{electrical}} \times 3600}{1000} \ \left(\text{KJ/h}\right) \tag{3}$$

Table 2. Instruments used.

Sr. No	Name of Instrument	Specification	Least Count
1	Measuring Flask	10 litres	25 ml
2	Pyronometer for global radiation	0 to 1800 W/m^2	1 W/m^2
3	Pyranometer with shading ring for diffuse radiation	0 to 1800 W/m^2	1 W/m^2
4	RTD for temperature measurement	0 to 300˚C	1˚C
5	Stop Watch		1 second
6	Pressure sensors	0 to 20 bar (gauge)	0.1 bar

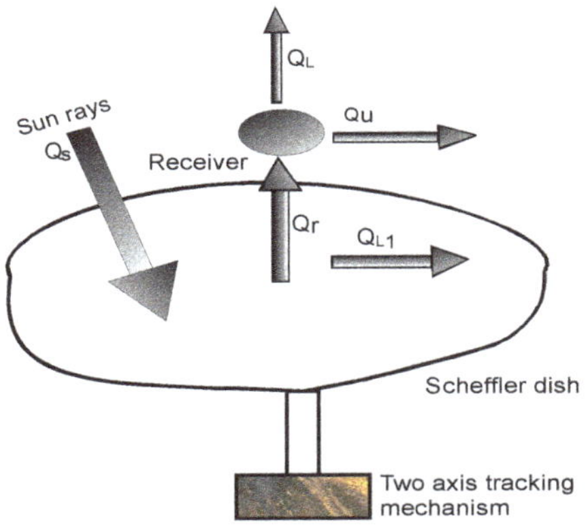

Figure 3. Energy balance on concentrator receiver system.

Useful heat energy gain rate at receiver during test period can be obtained as

$$Q_u = Q_{\text{sup}} - Q_{\text{electrical}} \quad (\text{KJ/h})$$

The quality of steam can be obtained as,

$$x = \frac{Q_u}{m_s h_{fg}} \tag{4}$$

where x is dryness fraction of steam and h_{fg} is latent heat for water at operating pressure, in kJ/kg.

The thermal efficiency of collector system is defined as ratio of useful energy on the receiver to the energy incident on the concentrator

Collector efficiency,

$$\eta_c = \frac{\text{Heat gain rate at receiver}}{\text{Heat incident rate on collector}} = \frac{Q_u}{Q_s} \tag{5}$$

Further, useful energy can also be expressed as difference of energy falling onto receiver, Q_r, and heat losses from the receiver, Q_L.

$$Q_L = Q_r - Q_u \tag{6}$$

The concentrated solar energy reaching on the receiver Q_r depends on the optical efficiency η_o of collector, which may be defined as

$$\eta_o = \frac{\text{Energy delivery rate on receiver}}{\text{Energy incident rate on concentrator's aperture}} = \frac{Q_r}{Q_s} \tag{7}$$

The optical efficiency depends on optical characteristic of material and geometry used for collector. It also accounts cosine loss, shading loss, reflection loss, transmission and absorption losses and energy spillage. Optical efficiency of most of collectors falls in range of 0.70 to 0.85 [13]. Further, the system efficiency can be defined as

$$\eta_r = \frac{\text{Useful energy gain rate by receiver}}{\text{Energy incident rate on receiver}} = \frac{Q_u}{Q_r} \tag{8}$$

Combining Equations (3) - (6), the collector efficiency can be interpreted as

$$\eta_c = \frac{Q_u}{Q_s} = \frac{Q_r}{Q_s} \times \frac{Q_u}{Q_r} = \eta_o \times \eta_r = \eta_o \left(1 - \frac{Q_L}{Q_r}\right) = \eta_o \left(1 - \frac{Q_L}{\eta_o Q_s}\right) = \eta_o - \frac{Q_L}{Q_s} \tag{9}$$

It is evident from Equation (9) that the thermal efficiency of collector is function of optical efficiency and total heat loss rate from the receiver.

4.5. Calculation of Heat Losses

The total heat loss rate Q_L from the receiver is sum of conductive, convective and radiative heat losses from the receiver surface. Mathematically;

$$Q_L = Q_{cond} + Q_{conv} + Q_{rad} \tag{10}$$

The outer surface of the receiver is covered with thick glass wool insulation to minimize the conductive heat loss and it is insignificant compare to convective and radiative losses [14]. Therefore, authors consider outer receiver wall adiabatic ($Q_{cond} = 0$) in this study.

The convection heat losses from receiver are most complicated phenomenon. It includes free and forced convections and contributes major portion of heat losses. The characteristic of convection heat losses is investigated by many researchers [15] and developed various laboratory models for estimation of natural convection heat losses. Paitoonsurikarn *et al.* [16] developed a angle dependent correlation for estimation of convection heat loss from receiver that is

$$Nu_L = 0.106 Gr_L^{1.3} \left(\frac{T_w}{T_\infty}\right)^{0.18} \left(\frac{40256 A_r}{A_w}\right)^s h(\varphi) \tag{11}$$

where Grashof number is $Gr_L = \frac{g\beta(T_w - T_\infty)L^3}{\nu^2}$, and $s = 0.56 - 1.01(A_r/A_w)^{0.5}$ is an angle dependent function $h(\varphi) = 1.1677 - 1.0762\sin(\varphi^{0.8324})$

In our experimental arrangement, the plain cylindrical receivers are mounted vertically, thus characteristic length is considered diameter of receiver. All properties of air are taken at film temperature; *i.e* average of receiver's surface temperature and ambient temperature.

The convective heat loss from receiver

$$Q_{conv} = hA_r(T_w - T_\infty) \tag{12}$$

The radiation heat loss from the receiver can be obtained as

$$Q_{rad} = A_r \varepsilon \sigma (T_w^4 - T_\infty^4) \tag{13}$$

4.6. Observation Table

A set of observations are presented in **Table 3**. Effective aperture area of 16 m^2 Scheffler reflector dish during the month of March is approximately 11.8 m^2.

5. Result and Discussion: Characterization for Plain Receiver

(a) Thermal Efficiency

The direct steam generating 16 m^2 solar concentrator is tested with the help of 0.5 m diameter plain receiver. The optical efficiency is assumed 85%. The effiiency varies from 52.38% to 26% with an error ±3.5% as shown in **Figure 4**. The efficiency is higher at loest operating pressure and it decreases at operating pressure increases. The radiation and convective heat losses bbecome dominating at higher operating temperatures.

Table 3. Recorded data of 16 m^2 direct steam generating Scheffler solar concentrator with plain receiver.

Sr. No.	Date	Start Time	Direct Beam Radiation (W/m^2)	Diffuse Beam Radiation (W/m^2)	Steam Pressure (bar)	Temperature of Superheated Steam (˚C)	Steam Generated in One Hour (kg/h)	Receiver Temp. (˚C)	Air Velocity (m/s) N-S dir.	Heater Input kWh
1	22/3/2015	10.53.00	990	251	2.40	153	8.15	155.7	1.2	19,944
2	22/3/2015	2.11.00	1012	257	3.50	161	7.65	195.1	1.2	20,232
3	23/3/2015	10.49.15	1061	283	4.40	167	6.90	200.0	1.3	20,592
4	23/3/2015	2.08.00	900	205	5.50	183	6.35	230.4	1.63	20,952
5	24/3/2015	11.02.00	1018	253	6.80	191	5.80	256.3	2.2	19,584
6	24/3/2015	1.58.00	1024	263	8.25	210	5.25	280.2	2.1	20,196
7	25/3/2015	11.05.00	1011	278	9.45	235	4.65	281.2	1.4	20,376
8	25/3/2015	2.08.00	1022	261	10.80	250	3.91	289.1	0.8	19,872
9	26/3/2015	10.50.00	1017	257	11.60	268	3.22	291.2	1.0	20,232
10	26/3/2015	1.58.00	1010	259	12.80	274	2.76	310	1.6	20,520

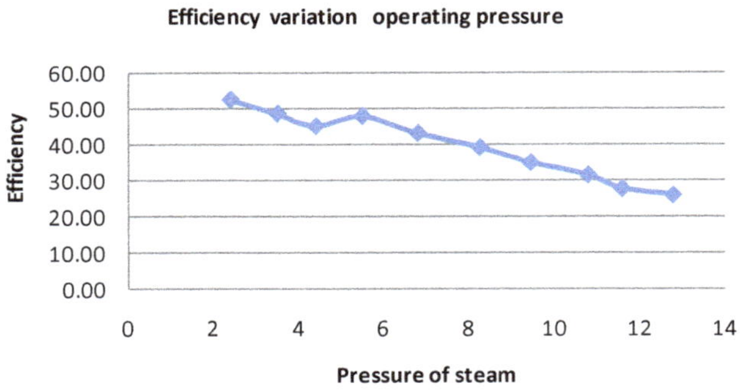

Figure 4. Efficiency of system decreases as operating pressure increases.

(b) Convective heal loss pattern

As wind direction angle increases, the convection heat loss increases and the heat loss reaches maximum value, when wind angle 90 degree with the surface of receiver as shown in **Figure 5**.

The heat losses from the receiver at different operating temperature are determined from Equation (6). The conduction heat losses are considered negligible and radiation heat loss from plain cylindrical receiver is calculated from Equation (14). The remaining heat loss is assumed convection heat loss, which is presented in **Table 1**. The heat transfer coefficient is obtained by using empirical relation Equation (11) and is used to obtain calculated values of convection heat losses.

Further, it is evident that the experimental and empirical values of convection losses closely agree, but as operating temperature increases, the error in estimation becomes widen from 0-30 degree receiver tilt and then it decreases.

6. Conclusions

The testing of 16 m^2 direct steam generating solar concentrator is done at constant pressure with change of phase of working substance. Therefore, only enthalpy of vaporization is required to consider in calculations, which is easy to obtain from steam tables. Further, during the evaporation, the temperature of working substance remains constant, which makes heat loss rate constant under the same ambient condition.

Proposed test standard also provides useful information to be reported to all stakeholders. Thermal performance tests are to be performed by the "Test Centers". The normalized parameters Q_n and η_n are the most im-

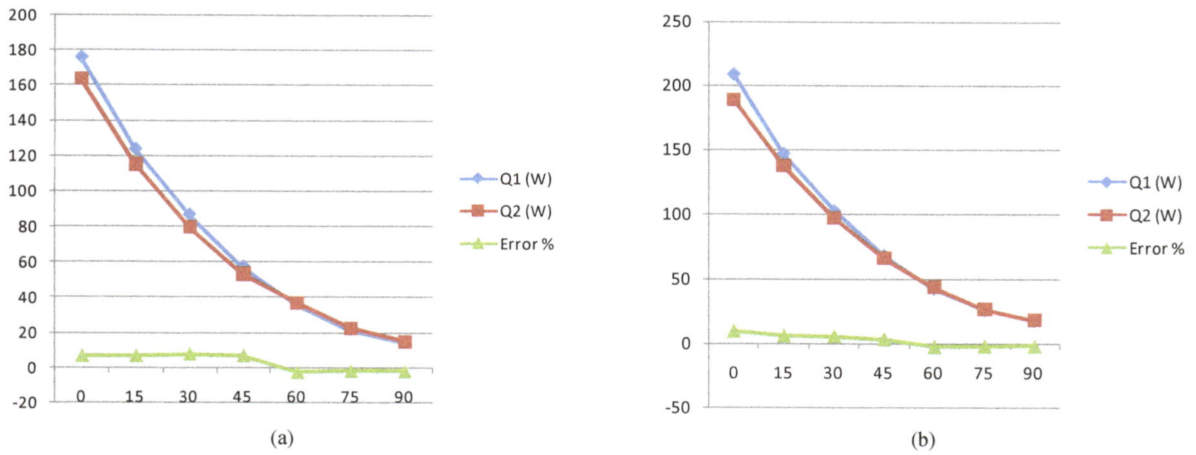

Figure 5. (a) Convection heat loss pattern with wind direction at 200˚C; and (b) Convection heat loss pattern with wind direction at 250˚C. Q1 = calculated value and Q2 = experimental measured value

portant parameters to bring in uniformity for comparing different solar concentrating cookers tested at different test centers in different climatic conditions.

The test standard provides important technical parameters which can be used for certification of the solar concentrating cookers. The convection heat and radiation losses from receiver reduce efficiency of system significantly. These losses must be estimated carefully. Further, at high operating temperature, the radiation heat loss is dominating over convection heat transfer. A rigorous work is required to develop a mathematical model for estimation of radiation losses.

The solar concentrators have huge potential for traditional fuel saving opportunity and cooking capability. Further, the technical date generated from the test will be useful for policy makers like GACC (Global Alliance for Clean Cookstoves), UNDP (United Nations Development Programs) and for governments especially in Asia and Africa. Data generated can be used for generation as well as validation of projects for CDM and similar carbon trading mechanisms.

References

[1] Rathore, M.M. and Warkhedkar, R.M. (2015) Development of Universal Test Standard for Concentrating Solar. *International Journal of Modern Trends in Engineering and Research* (*IJMTER*), **2**, 1655-1658.

[2] Ravi Kumar, K. and Reddy, K.S. (2009) Thermal Analysis of Solar Parabolic trough with Porous Disc Receiver. *Applied Energy*, **86**, 1804-1812. http://dx.doi.org/10.1016/j.apenergy.2008.11.007

[3] Reddy, K.S. and Kumar, N.S. (2008) Combined Laminar Natural Convection and Surface Radiation Heat Transfer in a Modified Cavity Receiver of Solar Parabolic Dish. *International Journal of Thermal Sciences*, **47**, 1647-1657. http://dx.doi.org/10.1016/j.ijthermalsci.2007.12.001

[4] Centre of Energy Studies, Ministry of Non-Conventional Energy Sources (2006) Draft Test Procedure Solar Cooker—Paraboloid Concentrator Type. 1-12.

[5] Mullick, S.C., Kandpal, T.C. and Kumar, S. (1991) Thermal Test Procedure for a Paraboloid Concentrator Solar Cooker. *Solar Energy*, **46**, 139-144. http://dx.doi.org/10.1016/0038-092X(91)90087-D

[6] Kumar, S., Kandpal, T.C. and Mullick, S.C. (1993) Heat Losses from a Paraboloid Concentrator Solar Cooker: Experimental Investigations on Effect of Reflector Orientation. *Renewable Energy*, **3**, 871-876. http://dx.doi.org/10.1016/0960-1481(93)90044-H

[7] Kumar, S., Kandpal, T.C. and Mullick, S.C. (1994) Effect of Wind on the Thermal Performance of a Paraboloid Concentrator Solar Cookers. *Renewable Energy*, **4**, 333-337. http://dx.doi.org/10.1016/0960-1481(94)90037-X

[8] ASTM International (2007) Standard Test Method for Determining Thermal Performance of Tracking Concentrating Solar Collectors. Designation: E 905-87, Reapproved 2007, 1-14.

[9] Shaw Shawn (2006) Development of a Comparative Framework for Evaluating the Performance of Solar Cooking Devices. Thesis submitted at Rensselaer Polytechnic Institute, USA, 1-63. http://www.solarcooker.org/Evaluating-Solar-Cookers.doc

[10] Kundapur, A. and Sudhir, C.V. (2009) Proposal for New World Standard for Testing Solar Cookers. *Journal of Engineering Science and Technology*, **4**, 272-281.

[11] Sardeshpande, V.R, Chandak, A.G. and Pillai, I.R. (2011) Procedure for Thermal Performance Evaluation of Steam Generating Point-Focus Solar Concentrators. *Solar Energy*, **85**, 1390-1398. http://dx.doi.org/10.1016/j.solener.2011.03.018

[12] Pillai, I.R., Chandak, A.G., Sardeshpande, V. and Somani, S.K. (2010) Methodology for Performance Evaluation of Fixed Focus Moving Solar Concentrators. *World Renewable Energy Congress XI*, Abu Dhabi, 25-30 September 2010, 1-6.

[13] Chandak, A. (2009) Apparatus for Testing Solar Concentrators' Patent Application No 2377/MUM/2009. Filed with Controller of Patents, Mumbai.

[14] Shuai, Y., Xia, X.-L. and Tan, H.-P. (2008) Radiation Performance of Dish Solar Concentrator/Cavity Receiver Systems. *Solar Energy*, **82**, 13-21. http://dx.doi.org/10.1016/j.solener.2007.06.005

[15] Leibfried, U. and Ortjohann, J. (1995) Convective Heat Loss from Upward and Downward Facing Cavity Receivers: Measurements and Calculations. *Journal of Solar Energy Engineering*, **117**, 75-84. http://dx.doi.org/10.1115/1.2870873

[16] Paitoonsurikarn, S., Taumoefolau, T. and Lovegrove, K. (2004) Estimation of Convection Loss from Paraboloidal Dish Cavity Receivers. *Proceedings of* 42*nd Conference of the Australia and New Zealand Solar Energy Society* (*ANZSES*), Perth, 30 November-3 December 2004, 1-7.

Nomenclature

A_p = aperture area of dish concentrator, m^2
A_a = area of absorber surface, m^2
A_w = cavity surface area, m^2
C = concentration ratio,
h = convective heat transfer coefficient, $W/m^2 \cdot K$
h_{fg} = enthalpy of vaporization of water, J/kg
h_{sup} = enthalpy of superheated steam, J/kg
I_{bn} = normal intensity of radiation, W/m^2
$\dot{I}_{av}$ = reference solar intensity, W/m^2
k_f= thermal conductivity of air, W/m·K
m_w = mass of water, kg
m_s = mass of steam, kg
Q_{cond} = conductive heat loss rate, W
Q_{conv} = convection heat loss rate, W
Q_L = heat loss rate from receiver, W
Q_r = concentrated heat rate on receiver, W
Q_s = energy incident rate on dish, W
Q_u = useful energy rate, W
Q_{un} = normalized useful energy rate, W
Q_{rad} = radiative heat loss rate, W
Q_{sup} = heat of superheated steam, W
T = surface absolute temperature, K
T_∞ = overall heat loss coefficient, K
T_{sat} = saturation temperature, K
T_{sup} = temperature of superheated steam, K
T_1 = initial temperature of water, K
U_L = overall heat loss coefficient, $W/m^2 \cdot L$
x = dryness fraction of steam.

Greek symbols
η_c = efficiency of concentrator,
η_o = optical efficiency of concentrator,
η_r = efficiency of receiver system,
ε = emissivity of surface,
σ = Stefan Boltzmann Constant,
Δ = difference in quantity,
φ = tilt angle of receiver, radian.

14

Preferred Economic Dispatch of Thermal Power Units

Selvaraj Durai, Srikrishna Subramanian, Sivarajan Ganesan

Department of Electrical Engineering, Annamalai University, Chidambaram, India

Email: duraiselvaraj86@gmail.com

Abstract

Economic Dispatch (ED) problem is one of the main concerns of the power generation operations which are basically solved to generate optimal amount of power from the generating units in the system by minimizing the fuel cost and by satisfying the system constraints. The accuracy of ED solutions is highly influenced by the fuel cost parameters of the generating units. Generally, the parameters are subjected to transform due to aging process and other external issues. Further the parameters associated with the transmission line modelling also change due to aforementioned issues. The loss coefficients which are the functions of transmission line parameters get altered from the original value over a time. Hence, the periodical estimation of these coefficients is highly essential in power system problems for obtaining ideal solutions for ED problem. Estimating the ideal parameters of the ED problem may be the best solution for this issue. This paper presents the Teaching Learning Based Optimization (TLBO) algorithm to estimate the parameters associated with ED problem. The estimation problem is formulated as an error minimization problem. This work provides a frame work for the computation of coefficients for quadratic function, piece-wise quadratic cost function, emission function, transmission line parameters and loss coefficients. The effectiveness of TLBO is tested with 2 standard test systems and an Indian utility system.

Keywords

Parameter Estimation, Cost Coefficients, Emission Coefficients, Error Estimation, Transmission Line Parameters, Teaching Learning Based Optimization

1. Introduction

1.1. Ideal Economic Dispatch

The economical operation of power system needs more accurate representation of fuel cost coefficients and transmission loss coefficients. Due to changing weather conditions and other external issues, the aforesaid coefficients are not ideal and hence the existing parameters with ED solutions also change. Therefore, the objective of this article is to revise the existing parameters for the Preferred Economic Dispatch (PED) solutions.

1.2. Literature Review

Various optimization techniques have been proposed by many researchers to deal with the parameter estimation problems. To obtain the exact parameters for ED problems different parameter estimation rehearsal has been projected to solve estimation problems in power systems. State estimation based technique such as Least Error Square (LES) and least absolute value methods have been demonstrated. Among the two techniques, LES technique has been the most famous static estimation technique and in use for a long time as the preferred technique for optimum estimation of parameters. In general, some limitations and disadvantages are associated with this approach. El-Hawary and Mansour conducted performance analysis of LES, Bard algorithm, Marquardt algorithm and Powell regression algorithm for estimating coefficients [1]. The methods based on the least absolute value approximations and curve fitting techniques have been reported for fuel cost coefficients estimation [2]. Two polynomial curve fitting methods, Gram-Schmidt orthonormalization and least square are also applied to evaluate the fuel cost coefficients [3]. Further Henry Y. K. Chen and Charles E. Postel applied sequential regression technique to online parameter estimation of input-output curves for thermal units [4].

Research endeavours indicate that the field of Transmission Line Parameters (TLP) estimation in power system studies is less focused. Traditionally, tower and conductor geometric parameters, conductor type, assumed ambient conditions etc., are utilized for estimating TLP [5]-[7]. The sending and receiving end voltage and current phasors are utilized to derive the TLP and propagation constant [8]. The synchronized phasor measurements at both ends of the transmission line emphasises the online parameter estimation [9]-[12]. Wagenaars *et al.*, have proposed the measurement method, based on a pulse response measurement, to determine the transmission line parameters of the shield-to-phase and phase-to-phase modes [13]. Researchers have reported Newton-Raphson based method for estimating TLP parameters utilizing the measurements of voltage, current and power [14].

A new category of classical optimization techniques has emerged to cope with some of the traditional algorithms in the field of power system estimation problems. The heuristic techniques such as Genetic Algorithm (GA) [15], Particle Swarm Optimization (PSO) [16] [17], Artificial Bee Colony (ABC) [18], Artificial Neural Network (ANN) [19] [20], Fuzzy Logic (FL) [21], Ant Colony Optimization (ACO) [22] and Bacterial Foraging Algorithm (BFA) [23]-[25] have been used for solving various estimation problems such as load forecasting, state estimation and induction motor parameters.

1.3. Optimization Techniques

The heuristic techniques outperform the mathematical methods but their solution quality is sensitive to the algorithmic controlling parameters like population size and number of generations. Besides common control parameters, different algorithms require their own algorithm specific control parameters. Major disadvantage of ABC, PSO and GA are the presence of various parameters that need to be carefully tuned to reach acceptable estimation performance. Recently, TLBO, a nature inspired algorithm that is based on the effect of influence of a teacher on the output of learners in a class [26] [27] is proposed. It is a powerful evolutionary algorithm that maintains a population of students, where each student represents a potential solution to an optimisation problem. Each searching generation includes initializing of class, teacher phase, learner phase and termination criterion. The advantage of TLBO are simple, easy implementation and necessitates few control parameters for tuning that make it suitable to implement for power system parameter estimation problems. TLBO is an algorithm-specific, parameter-less algorithm that does not require any algorithm-specific parameters to be tuned [28]. This algorithm can find the global solution for nonlinear constrained optimization problems with less computation effort and high consistency [29]. The authors have proposed the TLBO algorithm for estimating the accurate parameters of the input output characteristics of thermal units [30]. The adaptable properties of this algorithm encourage the authors to use TLBO as a parameter estimator.

1.4. Research Gap and Motivation

The most important issue in ED problem is to have an accurate estimate of parameters. The exact ED solution in power generation and the estimation of parameters in transmission systems is an important task in power systems.

The demonstration of literature review reveals that the estimation of accurate fuel cost coefficients and transmission line parameters are carried out independently. The impact of change in transmission line parameters on the loss coefficients of ED problem is seldom carried out. These points motivate us to contribute research in estimation of cost and emission coefficients, transmission line parameters and transmission loss coefficients for the exact ED solution. The TLBO algorithm is implemented for solving parameter and ED problems on different scale of test systems.

1.5. Highlights

- Accurate parameters of different generator cost functions and transmission lines are estimated.
- The preferred economic dispatch is carried out.
- A 19 unit practical Indian utility system is considered.
- A nature inspired TLBO is applied for both estimation of parameters and for solving ED problems.

1.6. Paper Organization

The rest of the paper is structured as follows: Problem formulation is explored in Section 2. Section 3 describes the implementation for preferred ED. The detailed discussions about numerical results achieved by various test systems are detailed in Section 4. Section 5 describes the potential verification of TLBO. Finally, Section 6 presents the conclusions of this article.

2. Problem Formulation

2.1. State of the Art Model

The state of the art multi objective ED model is presented as follows.

$$FC_i^{Exi} = W_1\left(\sum_{i=1}^{N} a_i^{Exi}P_i^2 + b_i^{Exi}P_i + c_i^{Exi} + \left|e_i^{Exi}\sin\left(f_i^{Exi}\left(P_{i,\min} - P_i\right)\right)\right|\right) + W_2\left(\sum_{i=1}^{N} e_{0i}^{Exi}P_i^2 + e_{1i}^{Exi}P_i + e_{2i}^{Exi}\right) + r_i \quad i = 1,2,\cdots,N \tag{1}$$

FC_i^{Exi} is a existing fuel cost function, N is the number of generating units and P_i is the power generated by ith generating unit.

a_i^{Exi}, b_i^{Exi}, c_i^{Exi} are the existing cost coefficients, e_i^{Exi}, f_i^{Exi} are the existing valve point coefficients and e_{0i}^{Exi}, e_{1i}^{Exi}, e_{2i}^{Exi} are the existing emission coefficients .

W_1 and W_2 are the weights of the function.

r_i is the error associated with the ith equation.

$$\text{Loss function: } P_L^{Exi} = \sum_{i=1}^{N}\sum_{j=1}^{N} P_i B_{ij}^{Exi} P_j + \sum_{i=1}^{N} B_{0i}^{Exi} P_i + B_{00}^{Exi} + r_i \quad i = 1,2,\cdots,N \tag{2}$$

P_L^{Exi} indicates existing transmission loss.

B_{ij}^{Exi}, B_{0i}^{Exi}, B_{00}^{Exi} are the existing transmission loss coefficients.

2.2. Proposed Parameter Estimation Problems

For preferred ED solution, the accurate parameters for fuel cost, emission, and transmission line and transmission loss are needed. Those parameters can be estimated by solving the following problems.

Cost coefficients Estimation, FC^{Est} = Estimate [a, b, c, e, f]

Emission coefficients Estimation, E^{Est} = Estimate [e_0, e_1, e_2]

Transmission line parameters Estimation, TLP^{Est} = Estimate [R, X, B]

Transmission loss coefficients Estimation, KC^{Est} = Estimate [B_{ij}, B_{0i}, B_{00}]

In general to estimate the accurate parameters, the following function are used

$$\text{Estimate } \left[FC^{Est}, E^{Est}, TLP^{Est}, KC^{Est}\right] \tag{3}$$

2.3. Accurate Model Using Estimated Parameters

The cost function of generating unit is expressed using estimated coefficients as follows,

$$\begin{aligned} FC_i^{Est} &= W_1\left(\sum_{i=1}^{N} a_i^{Est}P_i^2 + b_i^{Est}P_i + c_i^{Est} + \left|e_i^{Est}\sin\left(f_i^{Est}\left(P_{i,\min} - P_i\right)\right)\right|\right) \\ &+W_2\left(\sum_{i=1}^{N} e_{0i}^{Est}P_i^2 + e_{1i}^{Est}P_i + e_{2i}^{Est}\right) + r_i \quad i = 1,2,\cdots,N \end{aligned} \tag{4}$$

where FC_i^{Est} is an estimated fuel cost.

a_i^{Est} b_i^{Est}, c_i^{Est}, e_i^{Est}, f_i^{Est}, e_{0i}^{Est}, e_{1i}^{Est}, e_{2i}^{Est} are the estimated cost coefficients, valve point coefficients and emission coefficients respectively.

$$\text{Loss function: } P_L^{est} = \sum_{i=1}^{N}\sum_{j=1}^{N} P_i B_{ij}^{Est} P_j + \sum_{i=1}^{N} B_{0i}^{Est} P_i + B_{00}^{Est} \quad i = 1,2,\cdots,N \tag{5}$$

B_{ij}^{Est}, B_{0i}^{Est}, B_{00}^{Est} are the estimated transmission loss coefficients and P_L^{Est} is the estimated transmission loss.

2.4. Error Minimization

As detailed in [30], the error is estimated for fuel cost and emission coefficients as follows

The error at each step *i* can be calculated as,

$$\text{Error} = \left(FC^{Exi} - FC^{Est}\right) \text{ or } \left(E^{Exi} - E^{Est}\right) \tag{6}$$

$$\%\,\text{Error for Fuel cost} = \sum_{k=1}^{n}\left|\frac{FC_i^{Exi} - FC_i^{Est}}{FC_i^{Exi}}\right| * 100 \tag{7}$$

$$\%\,\text{Error for Emission} = \sum_{k=1}^{n}\left|\frac{E_i^{Exi} - E_i^{Est}}{E_i^{Exi}}\right| * 100 \tag{8}$$

The % error for each step *k* of transmission line parameter is expressed as,

$$\%\,\text{Error} = \sum_{k=1}^{n}\left|\frac{TLP_k^{Exi} - TLP_k^{Est}}{TLP_k^{Exi}}\right| * 100 \tag{9}$$

The objective function is minimised subjected to physical limits of each coefficients and TLP.

3. Implementation of TLBO for Preferred Economic Dispatch (PED)

This section details the computational flows for parameter estimation and PED problems.

3.1. Parameter Estimation Procedure

Step 1: Read the required data for estimating the parameters, population size, maximum number of iterations ($Iter_{\max}$), maximum and minimum limits of parameters ($PA^{\min}$, $PA_i^{\max}$).

Step 2:Randomly generate parameters using the following equation,

$$PA_i^j = PA_i^{\min} + rand_z * \left(PA_i^{\max} - PA_i^{\min}\right) \quad i = 1,\cdots,CG, \; j = 1,\cdots,PS \tag{10}$$

Step 3: Check for the constraint violation.

Step 4: If the limit is violated, then upgrade the parameters of the units using the following equation

$$P_i^j = \begin{cases} PA_i^{\min}; \left(PA_i^j < PA_i^{\min}\right) \\ PA_i^{\max}; \left(PA_i^j > PA_i^{\min}\right) \\ PA_i^j; \left(PA_i^{\max} \geq PA_i^j \geq PA_i^{\min}\right) \end{cases} \tag{11}$$

Step 5: Calculate FC^{Est} by using Equations (4).

Step 6: The error and % error associated with each measurement can be calculated by using Equations (6), (7), (8) and (9).

Step 7: Teacher phase: Compute the difference between existing mean result and best mean by utilizing teaching factor (ψ_F).

Step 8: Learner phase: Evaluate the learners' parameter values with the help of teacher's parameter values.

Step 9: Update the parameters subject to constraints.

Step 10: Stopping Criterion: If $Iter \geq Iter_{\max}$ print the optimal results, otherwise go to step 7.

3.2. Computational Flow of PED

Hence the estimated parameters are suitably achieved. The ED solution is carried out and the computational flow is as follows in **Figure 1**.

4. Numerical Simulation Results and Discussion

This section presents three different test systems such as 10 unit, IEEE 30 bus and a practical Indian Utility system to demonstrate the efficacy of the proposed solution technique for perfect estimate of parameters and preferred ED solutions. The algorithm is implemented in Matlab package and simulations are carried out on a personal computer having the configuration of Intel(R) core i3 CPU with 2 GB RAM. The following algorithmic parameters are chosen for all cases in order to validate the performance of the algorithm: Population size = 10;

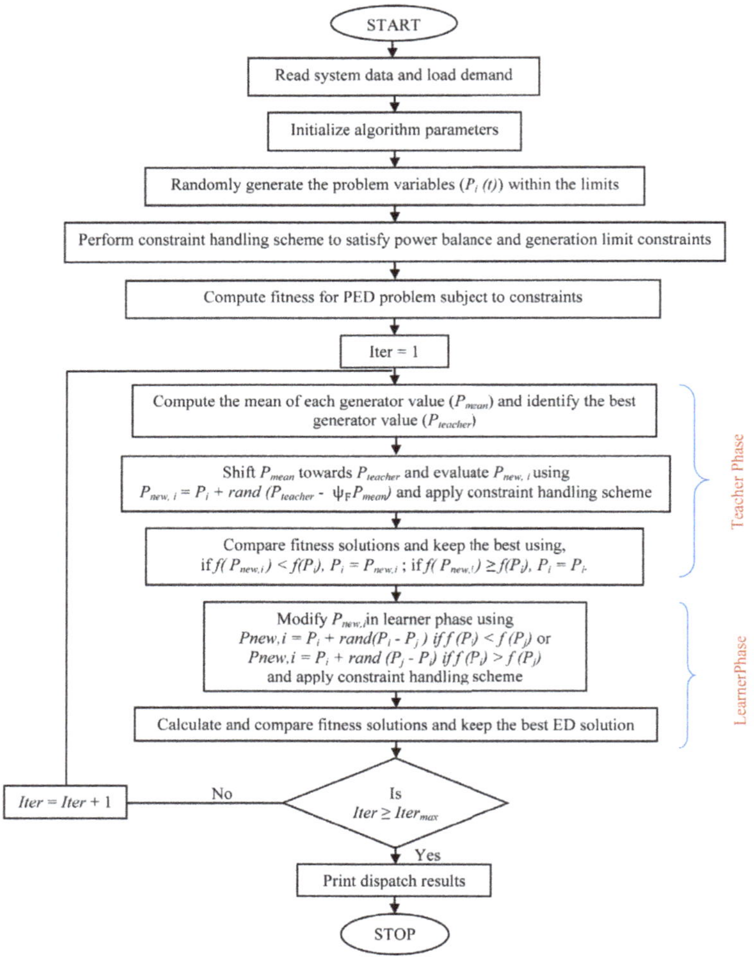

Figure 1. Flowchart for preferred economic dispatch using TLBO.

Maximum number of iterations = 120.

4.1. Test System 1: 10 Unit System with Piecewise Quadratic Functions

This test system consists of 10-generating units in which each unit has two or three fuel options like coal, oil and gas. As more than one fuel is used, the generator input-output characteristic is expressed as a piecewise quadratic function. The valve point effects are also considered. The test system data is available in [31].

The generator characteristic is expressed as a piecewise quadratic function including valve point effects increases the number of variables to be estimated. In this test system, the generator 1 has two fuel options and the remaining units have three fuel options. It is specified that the fuel 2 is uneconomical for generator 9. The simulation is carried out using the proposed TLBO which acquires new values for cost coefficients and are presented in the **Table 1**. The estimation process has been performed for each generation level and the results of existing

Table 1. Estimation of piecewise quadratic cost function coefficients using TLBO-10 unit system.

Unit	P (MW)		Fuel type	Coefficients estimated by TLBO					FC^{Exi} ($/h)		FC^{Est} ($/h)		Σ_{error}	
	Min	Max		a_i	b_i	c_i	e_i	f_i	Min	Max	Min	Max	TLBO	ABC
1	100	196	1	25.19938	−0.380846	0.0021609	0.025324	−3.028257	8.980000	32.6800	8.723994	33.59283	1.168802	3.98151
	196	250	2	20.03000	−0.305900	0.0018897	0.021337	−3.009000	32.66577	60.9879	32.67157	61.67932	0.697156	2.81373
2	50	114	2	1.816721	−0.037780	0.0011015	0.001901	−0.410800	22.54490	48.4386	23.48145	49.14125	1.639115	3.25496
	114	157	3	12.94016	−0.198000	0.0016731	0.013159	−1.880000	2.716000	12.1088	2.681891	11.82574	0.317203	2.75457
	157	230	1	120.6888	−1.282684	0.0042228	0.112869	−11.94430	12.13152	22.4996	12.12238	23.09567	0.605175	2.93069
3	200	332	1	39.04620	−0.305817	0.0014527	0.039875	−3.115954	35.75000	96.9444	35.98904	97.64218	0.936732	1.59272
	332	388	3	−2.97500	0.032890	0.0008075	−0.002976	0.336952	131.3536	187.056	131.7986	187.3240	0.712539	1.27426
	388	500	2	−58.1400	0.484892	1.196E−05	−0.058140	4.764000	96.94140	131.236	96.94769	131.3470	0.116458	1.66413
4	99	138	1	1.990265	−0.031040	0.0010434	0.001982	−0.310400	9.181389	17.6636	9.143991	17.57876	0.122282	1.15842
	138	200	2	52.94079	−0.633800	0.0027499	0.053850	−6.248000	17.77095	36.2506	17.84586	36.22169	0.103847	0.43500
	200	265	3	258.6681	−2.290143	0.0058688	0.269957	−22.39430	36.60000	64.2124	35.39341	64.15721	1.261771	4.91182
5	190	338	1	13.80777	−0.088330	0.0010722	0.013720	−0.870300	35.80990	106.192	35.73286	106.449	0.333505	1.15977
	338	407	2	99.87032	−0.515692	0.0015799	0.099573	−5.182580	106.2448	152.504	106.0639	151.7490	0.936775	5.70216
	407	490	3	−52.9900	0.445057	0.0001478	−0.053890	4.560945	152.4276	200.634	152.6311	200.6286	0.208988	0.41948
6	138	200	1	52.95000	−0.631800	0.0027309	0.052750	−6.308000	17.77095	36.2506	17.76962	35.88033	0.371620	1.59502
	85	138	2	1.976247	−0.031040	0.0010540	0.001973	−0.310400	6.915125	17.6642	6.952709	17.76567	0.139008	0.91532
	200	265	3	265.4849	−2.333976	0.0059502	0.261282	−23.28925	36.60000	64.2124	36.69646	64.94455	0.828617	2.44408
7	200	331	1	18.88226	−0.132500	0.0011050	0.018850	−1.355000	36.71000	96.3699	36.58365	96.11233	0.383942	2.96959
	331	391	2	44.00000	−0.220000	0.0011464	0.043769	−2.211000	96.37086	133.271	96.77999	133.2703	0.410536	1.94797
	391	500	3	−43.0000	0.356873	0.0002422	−0.043000	3.552032	133.3239	195.993	133.5603	196.0080	0.251265	2.66318
8	99	138	1	1.984912	−0.031000	0.0010294	0.001992	−0.310000	9.181389	17.6636	9.005129	17.31185	0.528055	1.15768
	138	200	2	52.76360	−0.630000	0.0027294	0.052870	−6.375000	17.77095	36.2506	17.80167	35.96774	0.313608	1.13761
	200	265	3	266.5000	−2.330000	0.0059150	0.260000	−23.00000	36.60000	64.2124	37.10000	64.53166	0.819262	4.66804
9	213	370	1	88.56713	−0.566963	0.0015544	0.088509	−5.684923	38.15592	91.3812	38.32396	91.61140	0.398192	2.68979
	130	213	3	14.20019	−0.018100	0.0006124	0.014100	−0.181000	22.21239	38.1385	22.19758	38.14001	0.016320	1.07008
	370	440	3	14.10104	−0.018100	0.0006111	0.014280	−0.181000	91.30359	124.739	91.06478	124.4491	0.529615	5.65214
10	200	362	1	13.99000	−0.099100	0.0011054	0.013887	−0.992500	38.17400	122.414	38.38720	122.9832	0.781727	1.87694
	362	407	3	46.65000	−0.201400	0.0011360	0.046801	−2.014000	152.6864	198.035	152.7299	198.0353	0.043682	1.10165
	407	490	2	−61.0900	0.508437	4.157E−05	−0.061100	5.074000	122.4382	152.677	122.6075	152.8768	0.368904	1.08521

fuel cost (FC^{Exi}) and estimated fuel cost (FC^{Est}) are presented independently for each fuel type in the **Table 1**. The error is also computed by comparing FC^{Exi} and FC^{Est} using Equation (6). The obtained error using the proposed algorithm is compared with ABC algorithm which clearly illustrates the superiority of TLBO over other algorithm.

The ED is carried out separately considering both existing and estimated fuel cost coefficients. The obtained results for various demands are presented in **Table 2** along with % error, which clearly illustrates the impact of estimated coefficients on ED problem.

4.2. Test System 2: IEEE 30 Bus System

The TLBO algorithm has been executed on the test system consists of six generating units interconnected with 41 branches of transmission lines. Generator data, emission characteristics and line data are obtained from [32]. TLBO algorithm is applied to estimate the cost, emission coefficients and transmission line parameters. For each unit, the fuel cost and emission coefficients are computed. **Table 3** and **Table 4** show the TLBO outcomes of the

Table 2. Economic dispatch with improved parameters for 10 unit system.

Demand (MW)	FC^{Exi} ($/h)	FC^{Est} ($/h)	% Error
2500	525.7539	532.2159	1.22090
2600	608.7376	615.3032	1.07856
2700	627.2440	634.3932	1.13978

Table 3. Estimation of quadratic cost function coefficients by TLBO-IEEE 30 bus system.

Unit	P (MW)	Coefficients			FC^{Exi} ($/h)	FC^{Est} ($/h)	Error	Total Error
			Existing	TLBO				
	50	a_i	0.00375	0.003709	109.375	109.6930	0.318013	
1	125	bi	2.00000	2.008420	308.594	309.0026	0.408860	0.763171
	200	c_i	0.00000	0.000000	550.000	550.0363	0.036298	
	20	a_i	0.01750	0.017377	42.0000	41.95063	0.049366	
2	60	b_i	1.75000	1.750000	168.000	167.5557	0.444297	1.283524
	80	ci	0.00000	0.000000	252.000	251.2101	0.789861	
	15	a_i	0.06250	0.061676	29.0625	29.55164	0.489144	
3	30	b_i	1.00000	1.044963	86.2500	86.85768	0.607683	1.286063
	50	c_i	0.00000	0.000000	206.250	206.4392	0.189236	
	10	a_i	0.00834	0.008000	33.3340	33.30000	0.034000	
4	20	b_i	3.25000	3.250000	68.3360	68.20000	0.136000	0.586500
	35	c_i	0.00000	0.000000	123.967	123.5500	0.416500	
	10	a_i	0.02500	0.022322	32.5000	33.03565	0.535646	
5	20	b_i	3.00000	3.080341	70.0000	70.53576	0.535763	1.071758
	30	c_i	0.00000	0.000000	112.500	112.5003	0.000349	
	12	a_i	0.02500	0.023081	39.6000	40.22570	0.625698	
6	25	b_i	3.00000	3.075169	90.6250	91.30487	0.679869	1.369166
	40	c_i	0.00000	0.000000	160.000	159.9364	0.063598	

Table 4. Estimation of emission function coefficients by TLBO-IEEE 30 bus system.

Unit	P (MW)		Coefficients		E^{Exi} (Kg/h)	E^{Est} (Kg/h)	Error	Total Error
			Existing	TLBO				
	50	e_{0i}	0.01260	0.012470	−0.5170	−0.01556	0.501437	
1	125	e_{1i}	−1.1000	−1.08064	82.3580	82.60794	0.249943	2.212806
	200	e_{2i}	22.9830	22.84106	306.983	305.5216	1.461426	
	20	e_{0i}	0.0200	0.019823	31.3130	31.64200	0.329003	
2	60	e_{1i}	−0.1000	−0.08000	91.3130	91.87402	0.561024	1.354070
	80	e_{2i}	25.3130	25.31300	145.313	145.7770	0.464043	
	15	e_{0i}	0.02700	0.026724	31.4300	31.29213	0.137871	
3	30	e_{1i}	−0.0100	−0.00900	49.5050	49.19559	0.309407	1.178847
	50	e_{2i}	25.5050	25.41431	92.5050	91.77343	0.731569	
	10	e_{0i}	0.02910	0.028212	27.7600	27.81122	0.051223	
4	20	e_{1i}	−0.0050	−0.00100	36.4400	36.26489	0.175107	1.073847
	35	e_{2i}	24.9000	25.00000	60.3725	59.52498	0.847516	
	10	e_{0i}	0.02900	0.030000	27.5600	26.67813	0.881871	
5	20	e_{1i}	−0.0040	−0.00100	36.2200	35.66813	0.551871	1.455612
	30	e_{2i}	24.7000	23.68813	50.6800	50.65813	0.021871	
	12	e_{0i}	0.02710	0.028000	29.1364	28.48005	0.656352	
6	25	e_{1i}	−0.0055	−0.00500	42.1000	41.88305	0.216952	1.541352
	40	e_{2i}	25.3000	24.50805	68.4400	69.10805	0.668048	

estimated coefficients. Existing fuel cost and existing emission for three different power generations are compared with its FC^{Est} and E^{Est} and it is found that those values are very close to each other. The new values of coefficients and their related total error are presented in the **Table 3** and **Table 4**. From the results the error obtained by TLBO for all specified output of generating units is less which shows the accurate estimation of the system parameters.

TLP are obtained using Equations (A.2-A.5) for each line and the estimated TLP results are listed in **Table 5**. The estimated TLP values are compared with base case values and the percentage error is computed using Equation (9). The % error values for transmission lines are almost close to zero except for the lines 24 and 29 for which the TLP values are above 1. The efficiency of transmission lines 11-16, 37 are high as compared to other transmission lines, due to the presence of three winding transformer in the transmission lines. The estimated *R*, *X*, *B* values for each transmission line are presented in **Table 5**. The estimated transmission loss coefficients using Equations (A.10-A.11) are detailed in the **Table 6**.

By using existing parameters, economic emission dispatch is carried out and the results are listed in **Table 7**. The dispatch results are obtained for three different load demands of 275 MW, 284.4 MW and 300 MW. With the new estimated parameters for fuel cost function, emission and loss, the multi-objective ED problem is solved and the results are compared with the existing results. From the results, the percentage deviation of fuel cost, emission and network loss for estimated parameters are slightly elevated as compared to the existing parameters. The % error of all these values reveals the impact of accurate parameter estimation for ED problems.

4.3. Test System 3: 62 Bus Indian Utility System

In this case, a 62 bus Indian utility system is considered for estimating the improved parameters by using the proposed algorithm. The chosen test system consists of nineteen power producers, eighty nine (220KV) lines

Table 5. Estimation of transmission line parameters for IEEE 30 bus system (in p.u).

Bus		Estimated			% Error with respect to given data			%Efficiency	
nl	nr	R	X	B	R_e	X_e	B_e	Base Case	TLBO
1	2	0.019100	0.057400	0.02640	0.52	0.17	0.00	97.0336	97.014
1	3	0.045143	0.185000	0.02040	0.13	0.11	0.00	96.7376	96.6021
2	4	0.056500	0.173500	0.01830	0.88	0.12	0.54	97.6514	97.5072
3	4	0.013200	0.037700	0.00416	0.00	0.53	0.95	99.0193	99.0112
2	5	0.047150	0.198100	0.02080	0.11	0.10	0.48	96.5316	96.4118
2	6	0.057894	0.176000	0.01860	0.35	0.17	0.53	96.8020	96.7192
4	6	0.011900	0.041194	0.00446	0.00	0.50	0.89	99.1441	99.1147
5	7	0.045600	0.114000	0.01010	0.87	0.60	0.98	98.9923	98.6166
6	7	0.026672	0.080000	0.00840	0.10	0.73	0.47	99.0280	99.0081
6	8	0.012000	0.041600	0.00470	0.00	0.95	0.67	99.6538	99.3394
6	9	0.000000	0.208000	0.00000	0.00	0.00	0.00	100.000	100.000
6	10	0.000000	0.553000	0.00000	0.00	0.54	0.00	100.000	100.000
9	11	0.000000	0.206000	0.00000	0.00	0.96	0.00	100.000	100.000
9	10	0.000000	0.109000	0.00000	0.00	0.91	0.00	100.000	100.000
4	12	0.000000	0.253000	0.00000	0.00	0.55	0.00	100.000	100.000
12	13	0.000000	0.139800	0.00000	0.00	0.14	0.00	100.000	100.000
12	14	0.123100	0.255600	0.00000	0.00	0.12	0.00	99.0801	99.0584
12	15	0.066196	0.130100	0.00000	0.01	0.23	0.00	98.8170	98.7090
12	16	0.094500	0.198500	0.00000	0.00	0.10	0.00	99.2953	99.2099
14	15	0.220121	0.199500	0.00000	0.40	0.10	0.00	99.6200	99.5142
16	17	0.082396	0.192017	0.00000	0.00	0.15	0.00	99.6856	99.5790
15	18	0.107300	0.218300	0.00000	0.00	0.09	0.00	99.3638	99.2110
18	19	0.063900	0.129090	0.00000	0.00	0.09	0.00	99.8418	99.7114
19	20	0.031000	0.068000	0.00000	1.18	0.00	0.00	99.7729	99.2677
10	20	0.093526	0.207000	0.00000	0.08	0.96	0.00	99.0979	99.0061
10	17	0.032285	0.084200	0.00000	0.35	0.36	0.00	99.7364	99.4325
10	21	0.034499	0.074500	0.00000	0.86	0.53	0.00	99.3010	99.1052
10	22	0.072300	0.149700	0.00000	0.55	0.13	0.00	99.3219	99.2125
21	22	0.011200	0.023600	0.00000	1.03	0.00	0.00	99.9496	99.8451
15	23	0.099858	0.200000	0.00000	0.14	0.99	0.00	99.3771	99.0043
22	24	0.114842	0.178100	0.00000	0.14	0.50	0.00	99.2247	99.1239
23	24	0.131505	0.268000	0.00000	0.37	0.74	0.00	99.6375	99.1248
24	25	0.188200	0.329000	0.00000	0.16	0.03	0.00	99.4132	99.0121
25	26	0.254154	0.378000	0.00000	0.10	0.53	0.00	98.7084	98.2095
25	27	0.109100	0.208700	0.00000	0.18	0.00	0.00	99.5007	99.4003
28	27	0.000000	0.393000	0.00000	0.00	0.76	0.00	100.000	100.000
27	29	0.219621	0.415000	0.00000	0.08	0.07	0.00	98.6190	98.6171
27	30	0.320200	0.602500	0.00000	0.00	0.03	0.00	97.7849	97.7771
29	30	0.239820	0.453000	0.00000	0.03	0.07	0.00	99.1302	99.0330
8	28	0.063600	0.197967	0.02140	0.00	1.02	0.00	100.000	99.9900
6	28	0.016900	0.059898	0.06480	0.00	0.00	0.31	99.6877	99.6425

nl: start bus of a line; nr: end bus of a line; Re, Xe and Be: Errors in TLP.

Table 6. Estimated B-Coefficients for IEEE 30 bus system.

B_{ij} =	0.021757	0.010710	−0.00084	−0.00059	0.000804	0.003558
	0.010710	0.018204	−0.00076	−0.00093	0.000581	0.002856
	−0.00084	−0.00076	0.020189	−0.00920	−0.00927	−0.00636
	−0.00059	−0.00093	−0.00920	0.017105	0.006700	0.003504
	0.000804	0.000581	−0.00927	0.006700	0.016060	−0.00011
	0.003558	0.002856	−0.00636	0.003504	−0.00011	0.025026
B_{i0} =	1.65E−05	0.002078	−0.00351	0.002386	0.001136	0.003069
B_{00} =	0.001409					

Table 7. Economic emission dispatch with improved parameters for IEEE 30 unit system.

Demand (MW)	FC^{Exi} ($/h)	FC^{Est} ($/h)	% Error	E^{Exi} (Kg/h)	E^{Est} (Kg/h)	% Error	P_L^{Exi} (MW)	P_L^{Est} (MW)	% Error
275.0	568.67	568.74	0.01231	331.71	337.88	1.86006	5.72	5.97	4.37063
283.4	591.51	591.94	0.07270	346.00	348.25	0.65029	5.74	6.06	5.57491
300.0	638.20	640.68	0.38859	390.59	394.16	0.91400	7.43	7.74	4.17227

with eleven tap changing transformers. The bus, line and generator data of the test system are taken from [32].

The proposed method of determining the improved parameters is applied for the chosen test system and the obtained results are listed in **Table 8**. For each unit the fuel costs using improved parameters for three different real power outputs are also specified. In comparison with base case values the error is calculated for each generating unit. The obtained error is small which reveals that the estimation of the coefficient is accurate.

Further the estimation of TLP is carried out using A.2-A.5. The obtained TLP values by TLBO algorithm are listed in **Table 9**. The % error values for each transmission line are close to zero, except for few lines, which shows that the obtained TLP values are accurate. With the improved TLP values, the transmission line efficiency is calculated and is listed in **Table 9**. It is also inferred from **Table 9** that the efficiency of each transmission line obtained using TLBO is nearly the same as compared to the base case.

The PED is carried out with improved cost coefficients of generators and improved TLP for three different load demands of 2900 MW, 2967 MW, 3200 MW are obtained and the results are presented in **Table 10**. The estimated transmission loss coefficients using Equations (A.10-A.11) are presented in the **Table 11**. The results obtained are compared with the existing fuel cost and loss and the % deviation of fuel cost and transmission loss is computed and presented. The obtained results clearly illustrate the impact of improved parameters on ED problem.

5. Potential Verification of TLBO

5.1. Solution Quality

Using the proposed algorithm several runs have been carried out by varying the population size.In this analysis there is no larger difference in the average fitness value above the population size of 10. Hence the population size of 10 is preferred for all estimation process. The simulations are performed to estimate the quadratic cost function coefficients, piecewise quadratic valve point coefficients, emission coefficients, transmission line parameters and transmission loss coefficients using three different test cases. For these cases the existing coefficients at the time of installation are only available, hence the obtained improved parameters are compared with only these values. From the obtained simulation results, the cost coefficients, emission coefficients and transmission line parameters are very close to the existing values and are presented in **Tables 2-6** and **Tables 8-10.** The comparisons of total error in each case clearly indicate that the proposed method provides the best estimate of accurate coefficients.

Table 8. Estimation of quadratic cost function coefficients using TLBO–62 Bus systems.

Unit	P (MW)	Coefficients			FC^{Exi} (Rs/h)	FC^{Est} (Rs/h)	Error	Total error
			Existing	TLBO				
	50	a_i	0.0070	0.007000	452.5000	451.9880	0.512035	
1	150	b_i	6.8000	6.805155	1272.500	1272.503	0.003454	1.292177
	300	c_i	95.000	94.23022	2765.000	2765.777	0.776688	
	50	a_i	0.0055	0.005473	243.7500	243.9042	0.154235	
2	200	b_i	4.0000	4.013154	1050.000	1051.133	1.133461	1.403032
	450	c_i	30.000	29.56281	2943.750	2943.865	0.115336	
	50	a_i	0.0055	0.005500	258.7500	258.2589	0.491087	
3	200	b_i	4.0000	4.002228	1065.000	1064.843	0.156884	1.048092
	450	c_i	45.000	44.39751	2958.750	2959.150	0.400120	
	0	a_i	0.0025	0.002500	10.00000	9.569746	0.430254	
4	50	b_i	0.8500	0.850000	58.75000	58.31975	0.430254	1.290761
	100	c_i	10.000	9.569746	120.0000	119.5697	0.430254	
	50	a_i	0.0060	0.006000	265.0000	264.6022	0.397802	
5	150	b_i	4.6000	4.600000	845.0000	844.6022	0.397802	1.193407
	300	c_i	20.000	19.60220	1940.000	1939.602	0.397802	
	50	a_i	0.0055	0.005500	303.7500	303.4052	0.344756	
6	200	b_i	4.0000	4.000000	1110.000	1109.655	0.344756	1.034268
	450	c_i	90.000	89.65524	3003.750	3003.405	0.344756	
	50	a_i	0.0065	0.006500	293.2500	292.8159	0.434064	
7	100	b_i	4.7000	4.700000	577.0000	576.5659	0.434064	1.302191
	200	c_i	42.000	41.56594	1242.000	1241.566	0.434064	
	50	a_i	0.0075	0.007503	314.7500	314.4751	0.274862	
8	250	b_i	5.0000	5.000000	1764.750	1764.653	0.096978	0.830749
	500	c_i	46.000	45.71773	4421.000	4421.459	0.458909	
	0	a_i	0.0085	0.008500	55.00000	54.58443	0.415572	
9	300	b_i	6.0000	6.000000	2620.000	2619.584	0.415572	1.246717
	600	c_i	55.000	54.58443	6715.000	6714.584	0.415572	
	0	a_i	0.0020	0.002000	58.00000	57.36662	0.633381	
10	50	b_i	0.5000	0.500000	88.00000	87.36662	0.633381	1.900144
	100	c_i	58.000	57.36662	128.0000	127.3666	0.633381	
	50	a_i	0.0045	0.004486	156.2500	155.9085	0.341495	
11	100	b_i	1.6000	1.600000	270.0000	269.5571	0.442877	1.396220
	150	c_i	65.000	64.69230	406.2500	405.6382	0.611848	
	0	a_i	0.0025	0.002259	78.00000	77.74735	0.252646	
12	25	b_i	0.8500	0.850000	100.8125	100.409	0.403531	1.512366
	50	c_i	78.000	77.74735	126.7500	125.8938	0.856189	

Continued

	50	a_i	0.0050	0.005000	177.5000	176.8509	0.649138	
13	150	b_i	1.8000	1.800000	457.5000	456.8509	0.649138	1.947415
	300	c_i	75.000	74.35086	1065.000	1064.351	0.649138	
	0	a_i	0.0045	0.004452	85.00000	83.87105	1.128948	
14	75	b_i	1.6000	1.617354	230.3125	230.2146	0.097931	1.618938
	150	c_i	85.000	83.87105	426.2500	426.6421	0.392060	
	0	a_i	0.0065	0.006494	80.00000	81.00000	1.000000	
15	250	b_i	4.7000	4.700862	1661.250	1662.108	0.858182	1.859955
	500	c_i	80.000	81.00000	4055.000	4055.002	0.001773	
	50	a_i	0.0045	0.004374	171.2500	170.5123	0.737748	
16	100	b_i	1.4000	1.435546	275.0000	275.0915	0.091535	1.118103
	150	c_i	90.000	87.80097	401.2500	401.5388	0.288821	
	0	a_i	0.0025	0.002468	10.00000	10.00000	0.000000	
17	50	b_i	0.8500	0.843902	58.75000	58.36492	0.385076	1.315574
	100	c_i	10.000	10.00000	120.0000	119.0695	0.930498	
	50	a_i	0.0045	0.004420	116.2500	115.6299	0.620105	
18	150	b_i	1.6000	1.631116	366.2500	367.1468	0.896771	1.698896
	300	c_i	25.000	23.02344	910.0000	910.1820	0.182020	
	100	a_i	0.0080	0.007983	720.0000	718.5346	1.465383	
19	300	b_i	5.5000	5.514826	2460.000	2460.146	0.146243	1.637241
	600	c_i	90.000	87.22119	6270.000	6270.026	0.025615	

5.2. Statistical Analysis

The reliability of the algorithm in finding the best solution can be viewed by performing statistical analysis. The convergence property and the robustness characteristics are focused primly for the statistical analysis. The standard statistical parameters of the objective functions such as best, worst and average values in addition to standard deviation and solution iter are also worked out and analyzed and are given in **Table 12**. The standard deviation value is small which indicates the accuracy of estimate.

Convergence: The TLBO algorithm is implemented on different scale of test systems and for all cases, the convergence characteristics are plotted. The convergence pattern illustrates the searching ability of algorithm.

Figure 2 shows the convergence characteristics of 10 unit multiple fuels system; the proposed approach converged to a good solution within the 50 iterations for the three different fuels. For the IEEE 30 bus and 62 bus Indian utility systems, the convergence graphs are shown in **Figure 3**. **Figure 2** and **Figure 3** show that the TLBO has good convergence property, thus resulting in good evaluation value and low error.

Robustness: Many trials with different initial population have been carried out to test the consistency of the TLBO algorithm. The algorithm is executed for 50 trials and the obtained objective function values are presented in **Figure 4**. The mean value is close to the optimal value for all the cases. This description clears that the TLBO provides great searching ability, higher solution quality and the best estimate.

5.3. Observations

- Accurate parameters are estimated for three different test systems involving a practical test system.
- For the first time in the literature, both the parameter estimation and economic dispatch are carried out

Table 9. Enhanced parameters for Indian Utility 62 Bus system (in p.u).

Bus		Estimated			% Error with respect to given data			%Efficiency	
nl	nr	R	X	B	R_e	X_e	B_e	Base case	TLBO
1	2	0.003040	0.015610	0.014450	0.33	0.26	0.00	99.6854	99.5225
1	4	0.007158	0.104834	0.033970	0.02	0.08	0.00	99.4529	99.4126
1	6	0.004080	0.070416	0.019500	0.73	0.09	0.05	99.7985	99.5990
1	9	0.002280	0.011710	0.010820	0.44	0.26	0.18	99.9574	99.7074
1	10	0.015690	0.015620	0.074427	0.00	0.03	0.00	98.1424	98.0420
1	14	0.005440	0.006240	0.103900	0.73	0.00	0.02	98.8023	98.5559
2	3	0.002875	0.014870	0.013710	0.52	0.00	0.15	99.7694	99.5076
2	6	0.001670	0.008590	0.007950	0.60	0.23	0.00	99.8778	99.6178
3	4	0.003780	0.048259	0.018070	0.79	0.10	0.00	99.8107	99.7705
4	5	0.007120	0.060220	0.033970	0.56	0.08	0.00	99.5645	99.4913
4	14	0.004090	0.055537	0.019510	0.50	0.09	0.00	99.8105	99.8012
4	15	0.004109	0.027387	0.019509	0.03	0.09	0.01	99.5702	99.4160
5	6	0.005747	0.010570	0.003090	0.05	0.06	0.00	99.7869	99.7711
5	8	0.005711	0.095449	0.003090	0.67	0.09	0.00	99.7041	99.7021
6	7	0.000297	0.001557	0.005766	1.00	0.82	0.24	99.9801	99.6785
7	8	0.000485	0.001680	0.086100	1.02	0.00	0.02	99.9619	99.8603
11	10	0.006868	0.020338	0.032514	0.00	0.06	0.02	99.3434	99.3427
11	16	0.014021	0.015620	0.066654	0.28	0.03	0.07	97.7699	97.7640
12	11	0.019010	0.006240	0.090326	0.21	0.03	0.00	98.3704	98.3658
12	13	0.015363	0.079900	0.072920	0.05	0.02	0.00	81.3413	81.3286
12	20	0.019810	0.062595	0.093950	0.00	0.03	0.00	98.1003	98.1001
13	14	0.013110	0.048259	0.062370	0.08	0.04	0.00	97.3945	97.3887
13	17	0.015600	0.060220	0.074125	0.06	0.00	0.03	97.7258	97.7219
14	15	0.005200	0.055537	0.024462	0.00	0.15	0.01	98.8763	98.8754
14	16	0.003920	0.027387	0.018777	1.01	0.20	0.07	99.6615	99.5581
14	18	0.001336	0.010570	0.025577	1.07	0.43	0.01	99.8328	99.6312
14	19	0.007020	0.095449	0.033500	0.71	0.04	0.09	99.0144	99.0070
16	17	0.003400	0.104834	0.065039	0.87	0.11	0.00	99.2834	99.2784
17	21	0.018500	0.070416	0.088153	0.00	0.03	0.01	97.7762	97.7624
20	23	0.020378	0.020338	0.088157	0.21	0.03	0.00	75.0000	74.9171
21	22	0.013707	0.015620	0.065040	0.02	0.02	0.00	98.3203	98.3186
22	23	0.003910	0.006240	0.075120	1.26	0.06	0.05	99.6443	99.4403
23	24	0.003010	0.079900	0.014446	1.31	0.19	0.03	99.5683	99.3626
23	25	0.001253	0.006423	0.005140	0.56	1.18	1.33	99.9154	99.1389
24	41	0.015527	0.048259	0.073710	0.09	0.04	0.00	99.3582	99.3572
24	45	0.012171	0.060220	0.057809	0.16	0.02	0.00	99.4619	99.4618
25	26	0.009390	0.055537	0.044590	0.10	0.04	0.00	98.7969	98.7914
25	27	0.011709	0.027387	0.055650	0.18	0.07	0.00	98.6864	98.6835
25	28	0.010589	0.010570	0.050370	0.29	0.01	0.00	99.3916	99.3901
27	29	0.005280	0.095449	0.025260	0.94	0.01	0.12	99.8378	99.8347
29	30	0.020540	0.104834	0.097580	0.19	0.03	0.05	99.3531	99.3506

Continued

30	31	0.009916	0.050900	0.047010	0.04	0.10	0.09	99.5128	99.5060
32	31	0.017830	0.091400	0.084765	0.22	0.44	0.01	98.8139	98.8117
32	34	0.003908	0.020301	0.075147	1.31	0.24	0.02	99.3724	99.1176
32	36	0.003050	0.015650	0.014430	0.13	0.00	0.14	99.9418	99.8413
32	37	0.022000	0.113006	0.104331	0.00	0.00	0.02	99.4598	99.1606
32	46	0.020920	0.107590	0.099370	0.14	0.02	0.00	99.5276	99.3413
33	32	0.016710	0.086090	0.079470	0.30	0.00	0.03	99.3724	99.3573
34	33	0.017370	0.089211	0.082500	0.00	0.01	0.00	99.8829	99.8826
34	35	0.007010	0.035500	0.033234	0.00	1.39	0.02	99.5225	99.0238
34	37	0.019900	0.010215	0.094371	0.00	0.05	0.01	99.4598	99.4168
35	32	0.000354	0.001834	0.006770	1.64	0.33	0.29	99.9683	99.1634
36	46	0.018278	0.093890	0.086720	0.01	0.02	0.00	99.5501	99.5405
37	46	0.001040	0.005354	0.019500	0.00	0.11	1.52	99.7916	99.7816
38	34	0.010720	0.055220	0.051020	0.37	0.05	0.00	99.3325	99.3124
38	37	0.010400	0.053590	0.049500	0.38	0.04	0.00	98.3801	98.3739
39	37	0.002278	0.011710	0.010828	0.52	0.26	0.11	99.7716	99.5176
39	42	0.006810	0.035190	0.032520	0.73	0.09	0.00	99.3313	99.1268
40	30	0.007136	0.036776	0.033940	0.33	0.01	0.09	99.7272	99.6971
40	41	0.006040	0.031000	0.028902	0.82	0.96	0.03	99.6259	99.1229
41	42	0.000756	0.003910	0.014434	0.48	0.00	0.11	99.9392	99.9381
41	45	0.003326	0.017120	0.015876	0.24	0.00	0.15	99.8795	99.8775
42	43	0.009110	0.046960	0.043320	0.11	0.00	0.09	99.7810	99.7710
42	44	0.014120	0.072750	0.067204	0.11	0.04	0.01	99.3497	99.3472
44	59	0.008840	0.045362	0.041910	0.00	0.06	0.00	98.6675	98.6424
46	44	0.016760	0.086064	0.079476	0.00	0.03	0.02	94.3014	94.2190
47	46	0.007920	0.040644	0.037550	0.00	0.14	0.08	98.1599	98.1283
47	48	0.013680	0.070390	0.065040	0.09	0.06	0.00	96.9166	96.7059
48	50	0.000658	0.003353	0.012406	0.30	0.50	0.12	99.8962	99.8812
48	54	0.012511	0.064380	0.059478	0.13	0.05	0.00	99.4306	99.4116
49	48	0.003646	0.018745	0.069340	0.38	0.19	0.06	99.4691	99.4208
49	50	0.006630	0.034390	0.031800	1.04	0.12	0.00	99.5397	99.1185
51	53	0.011825	0.061100	0.056440	0.63	0.03	0.00	97.9189	97.5027
51	54	0.004058	0.020800	0.019253	0.29	0.48	0.24	99.6324	99.5800
51	55	0.014110	0.072740	0.067210	0.14	0.05	0.00	98.9161	98.9033
52	53	0.011288	0.058100	0.053690	0.28	0.09	0.00	99.2357	99.1319
52	61	0.011200	0.057883	0.053650	0.62	0.05	0.07	99.6475	99.2441

Continued

55	58	0.006650	0.034390	0.031790	0.75	0.12	0.03	99.8991	99.4928
56	58	0.002572	0.013251	0.012290	0.69	0.37	0.00	99.6806	99.1762
57	56	0.001520	0.007814	0.007225	0.00	0.20	0.07	99.8051	99.6063
57	58	0.001830	0.009360	0.008670	0.00	0.32	0.00	99.4844	99.2851
58	12	0.012088	0.062180	0.057450	0.18	0.06	0.00	96.4591	96.3550
58	60	0.004110	0.021130	0.019505	0.00	0.00	0.02	99.4220	99.1219
58	61	0.003333	0.017184	0.063590	0.50	0.21	0.00	99.3535	99.1403
59	61	0.009180	0.047300	0.043720	0.43	0.11	0.00	98.6172	98.5121
60	12	0.013600	0.070090	0.064747	0.37	0.04	0.00	96.4907	96.3813
60	61	0.002415	0.012492	0.046250	1.04	0.23	0.00	99.7277	99.1250
61	62	0.014930	0.076997	0.071090	0.40	0.02	0.03	98.0600	98.0401
62	25	0.013800	0.071020	0.065620	0.22	0.06	0.00	99.0983	99.0711

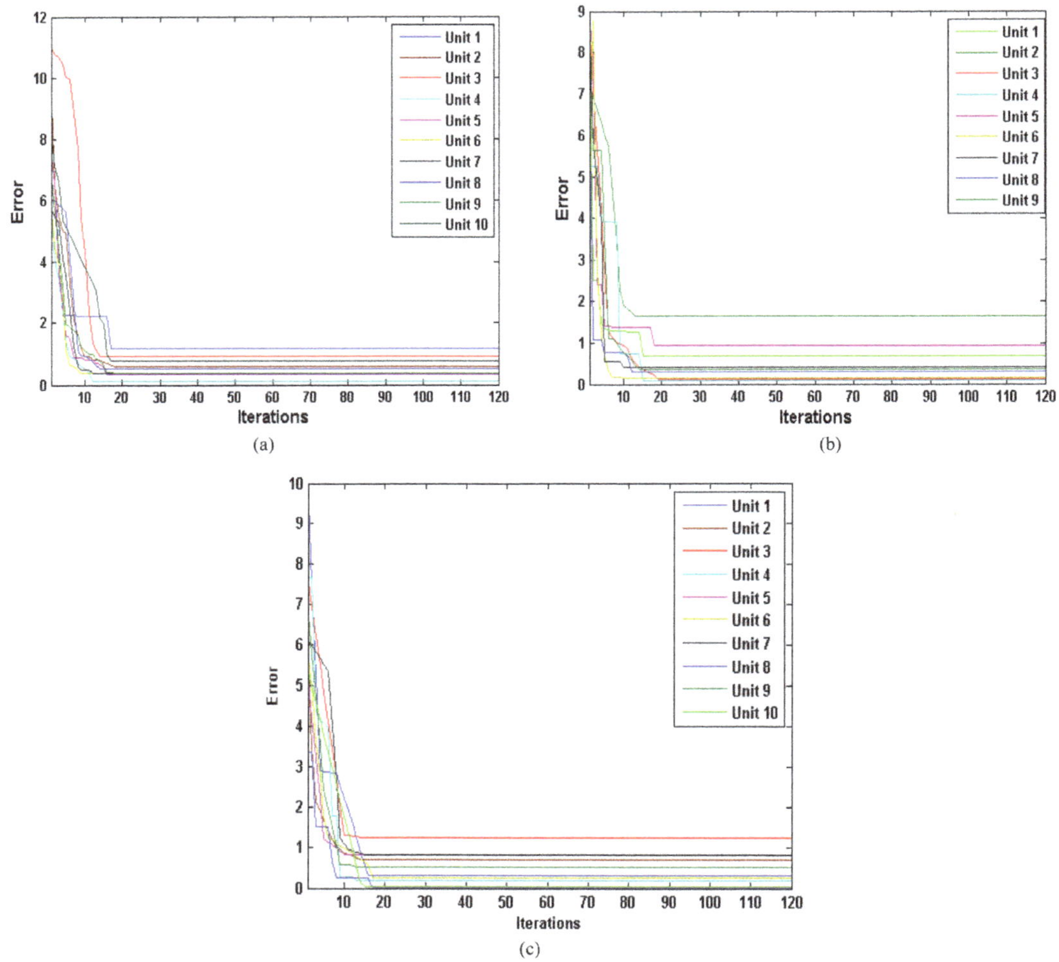

Figure 2. Cost coefficients convergence curve (a) fuel 1; (b) fuel 2 and (c) fuel 3.

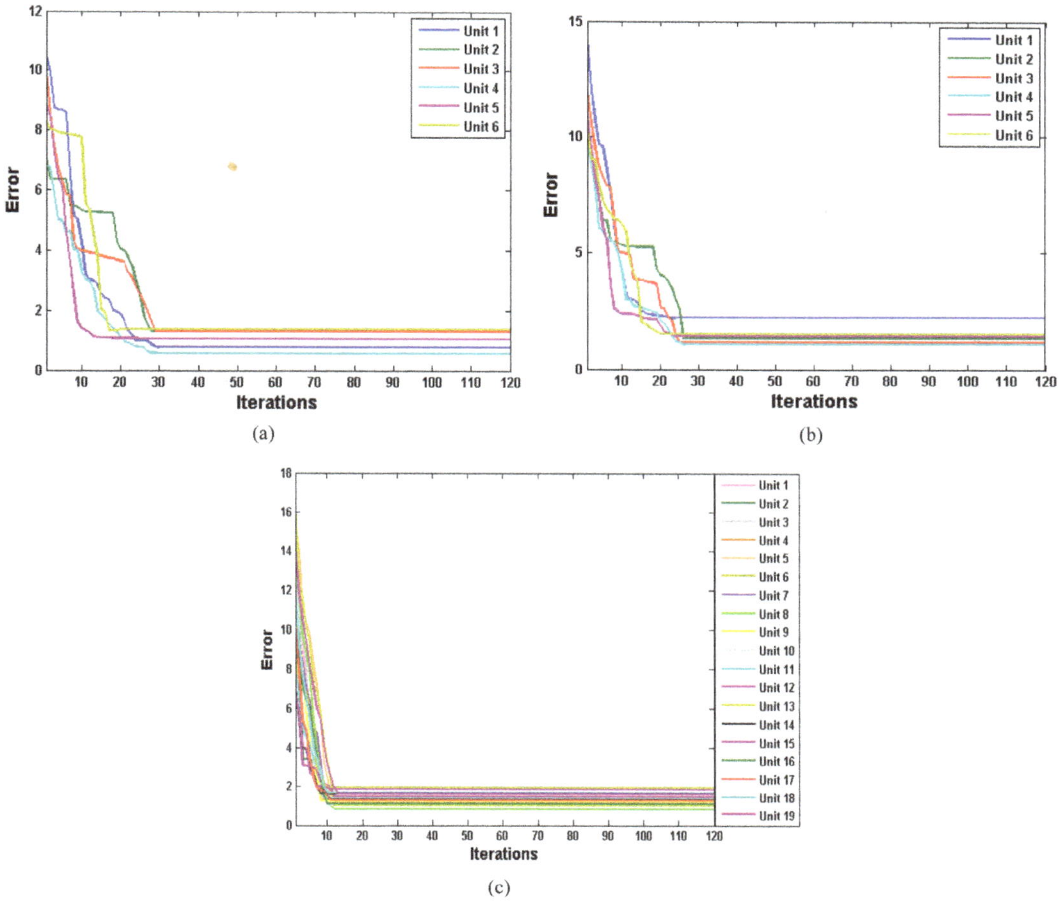

Figure 3. Convergence curve: (a) IEEE 30 bus cost coefficients; (b) IEEE 30 bus emission coefficients; (c) 62 bus cost coefficients.

Table 10. Economic dispatch with improved parameters for Indian utility 62 bus system.

Demand (MW)	FC^{Exi} (Rs/h)	FC^{Est} (Rs/h)	%Error	P_L^{Exi} (MW)	P_L^{Est} (MW)	%Error
2967	15651.73	15722.82	0.45420	94.61327	96.25311	1.73321
2900	15522.58	15890.59	2.37080	80.88011	83.48667	3.22275
3200	17493.59	18030.99	3.07198	92.90000	96.90000	4.30571

concurrently.

- The % error estimation of parameters reveals the rate of change in parameters over the years due to physical conditions like aging.
- The results clearly signify the impact of accurate parameters on ED problems on fetching the accurate dispatch results over existing results.

6. Conclusion

In this paper, the TLBO algorithm is successfully implemented for both the operation of parameter estimations and ED solution. The operational aspects of the generators such as valve-point loading, emission, multiple fuel options and transmission loss are considered. The three different test systems involving a practical test system are selected for this estimation and dispatch problem. The results obtained by satisfying all the constraints for all cases by the proposed algorithm are always comparable or better than the other methods. It is revealed that the

Table 11. Estimated B-Coefficients for 19 unit Indian utility system.

	0.00986	0.009897	0.00907	0.009803	0.007403	0.005591	−0.00375	−0.00357	−0.00712	−0.00739	−0.00685	−0.00524	−0.00402	−0.00361	−0.00332	−0.00245	−0.00345	−0.00078	−0.00209
	0.009897	0.012015	0.009786	0.009822	0.007155	0.005615	−0.00452	−0.00406	−0.0089	−0.00845	−0.0071	−0.00445	−0.00396	−0.00328	−0.00352	−0.00247	−0.00354	−0.00056	−0.00266
	0.00907	0.009786	0.011605	0.009023	0.007611	0.005583	−0.00344	−0.0034	−0.0063	−0.00695	−0.00683	−0.00576	−0.00415	−0.00387	−0.00332	−0.00254	−0.0035	−0.00097	−0.00191
	0.009803	0.009822	0.009023	0.011861	0.007375	0.005549	−0.00378	−0.00361	−0.00712	−0.00741	−0.00689	−0.00531	−0.00407	−0.00365	−0.00336	−0.00249	−0.00349	−0.0008	−0.00213
	0.007403	0.007155	0.007611	0.007375	0.008391	0.005861	−0.00242	−0.0027	−0.00409	−0.00558	−0.0064	−0.00651	−0.00408	−0.00411	−0.00294	−0.00237	−0.00325	−0.00106	−0.00114
	0.005591	0.005615	0.005583	0.005549	0.005861	0.008169	−0.00197	−0.002	−0.00446	−0.00527	−0.00539	−0.00468	−0.0032	−0.00306	−0.00242	−0.00182	−0.00261	−0.00058	−0.00097
	−0.00375	−0.00452	−0.00344	−0.00378	−0.00242	−0.00197	0.011636	0.008303	0.001709	0.001041	0.000789	0.000278	7.96E−5	−8.88 E−5	−0.00054	−0.00056	−0.00036	−0.00049	−5.79 E−5
	−0.00357	−0.00406	−0.0034	−0.00361	−0.0027	−0.002	0.008303	0.007873	0.001208	0.000746	0.000705	0.000677	0.000185	6.13E−5	−0.00035	−0.00037	−0.00021	−0.00039	0.000284
	−0.00712	−0.0089	−0.0063	−0.00712	−0.00409	−0.00446	0.001709	0.001208	0.026718	0.019649	0.012062	0.000424	0.002014	0.000593	0.000351	−0.00091	0.000828	−0.00046	−0.00347
B_{ij}=	−0.00739	−0.00845	−0.00695	−0.00741	−0.00558	−0.00527	0.001041	0.000746	0.019649	0.027091	0.013242	0.003525	0.002457	0.00157	0.000264	−0.00104	0.000842	−0.00107	−0.00264
	−0.00685	−0.0071	−0.00683	−0.00689	−0.0064	−0.00539	0.000789	0.000705	0.012062	0.013242	0.014385	0.006924	0.003204	0.00285	0.000609	−0.00071	0.001243	−0.00138	−0.00095
	−0.00524	−0.00445	−0.00576	−0.00531	−0.00651	−0.00468	0.000278	0.000677	0.000424	0.003525	0.006924	0.014532	0.005329	0.005915	0.001883	0.000655	0.002711	−0.00049	0.001063
	−0.00402	−0.00396	−0.00415	−0.00407	−0.00408	−0.0032	7.96E−5	0.000185	0.002014	0.002457	0.003204	0.005329	0.016443	0.014409	0.007107	0.003341	0.008906	0.001521	0.001778
	−0.00361	−0.00328	−0.00387	−0.00365	−0.00411	−0.00306	−8.88 E−5	6.13 E−5	0.000593	0.00157	0.00285	0.005915	0.014409	0.01593	0.006442	0.003321	0.008457	0.001908	0.000824
	−0.00332	−0.00352	−0.00332	−0.00336	−0.00294	−0.00242	−0.00054	−0.00035	0.000351	0.000264	0.000609	0.001883	0.007107	0.006442	0.011956	0.005508	0.010445	0.002346	0.004795
	−0.00245	−0.00247	−0.00254	−0.00249	−0.00237	−0.00182	−0.00056	−0.00037	−0.00091	−0.00104	−0.00071	0.000655	0.003341	0.003321	0.005508	0.01144	0.004947	0.003536	0.003576
	−0.00345	−0.00354	−0.0035	−0.00349	−0.00325	−0.00261	−0.00036	−0.00021	0.000828	0.000842	0.001243	0.002711	0.008906	0.008457	0.010445	0.004947	0.012607	0.002354	0.003416
	−0.00078	−0.00056	−0.00097	−0.0008	−0.00106	−0.00058	−0.00049	−0.00039	−0.00046	−0.00107	−0.00138	−0.00049	0.001521	0.001908	0.002346	0.003536	0.002354	0.010055	−0.0034
	−0.00209	−0.00266	−0.00191	−0.00213	−0.00114	−0.00097	−5.79E−5	0.000284	−0.00347	−0.00264	−0.00095	0.001063	0.001778	0.000824	0.004795	0.003576	0.003416	−0.0034	0.027605
	0.00986	0.009897	0.00907	0.009803	0.007403	0.005591	−0.00375	−0.00357	−0.00712	−0.00739	−0.00685	−0.00524	−0.00402	−0.00361	−0.00332	−0.00245	−0.00345	−0.00078	−0.00209

B_{i0}=	−0.0138	−0.01803	−0.01178	−0.01403	−0.00669	−0.00763	0.016537	0.011528	0.053561	0.037919	0.014398	−0.01547	−0.00371	−0.01247	−0.00268	−0.00539	−0.00256	−0.00102	−0.01581

B_{00}= 0.384294

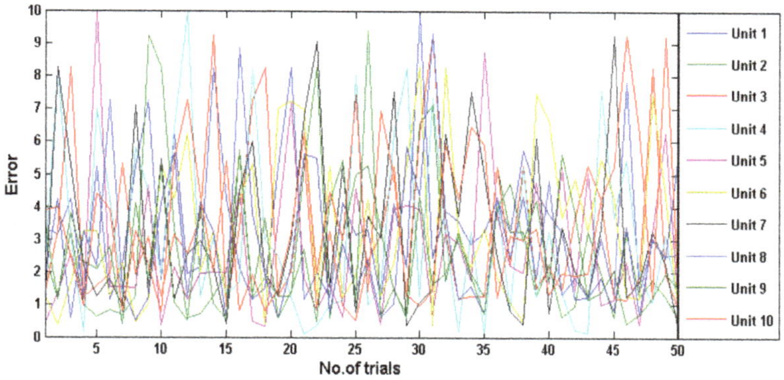
Error
No.of trials
Unit 1
Unit 2
Unit 3
Unit 4
Unit 5
Unit 6
Unit 7
Unit 8
Unit 9
Unit 10

(a)

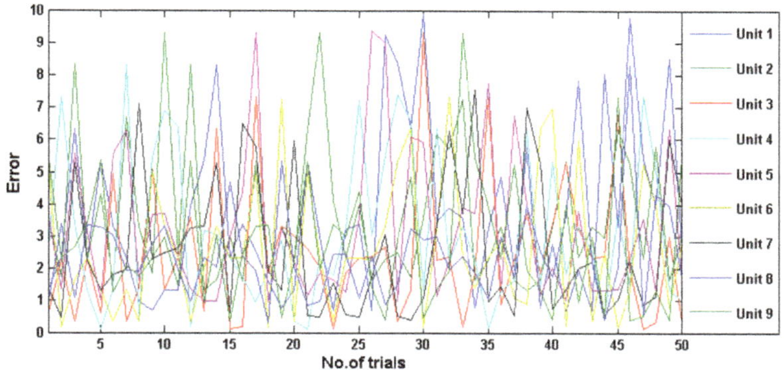
Error
No.of trials
Unit 1
Unit 2
Unit 3
Unit 4
Unit 5
Unit 6
Unit 7
Unit 8
Unit 9

(b)

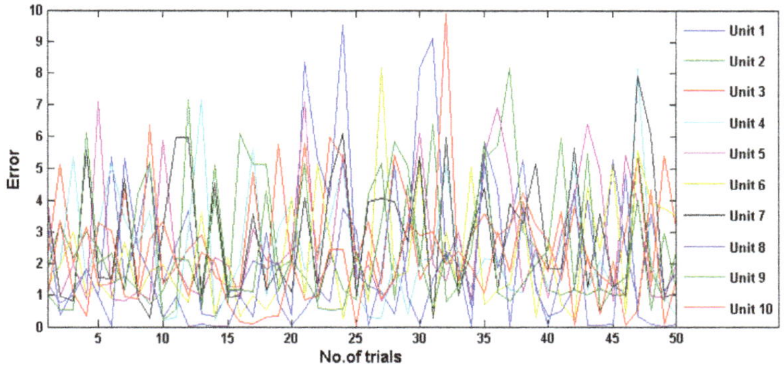
Error
No.of trials
Unit 1
Unit 2
Unit 3
Unit 4
Unit 5
Unit 6
Unit 7
Unit 8
Unit 9
Unit 10

(c)

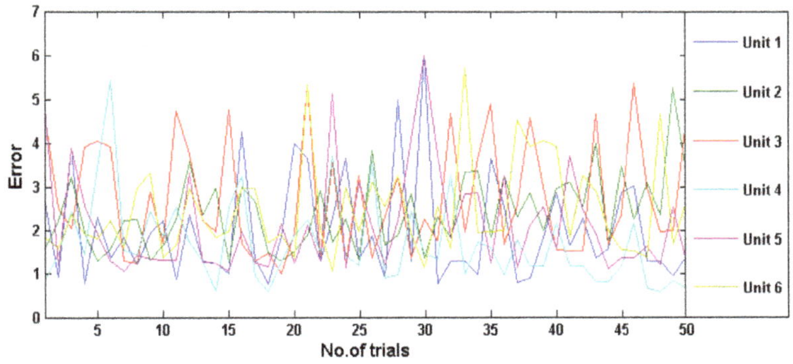
Error
No.of trials
Unit 1
Unit 2
Unit 3
Unit 4
Unit 5
Unit 6

(d)

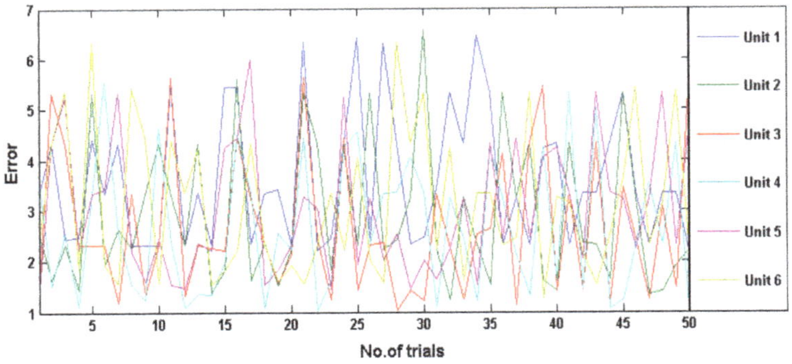

(e)

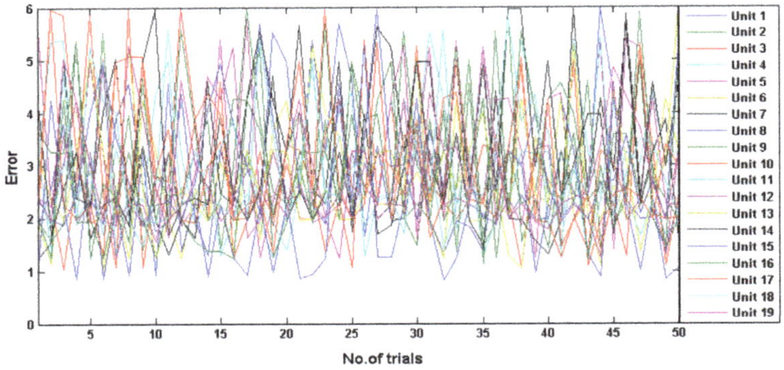

(f)

Figure 4. Robustness characteristics: (a) 10 unit system for fuel 1; (b) 10 unit system for fuel 2; (c) 10 unit system for fuel 3; (d) IEEE 30 bus system for cost function; (e) IEEE 30 bus system for emission function; (f) 62 bus system for cost function.

Table 12. Performance indices by TLBO for various test systems.

Test Systems	Units	Best	Worst	Mean	Solution Iter	Standard Deviation
	1	1.168802	9.657846	1.42197	8	1.3247
		0.697156	8.712287	1.51550	17	1.0365
		1.639115	7.88428	1.83870	16	1.9987
	2	0.317203	9.145168	0.18740	9	1.3658
		0.605175	8.510824	1.03250	7	0.6241
		0.936732	8.824971	0.18740	13	0.0874
	3	0.712539	7.533577	2.21470	17	2.1571
10-unit		0.116458	9.139696	0.28540	3	0.1547
		0.122282	8.834306	0.51020	8	0.3251
	4	0.103847	9.809413	1.89980	7	1.6624
		1.261771	5.867897	0.54770	18	0.3198
		0.333505	7.993594	1.69851	5	1.4873
	5	0.936775	9.220367	0.54102	21	0.2163
		0.208988	9.779851	1.80871	15	1.1024
	6	0.371620	9.690017	0.19546	6	0.1687

Continued

		0.139008	8.923714	2.10245	14	1.8547
		0.828617	8.869472	2.27845	11	1.9687
		0.383942	9.140136	1.15842	3	0.8554
	7	0.410536	9.799390	1.58751	16	1.3258
		0.251265	7.988590	1.19847	14	0.9962
		0.528055	6.047550	0.89870	17	0.5521
	8	0.313608	9.711690	3.18520	16	2.9385
		0.819262	7.804830	1.84680	11	1.6824
		0.398192	8.165560	1.71870	4	1.2365
	9	0.016320	8.000195	0.95230	6	0.5021
		0.529615	5.556480	1.94770	12	1.6324
		0.781727	6.100437	0.57450	20	0.3219
	10	0.043682	8.128070	1.53210	8	1.3278
		0.368904	6.657846	1.68970	8	1.3014
6-unit	1	0.763171	6.001400	1.50080	30	1.2017
	2	1.283524	5.254700	2.41872	28	1.3855
	3	1.286063	5.3606300	2.75281	29	1.3127
	4	0.586500	5.4471000	1.70420	30	0.9844
	5	1.071758	5.1210000	2.14751	25	1.1795
	6	1.369166	5.6981000	2.47828	20	1.4641
19-unit	1	1.292177	5.6987000	2.78649	12	1.2251
	2	1.403032	5.9874000	2.80534	9	1.2379
	3	1.048092	5.6987000	1.51007	10	1.1566
	4	1.290761	4.8746000	1.80493	9	1.3982
	5	1.193407	5.6856000	2.80835	11	1.3069
	6	1.034268	5.2587000	1.51712	12	1.1984
	7	1.302191	5.9655000	2.83551	10	1.4249
	8	0.830749	4.2387000	1.11428	12	1.2307
	9	1.246717	4.5214000	1.77940	8	1.2839
	10	1.900144	5.8945000	2.48302	10	2.2183
	11	1.396220	5.5854000	1.98359	12	1.4946
	12	1.512366	5.2399000	2.12035	12	1.6254
	13	1.947415	5.9658000	2.78960	11	2.0128
	14	1.618938	5.9574000	2.80584	8	1.7851
	15	1.859955	5.5304000	2.83611	11	1.9962
	16	1.118103	5.8512000	1.92709	10	1.2147
	17	1.315574	5.7415000	2.13565	11	1.4210
	18	1.698896	5.6987000	2.23442	11	1.8745
	19	1.637241	5.5718000	2.77713	13	1.7721

TLBO possesses better convergence characteristics and robustness. The proposed TLBO approach has shown merits such as better results, easy implementation for the accurate parameter estimation for PED problems. It can be concluded that the estimated parameters obtained show that the proposed method can be used as a very accurate tool for estimating the fuel cost, emission and loss coefficients.

Acknowledgements

The authors gratefully acknowledge the authorities of Annamalai University, Annamalai Nagar, Tamilnadu, India for the facilities provided to carry out this research work.

References

[1] El-Hawary, M.E. and Mansour, S.Y. (1982) Performance Evaluation of Parameter Estimation Algorithms for Economic Operation of Power Systems. *IEEE Transaction on Power Apparatus Systems*, **101**, 574-582. http://dx.doi.org/10.1109/TPAS.1982.317270

[2] Soliman, S.A., Emam, S.E.A. and Christensen, G.S. (1991) Optimization of the Optimal Coefficients of Non-Monotonically Increasing Incremental Cost Curves. *Electric Power System Research*, **21**, 99-106. http://dx.doi.org/10.1016/0378-7796(91)90023-G

[3] Liang, Z.X. and Glover, J.D. (1991) Improved Cost Functions for Economic Dispatch Computations. *IEEE Transactions of Power System*, **6**, 821-829. http://dx.doi.org/10.1109/59.76731

[4] Chen, H.Y.K. and Postel, C.E. (1986) On-Line Parameters Identification of Input Output Curves for Thermal Units. *IEEE Transaction Power System*, **1**, 221-224. http://dx.doi.org/10.1109/TPWRS.1986.4334933

[5] Grainger, J. and Stevenson, W. (1994) Power System Analysis. McGraw-Hill., New York.

[6] Chan, S.M. (1993) Computing Overhead Line Parameters. *Computer Applications and Power*, **6**, 43-45. http://dx.doi.org/10.1109/67.180436

[7] Dommel, H.W. (1985) Overhead Line Parameters from Handbook Formulas and Computer Programs. *IEEE Transactions on Power Apparatus and Systems*, **4**, 366-372. http://dx.doi.org/10.1109/TPAS.1985.319051

[8] Thorp, J.S., Phadke, A.G., Horowitz, S.H. and Begovic, M.M. (1988) Some Applications of Phasor Measurements To Adaptive Protection. *IEEE Transaction Power System*, **3**, 791-798. http://dx.doi.org/10.1109/59.192936

[9] Chen, C.S., Liu, C.W. and Jiang, J.A. (2002) A New Adaptive PMU Based Protection Scheme for Transposed/Untransposed Parallel Transmission Lines. *IEEE Transactions on Power Delivery*, **17**, 395-404. http://dx.doi.org/10.1109/61.997906

[10] Kim, I.-D., and Aggarwal, R.K. (2006) A Study on the On-Line Measurement of Transmission Line Impedances for Improved Relaying Protection. *Electric Power and Energy System*, **28**, 359-366. http://dx.doi.org/10.1016/j.ijepes.2006.01.002

[11] Liao, Y. and Kezunovic, M. (2009) Online Optimal Transmission Line Parameter Estimation for Relaying Applications. *IEEE Transactions on Power Delivery*, **24**, 96-102. http://dx.doi.org/10.1109/TPWRD.2008.2002875

[12] Sivanagaraju, G., Chakrabarti, S. and Srivastava, S.C. (2014) Uncertainty in Transmission Line Parameters Estimation and Impact on Line Current Differential Protection. *IEEE Transactions on Instrumentation and Measurement*, **63**, 1496-1504. http://dx.doi.org/10.1109/TIM.2013.2292276

[13] Wagenaars, P., Wouters, P.A.A.F., van der Wielen, P.C.J.M. and Steennis, E.F. (2010) Measurement of Transmission Line Parameters of Three-Core Power Cables with Common Earth Screen. *IET Science Measurement and Technology*, **4**, 146-155. http://dx.doi.org/10.1049/iet-smt.2009.0062

[14] Indulkar, C.S. and Ramalingam, K. (2008) Estimation of Transmission Line Parameters from Measurements. *Electric Power and Energy Systems*, **30**, 337-342. http://dx.doi.org/10.1016/j.ijepes.2007.08.003

[15] Al-Kandari, A.M. and El-Naggar, K.M. (2006) A Genetic-Based Algorithm for Optimal Estimation of Input-Output Curve Parameters of Thermal Power Plants. *Electrical Engineering*, **89**, 585-590. http://dx.doi.org/10.1007/s00202-006-0047-x

[16] El-Naggar, K.M. and Alrashidi, M.R. and Al-Othman, A.K. (2009) Estimating the Input-Output Parameters of Thermal Power Plants Using PSO. *Energy Conversion and Management*, **50**, 1767-1772. http://dx.doi.org/10.1016/j.enconman.2009.03.019

[17] Alrashidi, M.R., El-Naggar, K.M. and Al-Othman, A.K. (2009) Particle Swarm Optimization Based Approach for Estimating the Fuel-Cost Function Parameters of Thermal Power Plants with Valve Loading Effects. *Electric Power Components and Systems*, **37**, 1219-1230. http://dx.doi.org/10.1080/15325000902993589.

[18] Sonmez, Y. (2013) Estimation of Fuel Cost Curve Parameters for Thermal Power Plants Using the ABC Algorithm. *Turkish Journal of Electrical Engineering and Computer Science*, **21**, 1827-1841. http://dx.doi.org/10.3906/elk-1203-10

[19] Sinha, A.K. and Mandal, J.K. (1999) Hierarchical Dynamic State Estimator Using ANN-Based Dynamic Load Prediction. *IET Generation Transmission and Distribution*, **146**, 541-549. http://dx.doi.org/10.1049/ip-gtd:19990462

[20] Kumar, D.M.V. and Srivastava, S.C. (1999) Power System State Forecasting Using Artificial Neural Networks. *Electric Machine and Power System*, **27**, 653-664. http://dx.doi.org/10.1080/073135699269091

[21] Lin, J.M., Huang, S.J. and Shih, K.R. (2003) Application of Sliding Surface Enhanced Fuzzy Control for Dynamic State Estimation of a Power System. *IEEE Transaction Power System*, **18**, 570-577. http://dx.doi.org/10.1109/TPWRS.2003.810894

[22] Bhuvaneswari, R., Subramanian, S. and Madhu, A. (2008) A Novel State Estimation Based on Minimum Errors between Measurements Using Ant Colony Optimization Technique. *International Journal of Electric Engineering*, **15**, 457-568.

[23] Sakthivel, V.P., Bhuvaneswari, R. and Subramanian, S. (2010) Design Optimization of Three-Phase Energy efficient Induction Motor Using Adaptive Bacterial Foraging Algorithm. *International Journal of Computation and Mathematics in Electrical and Electronics Engineering*, **29**, 699-726.

[24] Sakthivel, V.P., Bhuvaneswari, R. and Subramanian, S. (2010) Non-Intrusive Efficiency Estimation Method for Energy Auditing and Management of In-Service Induction Motor Using Bacterial Foraging Algorithm. *IET Electric Power Applications*, **4**, 579-590. http://dx.doi.org/10.1049/iet-epa.2009.0313

[25] Sakthivel, V.P., Bhuvaneswari, R. and Subramanian, S. (2011) An Accurate and Economical Approach for induction motor Field Efficiency Estimation Using Bacterial Foraging Algorithm. *Measurement*, **44**, 674-684. http://dx.doi.org/10.1016/j.measurement.2010.12.008

[26] Rao, R.V., Savsani, V.J. and Vakharia, D.P. (2011) Teaching-Learning-Based Optimization: A Novel Method for Mechanical Design Optimization Problems. *Computer Aided Design*, **43**, 303-315. http://dx.doi.org/10.1016/j.cad.2010.12.015

[27] Rao, R.V. and Waghmare, G.G. (2014) Complex Constrained Design Optimisation Using an Elitist Teaching-Learning-Based Optimisation. *International Journal of Metaheuristics*, **3**, 81-102. http://dx.doi.org/10.1504/IJMHEUR.2014.058863

[28] Rao, R.V., Savsani, V.J. and Vakharia, D.P. (2012) Teaching-Learning-Based Optimization: An Optimization Method for Continuous Non-Linear Large Scale Problems. *Information Science*, **183**, 1-15. http://dx.doi.org/10.1016/j.ins.2011.08.006

[29] Rao, R.V., Savsani, V.J. and Vakharia, D.P. (2011) Teaching-Learning-Based Optimization: A Novel Method for Mechanical Design Optimization Problems. *Computer-Aided Design*, **43**, 303-315. http://dx.doi.org/10.1016/j.cad.2010.12.015

[30] Durai, S., Subramanian, S. and Ganesan, S. (2015) Improved Parameters for Economic Dispatch Problems by Teaching Learning Optimization. *Electric Power and Energy System*, **67**, 11-24. http://dx.doi.org/10.1016/j.ijepes.2014.11.010

[31] Chiang, C.-L. (2005) Improved Genetic Algorithm for Power Economic Dispatch of Units with Valve-Point Effects and Multiple Fuels. *IEEE Transaction Power System*, **20**, 1690-1699. http://dx.doi.org/10.1109/TPWRS.2005.857924

[32] Tamilnadu Electricity Board Statistics at a Glance (1999-2000), Compiled by Planning Wing of Tamilnadu Electricity Board, Chennai, India.

Appendix

Parameter Estimation Problems

The vector equation describing the relationships between the measured values *y*, the unknown parameters m, the system matrix *D* and the residual due to the change in values *r* are as follows,

$$y = Dm + r \tag{A.1}$$

The parameter estimation problem for estimating the cost coefficients *a*, *b*, *c*, *e*, *f*, emission coefficients e_0, e_1, e_2 and long line transmission parameters *R*, *X*, *B* are formulated as follows:

Let, *k*-Number of measurements, *y*(*k*)-Measured value of the *k*th measurement, *D*-System matrix

$$m = [a,b,c,e,f]^{\mathrm{T}}, [e_0,e_1,e_2]^{\mathrm{T}}, [R,X,B]^{\mathrm{T}}, \begin{bmatrix} B_{ij} & B_{0i} \\ B_{0i} & B_{00} \end{bmatrix}^{\mathrm{T}}$$

r-Error vector that relates *y*(*k*) to *m*.

There are *k* equations available to represent the parameter estimation that forms measurement matrix *y*(*k*).

TLBO is used as an estimator to find the unknown values of [*m*]. These estimates are used to recalculate parameters using consequent equations at each time step. The calculation procedure for the evaluation of parameters such as *a*, *b*, *c*, *e*, *f*; e_0, e_1, e_2 and B_{ij}, B_{0i}, $B_{00}$are detailed in Section 2.2, and the transmission line parameters such as *R*, *X*, *B* are presented in (A.2-A.5).

$$v_s \cos\delta = V_r - B^{Est} X^{Est} V_r + R^{Est} I_r \cos\varphi_r + X^{Est} I_r \sin\varphi_r, \Delta a, say \tag{A.2}$$

$$v_s \sin\delta = B^{Est} X^{Est} V_r + X^{Est} I_r \cos\varphi_r - R^{Est} I_r \sin\varphi_r, \Delta b, say \tag{A.3}$$

$$I_s(\cos(\delta - \varphi_s)) = -R^{Est} B^{2Est} V_r + I_r \cos\varphi_r - X^{Est} B^{Est} I_r \cos\varphi_r - R^{Est} B^{Est} Ir \sin\varphi_r, \Delta c, say \tag{A.4}$$

$$I_s(\sin(\delta - \varphi_s)) = 2B^{Est} V_r - X^{Est} B^{2Est} V_r - I_r \sin\varphi_r + R^{Est} B^{Est} I_r \cos\varphi_r + X^{Est} B^{Est} I_r \sin\varphi_r, \Delta d, say \tag{A.5}$$

Next, using the Newton-Raphson (NR) method, the following four non-linear equations of the form $F(x) = 0$, where $F = (f_1, f_2, f_3, f_4)^{\mathrm{T}}$ and $x = (X, R, B)^{\mathrm{T}}$ are solved for the unknown *X*, *R*, *B*,

$$f_1(x) = -V_s^2 + a^2 + b^2 = 0 \tag{A.6}$$

$$f_2(x) = -\tan\delta + b/a = 0 \tag{A.7}$$

$$f_3(x) = -I_s^2 + c^2 + d^2 = 0 \tag{A.8}$$

$$f_4(x) = -\tan(\delta - \varphi_s) + d/c = 0 \tag{A.9}$$

Transmission Line Loss Parameters

The network loss is a function of transmission line parameters and network configurations. Thus accuracy in the transmission line model and B coefficients are necessary. The transmission line model consists of *R*, *X* and *B* parameters and to determine the existing network loss these parameters must be accurate. The estimation of transmission line parameters is mathematically formulated as an optimization problem (Indulkar and Ramalingam, 2008) and the accurate TLP can be determined by using an optimization technique.

$$TLP_i^{Est} = \left[R_i^{Est}, X_i^{Est}, B_i^{Est}\right] \quad i = 1, 2, \cdots, TL \tag{A.10}$$

$R_i^{Est}, X_i^{Est}, B_i^{Est}$ are the estimated transmission line parameters

The Kron's coefficients (B_{ij}^{Est}, B_{0i}^{Est} and B_{00}^{Est}) are dependent of network parameters and configuration.

$$KC_i^{Est} = \begin{bmatrix} B_{ij}^{Est} & B_{0i}^{Est} \\ B_{0i}^{Est} & B_{00}^{Est} \end{bmatrix} = f\left(TLP_i^{Est}\right) \tag{A.11}$$

KC_i^{Est} —Estimated loss coefficients

The ideal transmission loss can be determined using KC_i^{Est} by Equation (5).

15

Electric-Field-Assisted Growth of Ga-Doped ZnO Nanorods Arrays for Dye-Sensitized Solar Cells

Jinxia Duan[1], Qiu Xiong[1], Jinghua Hu[2], Hao Wang[1*]

[1]Hubei Collaborative Innovation Center for Advanced Organic Chemical Materials, Hubei Key Laboratory of Ferro & Piezoelectric Materials and Devices, Faculty of Physics and Electronic Science, Hubei University, Wuhan, China
[2]School of Science, Wuhan University of Technology, Wuhan, China
Email: nanoguy@126.com

Abstract

A photoanode with Ga-doped ZnO nanorods has been prepared on F-doped SnO_2 (FTO) coated glass substrate and its application in dye-sensitized solar cells (DSSCs) has been investigated. Ga-doped ZnO nanorods have been synthesized by an electric-field-assisted wet chemical approach at 80˚C. Under a direct current electric field, the nanorods predominantly grow on cathodes. The results of the X-ray photoelectron spectroscopy and photoluminescence verify that Ga dopant is successfully incorporated into the ZnO wurtzite lattice structure. Finally, employing Ga-doped ZnO nanorods with the length of ~5 μm as the photoanode of DSSCs, an overall energy conversion efficiency of 2.56% is achieved. The dramatically improved performance of Ga-doped ZnO based DSSCs compared with that of pure ZnO is due to the higher electron conductivity.

Keywords

Ga-Doped ZnO, Electric-Field-Assisted, Wet Chemical Method, Dye-Sensitized Solar Cells

1. Introduction

Dye-sensitized solar cells (DSSCs) have drawn much attention as a promising renewable energy technology because of their low cost, environmental friendliness and large-scale solar energy conversion [1] [2]. As a result, energy conversion efficiency (PCE) exceeding 12% has been achieved using mesoporous films of sintered TiO_2

*Corresponding author.

nanoparticles as the photoanode [3] [4]. The photovoltaic performance of the typical DSSCs with TiO_2 nanocrystalline photoanodes has been hampered by electron loss during percolation in the nanoparticle network [5] [6]. However, one-dimensional (1D) nanostructures, such as nanowires (NWs), nanorods (NRs) and nanotubes (NTs), can provide direct electric pathways and reduce the electron recombination, which may improve the performance of photovoltaic devices such as DSSCs [7] [8]. However, most of 1D nanostructured DSSCs have been limited to single-crystalline ZnO NRs/NWs or TiO_2 NWs/NTs by available synthetic techniques [7]-[12]. Compared to TiO_2, ZnO shows higher electron mobility with similar bandgap and conduction band energies and more abundant morphologies [13]-[15]. Driven by this, ZnO is regarded as an alternative for high efficient DSSCs. Among the various nanostructures, the application of highly ordered 1D ZnO nanostructures in DSSCs is of prime interest [11] [16].

It is known that nominally undoped ZnO reveals n-type conduction with a typical carrier concentration of $10^{17}/cm^3$, which is smaller than the carrier concentration of 10^{18} to $10^{20}/cm^3$ in optoelectronic applications [17]. The n- or p-type doping of ZnO to increase the carrier concentration or decrease the resistivity remains a significant challenge. Controllable n-type doping in ZnO with high conductivity can be easily achieved by substituting Zn or O with group III elements such as Al, Ga, In or group VII elements such as Cl, Br and I [18]-[24]. Among these elements, Ga is the most effective n-type dopant in ZnO since the covalent bond length of Ga-O (1.92 Å) is nearly equal to that of Zn-O (1.97 Å). Moreover, the group III elements doped ZnO are believed to have more potential for diverse applications, including microelectronics, chemical and biological sensor, energy conversion and storage, light-emitting displays, catalysis, drug delivery, and optical storage [18] [19] [23].

Most techniques to obtain 1D ZnO require the temperature above 500˚C. The successful growth of 1D ZnO via an aqueous solution route suggests an opportunity for low temperature doping [24]-[28]. However, simple addition of group III metal compounds to the growth solution does not result in incorporation of dopants into ZnO [24]. In this study, we extend the solution method to an electrochemical process by applying a negative potential to the substrate. Ga-doped ZnO NRs can be obtained at a growth temperature of 80˚C. The optical and electrical properties of Ga-doped ZnO NRs were discussed in detail associated with the results of X-ray photoelectron spectroscopy (XPS) and photoluminescence (PL) measurement. Using 1% Ga-doped ZnO NRs as photoanode for DSSC, the photovoltaic performance of DSSC was examined. Therefore, this work will help to develop flexible DSSCs and provide a novel strategy of using solar cells for power nanodevices.

2. Experimental Section

2.1. Synthesis of Ga-Doped and Pure ZnO Nanorods

The synthesis of Ga-doped ZnO nanorods (NRs) was conducted in an aqueous solution including 9.5 mM $Zn(NO_3)_2 \cdot 6H_2O$, 0.5 mM $Ga(NO_3)_3 \cdot 6H_2O$, and the proper amount of hexamethylenetetramine (HMT). The pH value of the solution was adjusted to about 7 by addition of appropriate amounts of ammonia. An Electric field was applied between a Zinc sheet anode and a FTO glass substrate cathode in a common two-electrode plating cell at a temperature of around 80˚C for 1 h. The applied direct current (dc) voltage was −0.8 V. The distance between the electrodes was 4 cm. After deposition, the samples were washed carefully with deionized water and absolute ethanol before further characterization. For comparison, pure ZnO NRs were grown by the same procedure.

To grow ZnO:Ga nanorod arrays on FTO glass substrates, a ZnO buffered layer was first deposited on the surface of FTO substrates by radio frequency (rf) magnetron sputtering using a ZnO target with 99.99% purity at room temperature.

2.2. Fabrication of Nanorod-Based DSSCs

Prior to the solar cell assembly, the NRs photoanodes were annealed at 450˚C for 30 min. Annealed electrodes were soaked in 0.3 mM of ruthenium (II) dye (known as N719, Dyesol) in a tert-butanol/acetonitrile (1:1, V/V) solution for 2 h. The electrodes were washed with acetonitrile, dried and immediately used for assembling the DSSCs. The cells, whose active area was 0.16 cm^2, were fabricated by using the dye-adsorbed photoanode and a platinized counter electrode. The two electrodes were clipped together and a Himilan film (Solaronix), with 10 μm thickness, was used as the spacer. The internal space of the cells was filled with liquid electrolyte by capillary action. The electrolyte was composed of 0.5 M butylmethylimidazolium iodide (BMII), 0.05 M iodide (I_2)

and 0.5 M 4-tert-butylpyridine and 0.1 M guanidinium thiocyanate (GuNCS) in a solvent mixture of 85% acetonitrile with 15% valeronitrile by volume. The DSSCs were then heated to 120°C to soften the spacer and seal the edges to prevent the leakage of the electrolyte.

2.3. Characterization and Measurements

The size and morphology of as-prepared produces were observed by field-emission scanning electron microscopy (FE-SEM, JSM 6700F). Transmission electron microscope (TEM), high-resolution TEM (HRTEM) images and selected-area electron diffraction (SAED) patterns were obtained on a transmission electron microscope (TEM, FEI Tecnai G2, accelerating voltage: 200 kV). The chemical compositions of Ga-doped ZnO NRs were measured by X-ray photoelectron spectroscopy (XPS) at room temperature. The photoluminescence (PL) spectra of the samples were characterized using a 325 nm He-Cd laser. An electrical transport measurement (two-probe) was conducted in a home-built system. A Keithley millimeter was used to determine nanoscale electrical characteristics in nanoscale materials. Photovoltaic measurements were recorded employing an Oriel solar simulator system (100 mW/cm^2) with an AM 1.5 spectrum distribution.

3. Results and Discussion

Figure 1 presents (a) plan view and (b) cross-sectional view FE-SEM images of the grown Ga-doped ZnO NRs. It is observed that densely packed arrays of Ga-doped ZnO NRs have been perpendicularly grown on the surfaces of the FTO substrates. The size of the grown NRs is ~ 100 - 200 nm in width and ~5 um in length. The Ga-doped ZnO NRs are uniform, vertically oriented and well-aligned over the entire substrate. The layer appears to be homogeneous and uniform. There is excellent adherence and connection between Ga-doped ZnO NRs and FTO substrate. The morphology of the NRs shows a hexagonal structure, which is due to the wurtzite ZnO structure.

Transmission electron microscopy (TEM) was employed to characterize the Ga-doped ZnO NRs. The TEM image in **Figure 2** depicts the crystal structure of ZnO NRs and the diameter of ~ 100 - 200 nm. The SAED (the upper inset in **Figure 2(a)**) demonstrates that the nanorod is a single crystallite. The indexed diffraction spots confirm that it is hexagonal ZnO with a growth direction along [001] and Ga has been successfully incorporated

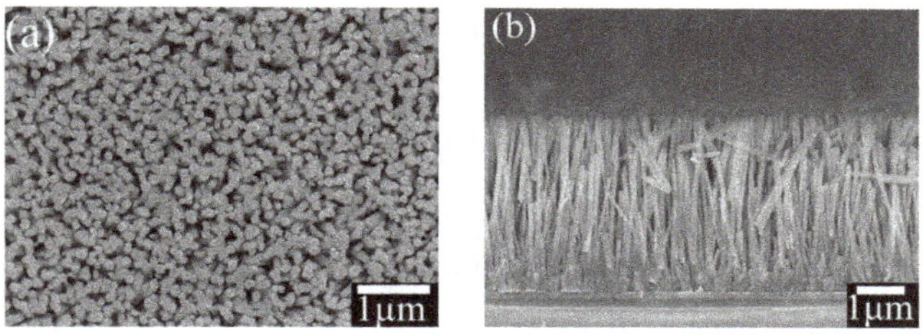

Figure 1. SEM images of Ga-doped ZnO nanorod arrays on FTO glass: (a) plan view; (b) cross-sectional view.

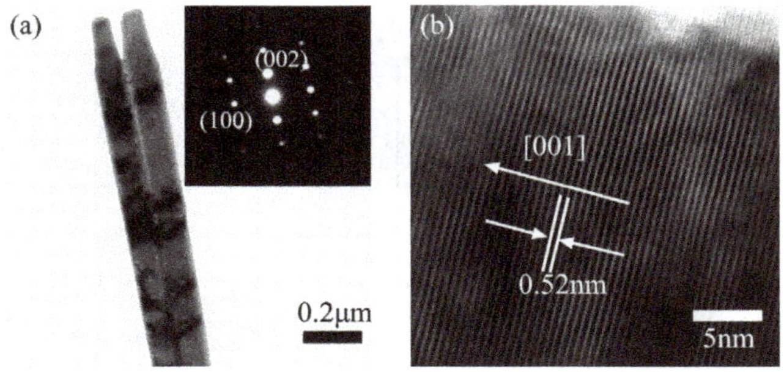

Figure 2. (a) TEM image of Ga-doped ZnO nanorods, Inset shows selected electron diffraction patterns (SAED); (b) HRTEM image showing individual nanorod with [002] growth direction. Both SAED and HRTEM images are indicative of high crystalline quality for most Ga-doped ZnO.

into ZnO NRs. The HRTEM image in **Figure 2(b)** reveals the clear fringes with a spacing of 5.2 nm corresponding to the (002) plane of hexagonal ZnO. It also demonstrates the [001] direction is the preferential growth direction.

To further confirm the purity and composition of Ga-doped NRs, XPS analysis has been performed. The XPS spectra of Ga 2p and Zn 2p are shown in **Figure 3**. The spectrum in **Figure 3(a)** exhibits two significant binding energy peaks at 1117.88 and 1144.35 eV corresponding to the electronic states of Ga 2p3/2 and Ga2p1/2 respectively. Two small satellite peaks come out along with Ga2p peaks, which confirm the successful substitution of Ga into ZnO NRs. Two peaks at 1021.04 and 1044.03 eV corresponding to Zn2p3/2 and Zn2p1/2 appear in **Figure 3(b)**. The energy difference is 26.47 eV between two Ga2p peaks and 22.99 eV between two Zn2p peaks, which agrees with the standard value of 26.84 and 22.97 eV, respectively [29] [30]. From close observation, it is learned that binding energy of Ga2p peaks exhibit a positive shift in comparison to standard value of Ga, probably caused by electron transfer from ZnO to Ga due to the strong electronic interaction between Ga and oxide support. In contrast, the binding energy of Zn2p peaks exhibits a negative shift due to the electronegativity (χ) difference between Zn (χ = 1.65) and Ga (χ = 1.81). The scan of O1s spectrum is not shown here, exhibiting a peak at 530.3 eV attributed to oxidized metal ions in the NRs, viz, O-Ga and O-Zn, in the ZnO lattice. This implies that Ga doping can significantly influence the structure of valence band states. Therefore, XPS analysis again confirms the successful incorporation of Ga into ZnO NRs. Samples grown without applied potential were measured no dopant, suggesting the electric field plays a key role in the doping process.

The above electric-field-assisted growth of Ga-doped ZnO NRs was reproducible. The growth behaviors of the NRs were likely attributed to the interaction of electric field and chemical reactivity. When a dc power was applied, the chemical reaction in the solution was adjusted by applied potential, which determined the interfacial free energy. The formation process of ZnO:Ga deposits was investigated here. When the electric field was carried out, Ga^{3+} and Zn^{2+} ions will be transferred from bulk solution to cathode, obtaining electrons on the surface of the cathode to form Zn and Ga atoms (Equations (1)-(2)). These newly formed Ga and Zn atoms are very active, and rapidly react with O_2 and H_2O in solution to form oxides via reactions. Furthermore, high-temperature (80˚C) can also promote the corrosion of electro-induced Zn and Ga atoms to form the stable passive phase ZnO and Ga_2O_3. In addition, the negative potential at the cathode surface resulted in the concentration of hydroxide ions and the consequent formation of ZnO and Ga_2O_3 (Equation (3)-(5)). The produced ZnO and Ga_2O_3 will mix at molecular level, and Ga_2O_3 can uniformly enter into the crystal lattices of ZnO, finally leading to the formation of ZnO:Ga^{3+} [31].

$$Ga^{3+} + 3e \rightarrow Ga \quad (1)$$

$$Zn^{2+} + 2e \rightarrow Zn \quad (2)$$

$$NO_3^- + H_2O + 2e \rightarrow NO_2^- + 2OH^- \quad (3)$$

$$Zn^{2+} + 2OH^- \rightarrow ZnO + H_2O \quad (4)$$

$$Ga^{3+} + 3OH^- \rightarrow Ga_2O_3 + H_2O \quad (5)$$

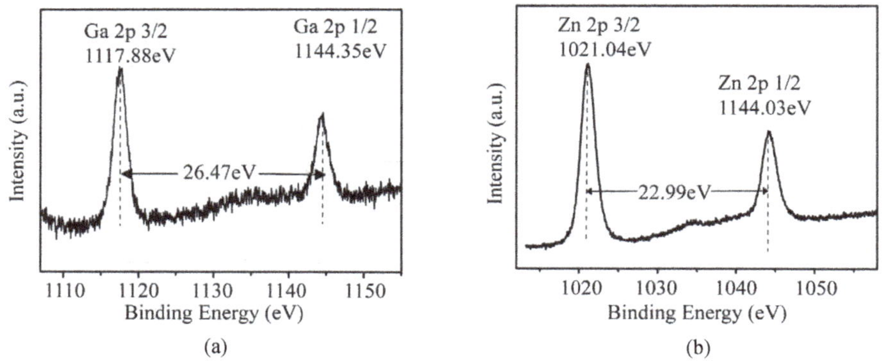

Figure 3. XPS spectra of Ga-doped ZnO nanorods corresponding to (a) Ga2p and (b) Zn2p.

Room-temperature PL curves in **Figure 4(a)** reveal that the near-band edge (NBE) emission peak shifts from 376.3 nm (3.29 eV) in pure ZnO to 382.5 nm (3.25 eV) in Ga-doped ZnO, while the emission intensity decreases with Ga doping. A small red shift in UV emission and the apparent band gap narrowing related to Ga doping are attributed to semiconductor-to-metal transition [18] [32] [33]. The band gap of a semiconductor increases when the impurity concentration is below Mott's critical density. However, an abrupt decrease in band gap would occur when the impurity concentration is above Mott's critical density. In the present study, the red shifts in the emission peak of Ga-doped ZnO NRs (6.2 nm) suggest that the impurity concentration is higher than Mott's critical density. Actually, Ga heavily doped ZnO shows metallic behavior [18]. The intensity of NBE emission decreases slightly in the Ga-doped sample as compared to the pure one, and all the PL spectra are normalized to maximum emission peaks. A significant difference is observed in the intensity ratio of NBE emission to the defect emission centered at ~580 nm, which is usually ascribed to the O-related native defect states in ZnO. In fact, such defects are clearly suppressed in the Ga-doped sample. One should note that Ga doping in ZnO modify the electronic structure of ZnO. At low doping concentrations (below $10^{20}/cm^3$), the substitutional Ga atoms on Zn sites form a shallow donor level, which will increase the free electron density, compensate the positive O vacancy defects, and suppress O vacancy formation in the ZnO lattice [34]. This is consistent with our experimental observation of depressed O defect emission in room-temperature PL spectrum of the low doping concentration of Ga.

The schematic illustration of NRs/buffer-layer/FTO glass substrate and its external circuit for I-V measurement are shown in **Figure 4(b)**. For ohmic contact, an Au/Ti bilayer was deposited by magnetron sputtering on the surface of Ga-doped ZnO NRs and FTO substrate, and patterned with a shadow mask. I-V spectra on undoped ZnO and Ga-doped ZnO NRs as well as calibration experiments were performed to ensure reproducibility. Black curves and red curves correspond to the spectra of undoped and Ga-doped ZnO NRs, respectively. A significantly steeper rising slope in I-V spectra is clear in Ga-doped ZnO NRs in comparison with those in the undoped NRs. It confirms the conductivity enhancement in NRs due to Ga doping, which is similar to the reports [32] [35].

In a preliminary attempt, Ga-ZnO NRs with a thickness of approximately 5 μm were fabricated for use as photoanode in DSSC tests (see **Figure 5**). For comparison, an undoped ZnO NRs photoanode with the same thickness was also constructed. **Figure 6** demonstrates the characteristic photocurrent-voltage (J-V) curves under AM 1.5 sunlight illumination (100 mW·cm^{-2}) and their photovoltaic parameters derived from the J-V curves are summarized in **Table 1**. An overall solar to electrical PCE (η) of 2.56% was achieved for Ga-ZnO NRs based DSSC with a current density (Jsc) of 8.88 mA·cm^{-2}, opencircuit voltage (V_{OC}) of 0.59 V, and fill factor (FF) of 0.49, much higher than the performance of ZnO NRs based DSSC (η = 1.07%, Jsc = 4.50 mA·cm^{-2}, V_{OC} = 0.55 V, FF = 0.43). The data demonstrates that PCE can be enhanced more than 2 times by doping Ga into ZnO. The improvement in PCE value of Ga-ZnO NRs DSSC is obviously caused by the higher Jsc. The significantly enhanced PCE and Jsc in Ga-doped ZnO based DSSCs could be attributed to the improved conductivity

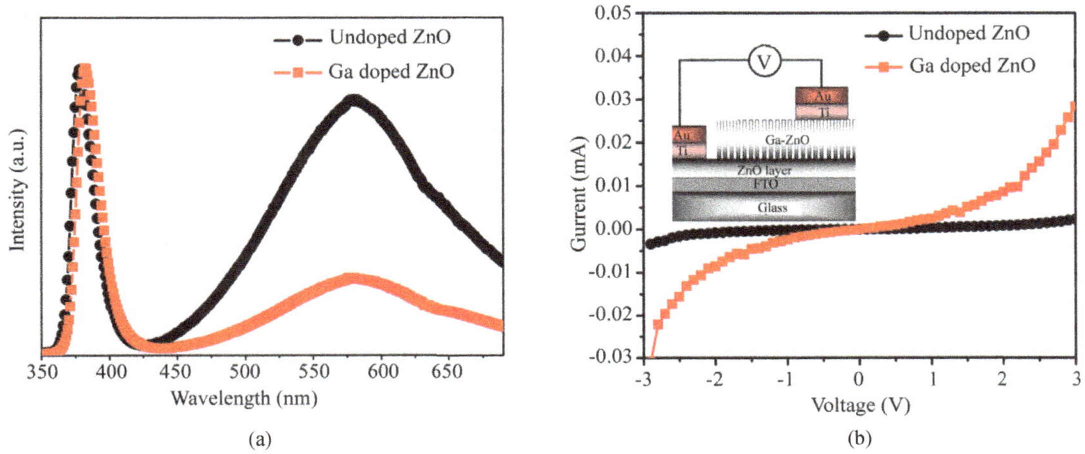

Figure 4. (a) Room-temperature PL spectra of undoped and Ga-doped ZnO NRs. (b) I-V characteristics of undoped and Ga-doped ZnO nanorods. The inset shows the schematic illustration of the nanorods/buffer-layer/FTO glass and its external circuit for I-V measurement.

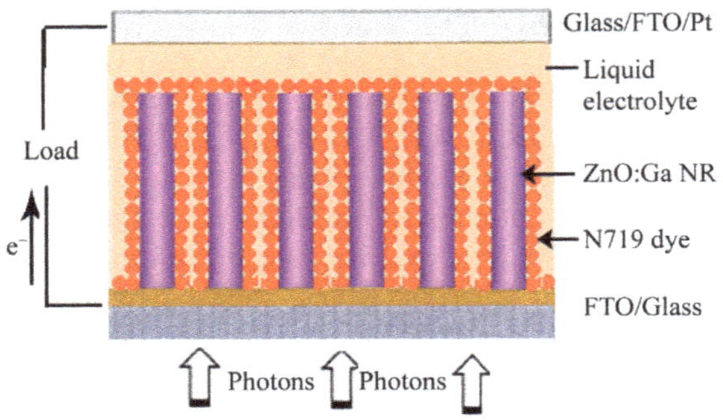

Figure 5. Schematic illustration of the configuration of a dye sensitized Ga-doped ZnO nanorods (ZnO:Ga NRs) Solar Cell.

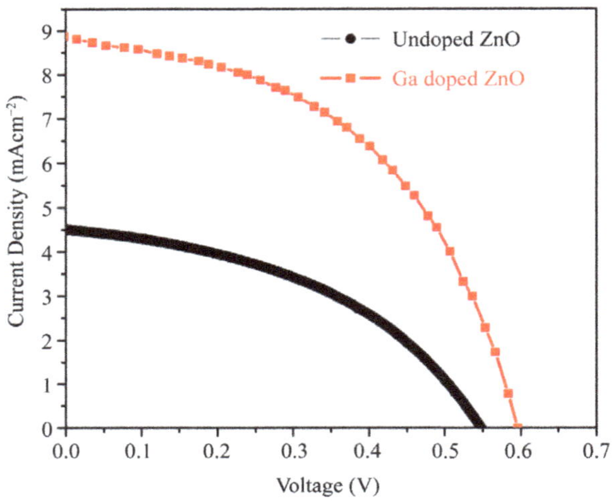

Figure 6. J–V characteristics (full line) for ZnO nanowire arrays electrodeposited on FTO electrode (0.25 cm^2) sensitized with N719 (simulated AM1.5G illumination with intensity 100 mW/cm^2).

Table 1. Photovoltaic parameters of the Ga-ZnO NRs and ZnO NR based DSCs under AM 1.5 sunlight illumination (100 $mW \cdot cm^{-2}$).

Sample	J_{sc} ($mA \cdot cm^{-2}$)	V_{oc} (V)	FF	Efficiency (%)
Ga-ZnO	8.88	0.59	0.49	2.56
ZnO	4.50	0.55	0.43	1.07

of Ga-doped ZnO layer. The slightly higher V_{OC} of Ga-ZnO based DSSCs should be a result of the addition of Ga into the ZnO to decrease charge recombination [36].

The thickness of ZnO thin films in our device is about 5 μm, much thinner than that in the previous work, resulting in a decrease of PCE due to deterioration of the transparency [11] [37]. We can further improve the PCE of our DSSCs with either pure ZnO or Ga-doped ZnO by controlling thicknesses of a nanostructured ZnO layer. We believe that the current PCE (2.56% for Ga-doped ZnO based DSSCs and 1.07% for pure ZnO based DSSCs) is enough to evaluate Ga-doping effect on the active layer in DSSCs.

4. Conclusion

Ga-doped ZnO NRs were successfully prepared on FTO substrates by the electric-field-assisted wet chemical method. XPS and PL results manifested that Ga atoms had been doped effectively into the ZnO lattices. A 5 μm thick Ga-doped ZnO NRs photoanode based DSSC showed a solar to electrical energy conversion efficiency

(PCE) of 2.56%, while the efficiency of a cell with ZnO NRs photoanode was 1.07%. The dramatic enhancement of the PCE of the Ga-doped ZnO-based DSSCs is due to the higher electron conductivity. Further experiments are underway to investigate the effect of the fundamental geometrical features of Ga-ZnO NRs on solar cell conversion efficiency. These will lead to new strategies for improvement.

Acknowledgements

This work was partially supported by the National Nature Science Foundation of China (Nos. 51072049 and 11204070), ED of Hubei Province (Nos. 2009CDA035, 2010BFA016, Z20091001 and Q20120106).

References

[1] O'Regan, B. and Grätzel, M. (1991) A Low-Cost, High-Efficiency Solar Cell Based on Dye-Sensitized Colloidal TiO_2 Films. *Nature*, **353**, 737-740. http://dx.doi.org/10.1038/353737a0

[2] Chen, C.Y., Wang, M.K., Li, J.Y., Pootrakulchote, N., Alibabaei, L., Ngoc-le, C., Decoppet, J., Tsai, J., Grätzel, C., Wu, C., Zakeeruddin, S.M. and Grätze, M. (2009) Highly Efficient Light-Harvesting Ruthenium Sensitizer for Thin-Film Dye-Sensitized Solar Cells. *ACS Nano*, **3**, 3103-3109. http://dx.doi.org/10.1021/nn900756s

[3] Yella, A., Lee, H.W., Tsao, H.N., Yi, C.Y., Chandiran, A.K., Nazeeruddin, M.K., Diau, E.W.G., Yeh, C.Y., Zakeeruddin, S.M. and Grätzel, M. (2011) Porphyrin-Sensitized Solar Cells with Cobalt (II/III)—Based Redox Electrolyte Exceed 12 Percent Efficiency. *Science*, **334**, 629-634. http://dx.doi.org/10.1126/science.1209688

[4] Hagfeldt, A. (2012) Brief Overview of Dye-Sensitized Solar Cells. *Ambio*, **41**, 151-155.

[5] Bazzan, G., Deneault, J.R., Kang, T., Taylor, B.E. and Durstock, M.F. (2011) Nanoparticle/Dye Interface Optimization in Dye-Sensitized Solar Cells. *Advanced Functional Materials*, **21**, 3268-3274. http://dx.doi.org/10.1002/adfm.201100595

[6] Tétreault, N., Arsenault, É., Heiniger, L.P., Soheilnia, N., Brillet, J., Moehl, T., Zakeeruddin, S., Ozin, G.A. and Grätzel, M. (2011) High-Efficiency Dye-Sensitized Solar Cell with Three-Dimensional Photoanode. *Nano Letters*, **11**, 4579-4584. http://dx.doi.org/10.1021/nl201792r

[7] Desai, U.V., Xu, C.K., Wu, J.M. and Gao, D. (2013) Hybrid TiO_2-SnO_2 Nanotube Arrays for Dye-Sensitized Solar Cells. *Journal of Physical Chemistry C*, **117**, 3232-3239. http://dx.doi.org/10.1021/jp3096727

[8] Lv, M., Zheng, D., Ye, M., Sun, L., Xiao, J., Guo, W. and Lin, C. (2012) Densely Aligned Rutile TiO_2 Nanorod Arrays with High Surface Area for Efficient Dye-Sensitized Solar Cells. *Nanoscale*, **4**, 5872-5879. http://dx.doi.org/10.1039/c2nr31431b

[9] Irene, G.V. and Monica, L.C. (2009) Vertically-Aligned Nanostructures of ZnO for Excitonic Solar Cells: A Review. *Energy & Environmental Science*, **2**, 19-34. http://dx.doi.org/10.1039/B811536B

[10] Wang, X.N., Zhu, H.J., Xu, Y.M., Wang, H., Tao, Y., Hark, S., Xiao, X.D. and Li, Q. (2010) Aligned ZnO/CdTe Core-Shell Nanocable Arrays on Indium Tin Oxide. *ACS Nano*, **4**, 3302-3308. http://dx.doi.org/10.1021/nn1001547

[11] Xu, C.K., Wu, J.M., Desai, U.V. and Gao, D. (2011) Multilayer Assembly of Nanowire Arrays for Dye-Sensitized Solar Cells. *Journal of the American Chemical Society*, **133**, 8122-8125. http://dx.doi.org/10.1021/ja202135n

[12] Wang, H., Wang, T., Wang, X.N., Liu, R., Wang, H.B., Xu, Y., Zhang, J. and Duan, J.X. (2012) Double-Shelled ZnO/CdSe/CdTe Nanocable Arrays for Photovoltaic Applications: Microstructure Evolution and Interfacial Energy Alignment. *Journal of Materials Chemistry*, **22**, 12532-12537. http://dx.doi.org/10.1039/c2jm32253f

[13] Son, D.Y., Im, J.H., Kim, H.S. and Park, N.G. (2014) 11% Efficient Perovskite Solar Cell Based on ZnO Nanorods: An Effective Charge Collection System. *Journal of Physical Chemistry C*, **118**, 16567-16573. http://dx.doi.org/10.1021/jp412407j

[14] Duan, J.X., Huang, X.T., Wang, E.K. and Ai, H.H. (2006) Synthesis of Hollow ZnO Microspheres by an Integrated Autoclave and Pyrolysis Process. *Nanotechnology*, **17**, 1786-1790. http://dx.doi.org/10.1088/0957-4484/17/6/040

[15] Zeng, H.B., Duan, G.T., Li, Y., Yang, S.K., Xu, X.X. and Cai, W.P. (2010) Blue Luminescence of ZnO Nanoparticles Based on Non-Equilibrium Processes: Defect Origins and Emission Controls. *Advanced Functional Materials*, **20**, 561-572. http://dx.doi.org/10.1002/adfm.200901884

[16] Law, M., Greene, L.E., Johnson, J.C., Saykally, R. and Yang, P.D. (2005) Nanowire Dye-Sensitized Solar Cells. *Nature Materials*, **4**, 455-459. http://dx.doi.org/10.1038/nmat1387

[17] Xu, C.K., Chun, J.W., Kim, D.E., Kim, J., Chon, B. and Joo, T. (2007) Electrical Properties and Near Band Edge Emission of Bi-Doped ZnO Nanowires. *Applied Physics Letters*, **90**, Article ID: 083113. http://dx.doi.org/10.1063/1.2431715

[18] Yao, Y.F., Tu, C.G., Chang, T.W., Chen, H.T., Weng, C.M., Su, C.Y., Hsieh, C., Liao, C.H., Kiang, Y.W. and Yang, C.C. (2015) Growth of Highly Conductive Ga-Doped ZnO Nanoneedles. *ACS Applied Materials & Interfaces*, **7**, 10525-10533. http://dx.doi.org/10.1021/acsami.5b02063

[19] Ahmad, M., Sun, H. and Zhu, J. (2011) Enhanced Photoluminescence and Field-Emission Behavior of Vertically Well Aligned Arrays of In-Doped ZnO Nanowires. *ACS Applied Materials & Interfaces*, **3**, 1299-1305. http://dx.doi.org/10.1021/am200099c

[20] Yuan, G.D., Zhang, W.J., Jie, J.S., Fan, X., Tang, J.X., Shafiq, I., Ye, Z.Z., Lee, C.S. and Lee, S.T. (2008) Tunable *n*-Type Conductivity and Transport Properties of Ga-Doped ZnO Nanowire Arrays. *Advanced Materials*, **20**, 168-173. http://dx.doi.org/10.1002/adma.200701377

[21] Onwona-Agyeman, B., Nakao, M., Kohno, T., Liyanage, D., Murakam, K.I. and Kitaoka, T. (2013) Preparation and Characterization of Sputtered Aluminum and Gallium Co-Doped ZnO Films as Conductive Substrates in Dye-Sensitized Solar Cells. *Chemical Engineering Journal*, **219**, 273-277. http://dx.doi.org/10.1016/j.cej.2013.01.006

[22] Du, S.F., Liu, H. and Chen, Y. (2009) Large-Scale Preparation of Porous Ultrathin Ga-Doped ZnO Nanoneedles from 3D Basic Zinc Carbonate Superstructures. *Nanotechnology*, **20**, Article ID: 085611. http://dx.doi.org/10.1088/0957-4484/20/8/085611

[23] Yoo, J., Lee, C., Joo Doh, Y.H., Jung, S. and Yi, G.C. (2009) Modulation Doping in ZnO Nanorods for Electrical Nanodevice Applications. *Applied Physics Letters*, **94**, Article ID: 223117. http://dx.doi.org/10.1063/1.3148666

[24] Wang, H., Baek, S., Song, J., Lee, J. and Lim, S. (2008) Microstructural and Optical Characteristics of Solution-Grown Ga-Doped ZnO Nanorod Arrays. *Nanotechnology*, **19**, Article ID: 075607. http://dx.doi.org/10.1088/0957-4484/19/7/075607

[25] Wang, H., Wang, H.B., Yang, F.J., Chen, Y., Zhang, C., Yang, C.P., Qi, L. and Wong, S.P. (2006) Structure and Magnetic Properties of $Zn_{1-x}Co_xO$ Single-Crystalline Nanorods Synthesized by a Wet Chemical Method. *Nanotechnology*, **17**, 4312-4316. http://dx.doi.org/10.1088/0957-4484/17/17/005

[26] Duan, J.X., Wang, H., Wang, H.B., Zhang, J., Wu, S. and Wang, Y. (2012) Mn-Doped ZnO Nanotubes: From Facile Solution Synthesis to Room Temperature Ferromagnetism. *CrystEngComm*, **14**, 1330-1336. http://dx.doi.org/10.1039/C1CE06221B

[27] Wang, H., Chen, Y., Wang, H.B., Zhang, C., Yang, F.J., Duan, J.X., Yang, C.P., Xu, Y.M., Zhou, M.J. and Li, Q. (2007) High Resolution Transmission Electron Microscopy and Raman Scattering Studies of Room Temperature Ferromagnetic Ni-Doped ZnO Nanocrystal. *Applied Physics Letters*, **90**, Article ID: 052505.

[28] Zhou, H., Fang, G.J., Liu, N. and Zhao, X.Z. (2011) Effects of Thermal Annealing on the Performance of Al/ZnO Nanorods/Pt Structure Ultraviolet Photodetector. *Materials Science and Engineering B*, **176**, 740-744. http://dx.doi.org/10.1016/j.mseb.2011.03.003

[29] Wagner, C.D., Riggs, W.M., Davis, L.E., Monlder, J.I. and Muilenberg, G. E. (1979) In Handbook of X-Ray Photoelectron Spectroscopy. Perkin-Elmer Corporation, Eden Prarie, 171-174.

[30] Bae, S.Y., Na, C.W., Kang, J.H. and Park, J. (2005) Comparative Structure and Optical Properties of Ga-, In-, and Sn-Doped ZnO Nanowires, Synthesized via Thermal Evaporation. *The Journal of Physical Chemistry B*, **109**, 2526-2531. http://dx.doi.org/10.1021/jp0458708

[31] Li, G.R., Lu, X.H., Su, C.Y. and Tong, Y.X. (2008) Low-Temperature Growth and Characterization of Cl-Doped ZnO Nanowire Arrays. *Journal of Physical Chemistry*, **112**, 2927-2933.

[32] Zhou, M.J., Zhu, J.H., Jiao, Y., Rao, Y.Y., Hark, S., Liu, Y., Peng, L.M. and Li, Q. (2009) Optical and Electrical Properties of Ga-Doped ZnO Nanowire Arrays on Conducting Substrates. *Journal of Physical Chemistry*, **113**, 8945-8947.

[33] Mott, N.F. (1974) Metal-Insulator Transitions. Taylor and Francis, London.

[34] Matsui, H., Saeki, H., Tabata, H. and Kawai, T. (2003) Role of Ga for Co-Doping of Ga with N in ZnO Films. *Japanese Journal of Applied Physics*, **42**, 5494-5499. http://dx.doi.org/10.1143/JJAP.42.5494

[35] Yang, P.Y., Wang, H., Wang, X.N., Zhang, J. and Jiang, Y. (2010) Optical and Electrical Properties of Ga-Doped ZnO Nanorod Arrays Fabricated by Catalyst-Free Thermal Evaporation. *Proceedings of the 3rd International Nanoelectronics Conference*, Hong Kong, 3-8 January 2010, 1187-1188. http://dx.doi.org/10.1109/inec.2010.5424956

[36] Shin, K.S., Lee, K.H., Lee, H.H., Choi, D. and Kim, S.W. (2010) Enhanced Power Conversion Efficiency of Inverted Organic Solar Cells with a Ga-Doped ZnO Nanostructured Thin Film Prepared Using Aqueous Solution. *Journal of Physical Chemistry C*, **114**, 15782-15785. http://dx.doi.org/10.1021/jp1013658

[37] Zhang, Q.F., Dandeneau, C.S., Zhou, X.Y. and Cao, G.Z. (2009) ZnO Nanostructures for Dye-Sensitized Solar Cells. *Advanced Materials*, **21**, 4087-4108. http://dx.doi.org/10.1002/adma.200803827

16

Modeling of Solar Drying Economics Using Life Cycle Savings (L.C.S) Method

Oyetunde Adeoye Adeaga[1*], Ademola Adebukola Dare[2], Kamilu Moradeyo Odunfa[2], Olayinka Soledayo Ohunakin[3]

[1]Department of Mechanical Engineering, The Polytechnic, Ibadan, Adeseun Ogundoyin Campus, Eruwa, Nigeria
[2]Department of Mechanical Engineering, University of Ibadan, Ibadan, Nigeria
[3]Department of Mechanical Engineering, Covenant University, Otta, Nigeria
Email: [*]engr.adeaga@gmail.com

Abstract

Major goals of industrialization include but are not limited to provision of employment, establishing a platform for overall national development and improving the capital income of whoever is involved, which invariably improve the overall standard of living. A better pre-visibility study must encompass a well analyzed economic appraisal of the plan. The law of mass conservation was applied to develop computer software with a view to analyzing the major preliminary economic indexes of industrial solar drying in both developed and rapidly developing economy. The present work used the life cycle cost method to investigate the solar process economics. In the paper three major geographical locations in Nigeria (*i.e.* Ibadan, Kano and Port Harcourt) were selected and their respective economic appraisal was investigated. Sample simulations revealed that, at a realistic initial moisture content of 30 (% wet basis) of the agricultural produce, economic analysis of over 20 years shows that recommended solar collector area of 85.46 m^2, 80.71 m^2 and 75.96 m^2 supplied about 67%, 88% and 55.8% of the annual energy needed for Ibadan, Kano and Port-Harcourt respectively.

Keywords

Industrialization, Platform, Pre-Visibility, Software, Economic

1. Introduction

The Sun is the largest source of energy in the solar system and it has the potential to supply all the energy

[*]Corresponding author.

requirement of the earth. Its economic potential for any country however depends on a specific location and locality. Solar energy is the most abundant energy source in the solar system. Despite the abundance of this energy, little use is being made of it in most part of the world. This could be attributed to the initial high cost of solar energy technologies, although on a life cycle costing basis, it is generally competitive with other energy technologies where a level playing field is provided and environmental cost is considered. Solar system applications are found in different facet of life. These include space heating, water heating, industrial/domestic cooking, drying of agricultural products, solar cooling and photovoltaic generation of electricity [1].

1.1. Why Solar Drying?

The energy from the sun reaching the earth's atmosphere amounts to about 1.395 kW/m^2. This amount is only $1/10^{10}$ of the actual energy released by the sun. Out of this energy, 23% are used as source of hydrological cycles and photosynthesis in plants, 47% are absorbed by the atmosphere, land and ocean and are converted to long wave radiation (terrestrial radiation) and 30% are reflected and scattered back into space [2]. The use of solar dryers represents an alternative to the traditional open sun drying in developing countries. It satisfies several conditions such as fast processing, better quality of product, low energy demand and non-contaminating energy source. The main disadvantages of solar dryers are the limited time of solar radiation and the short season of harvesting of many agricultural products. Several designs of solar dryers have been proposed for use in developing countries. It has been concluded that to meet the increasing demands for food preservation in developing countries, simple, cheap but efficient solar dryers should be developed where forced convection and supplementary heat are applied [3].

The drying potential in a cabinet drying bed can be employed when air is first dehumidified and then employed for drying of agricultural produce in an attached dryer. The proposed solar drying installation in this work is a coupling of solar collector, auxiliary energy source, and solar dryer of forced-convection type. The processes of mass and heat transfer in these units are simulated. The drying kinetics in a fixed-bed assumes a non-isothermal non-trace plug flow system with some basic variables [4]. One main reason for considering solar is due to its environmental friendliness, as it does not give out any form of environmental pollution, like smoke which characterizes the conventional fossil fuel heater. It also runs smoothly and quietly. This is because it has no mechanical moving part [5]. It also means that wear and tear in solar systems is relatively small, if not totally eliminated [4]. The environmental benefit of solar application also includes no global-warming potential associated as in conventional drying systems with fossils fuels with increasing fossil fuel prices opting for solar system in order to meet the heat energy requirement will save fuel costs and also, it is economically competitive on a life cycle costing basis [4].

1.2. Significance of Life Cycle Savings Method

Solar energy devices and application are generally considered to be relatively new to the underdeveloped and developing nations of the world. Solar processes are generally characterized by high investment cost and low operating costs. Thus the basic economic problem is one comparing initial known investment with estimated future operating costs. Most solar energy processes require an auxiliary (*i.e.*, conventional) energy source so that the system includes both solar and conventional equipment and the annual loads are met by a combination of the sources [2]. The Life Cycle Savings Method is a type of solar economic analysis of approach that takes the following into account 1) time value of money 2) detailed consideration of the complete range of costs 3) design criteria and variation in design factors.

1.3. Cost of Solar Energy Delivery

The cost of any energy delivering process includes all items of hardware and labour that are involved in installing the equipment, plus the operating expenses [6]. Factors that may be taken into consideration includes interest on borrowed money, property tax and income tax. Property tax and income tax may not be applicable in a country like Nigeria also the equipment resale value, maintenance insurance, fuel and other operating expenses should be taken into consideration.

Installed cost of solar equipment can be shown to be sum of two terms [3];

1) C_A = total area dependent cost (N)

2) C_E = total cost of equipment which is independent of the collector area (N)

Therefore $C_s = C_A \times A_C + C_E$
where, C_s = total cost of installed solar energy equipment (N), A_c = collector area (m^2)

The total area dependent cost, C_A includes costs such as the purchase and installation of collector and a portion of storage cost. The area independent cost C_E includes items like controls and bringing the construction erection equipment to site. Operating cost that are associated with solar process include cost of auxiliary energy, energy cost for operating fans/blower (this energy is often termed parasitic energy and should be minimized by careful design, extra insurance exists on solar equipment, maintenance etc.

1.4. Economic Figures of Merit

Some of the criteria proposed and used for evaluating and optimizing economics of solar energy systems are:

1.4.1. Least Cost Energy (LCE)

This is a reasonable figure of merit if solar energy is the only energy resource. The system with the least cost can be defined as that showing the minimum owing cost over the life of the system.

1.4.2. Life Cycle Cost (L.C.C.)

This is the sum of all the cost associated with an energy delivering system over its lifetime or over a selected period of analysis. This method includes inflation when estimating the future expenses.

1.4.3. Life Cycle Savings (L.C.S)

It is also known as the net present worth and it is defined as the difference between the life cycle of conventional fuel (only system and life cycle cost of the solar plus auxiliary energy system).

1.4.4. Annualized Life Cycle Cost (ALCC)

This is the average yearly outflow of money (cash flow).

1.4.5. Pay-Back Time

This have many definitions but in this paper, it is taken to be the time needed for the cumulative savings to equal the total initial investment, *i.e.*, how long it takes to get investment back by saving fuel.

2. Methodology Solar System Cost Analysis

In the mathematical model formed, the annual cost for both solar and non-solar system to meet energy need can be expressed as [2];

$$\begin{aligned}\text{Yearly Cost} = {} & \text{Mortgage payment} + \text{Fuel expense} + \text{Maintenance and Insurance} \\ & + \text{Parasitic energy cost} + \text{Property tax} - \text{Income tax savings}\end{aligned} \quad (1)$$

For income producing installation [2];

$$\begin{aligned}\text{Income tax savings} = {} & \text{Effective tax ratex}\left(\text{Interest payment}\right) + \text{Property tax} + \text{Fuel expense} \\ & + \text{Maintenanceand insurance} + \text{Parasitic energy cost} + \text{Depreciation}\end{aligned} \quad (2)$$

$$\text{Effective taxrate} = \text{Federal rate} \times \text{State tax} - \text{Federal tax-statetax} \quad (3)$$

$$\text{Solar savings} = \text{Cost of conventional energy} - \text{Cost of solar energy} \quad (4)$$

With this savings concept, it is only necessary to estimate the incremental cost of installing solar system because the solar system may have some equipment which is also common to the conventional non solar system. For example the auxiliary furnace and much of the duct work or plumbing in solar system are often the same as would be for a non-solar system. Therefore, solar savings can be rewritten as expressed below [2];

$$\begin{aligned}\text{Solar savings} = {} & \text{Fuel saving} - \text{Incremental Mortgage payment} - \text{Incremental insurance andmaintenance} \\ & - \text{Incremental parasitic energy cost} + \text{Tax saving}\end{aligned} \quad (5)$$

For income producing system [2];

$$\text{Income tax rate} = \text{Effectiveness tax rate} \times (\text{incremental interest payment} + \text{Incremental property tax} + \text{Incremental maintenanceand insurance} + \text{Incrementalparasitic energy cost} - \text{Valueof fuel saved}) \tag{6}$$

Fuel saved is a negative tax deduction since a business already deducts fuel expenses, therefore, value of fuel saved is a taxable income [2].

2.1. Discounting of Future Cost: Inflation

An approach to solar process economics is to use life cycle cost method that takes into account all future costs. The method provides a means of comparison of future costs with present costs. This can be done by discounting all anticipated costs to the common basis of present worth (or present value), that is, what would have been invested today, at the best alternative investment rate to have the funds available in the future to meet all anticipated expenses. The reason that cash flow must be discounted lies in the time value of money [2].

2.1.1. Present Worth (PW)

The relationship for determining the present worth of an amount "*A*" needed "*N*" (usually years) in future, with a market discount rate of "*d*" (present per time period) is [2];

$$\text{PW} = \frac{A}{(1+d)^N} \tag{7}$$

2.1.2. Present Worth Factor (PWF)

If obligation reoccurs each year and inflate at a rate "*i*" per period, a present worth factor, PWF, of "*N*" such payment can be found by using the following relationship [7];

$$\text{PWF}(N,i,d) = \begin{cases} \dfrac{1}{(d-i)}\left[1-\left[\dfrac{1+i}{1+d}\right]^N\right] & \text{for } i \neq d \\ N(1+i), & \text{for } i = d \end{cases} \tag{8}$$

2.2. The Computer Program

The computer program makes use of the metrological data to design the solar collector. The program attempts to obtain a collector area which is capable of supplying the whole annual air heating load based on the size of a single solar module which is also a variable. Although the attainment of this state might not be practically possible as there will always be some period of cloudiness, but as the number of solar modules increases the annual fraction by solar also increases. The economic analysis is performed using the life cycle savings method. By considering the life cycle saving of the different collector size (or area) and their corresponding annual solar fraction by solar, the economically optimum collector size can therefore be selected. If reducing cost is to be considered, the optimum collector is often the one with the highest solar savings. The database contains the global radiation and the extraterrestrial solar radiation, average sunshine hours, and average relative humidity of 10 different locations in Nigeria. The database also contains properties of air as the working fluid and also that of steam. Other data in the data base are monthly averaged ambient temperature of each location, the geographical position and their monthly averaged wind speed. The database also contains data on the materials that could be used in constructing the solar collector, thereby providing the users with a choice of materials and hence cost flexibility. The program gives room for adding, deleting or editing the data concerning any location. However, additional materials which could be used in constructing the collector can also be added to the database but with the required material properties.

3. Results and Discussions

The simulation software was used for investigation with different thermo properties of air as fluid, locations etc., and results were obtained for 50,000 kg of agricultural produce per month. **Table 1** presents the input data/parameters needed for simulation of the selected locations *i.e.* Ibadan, except that the location name needs to be edited each time there is change in location. The input data are kept in input data file within the computer program. **Table 2** shows how monthly heat load and the total fraction of 0.667 supplied by solar energy varies in a year for Ibadan location. **Table 3** presents the input necessary for the cost analysis for the three locations since cost is cost everywhere except cost differ in value. **Table 4** gives the analysis of solar savings of 1,146,112,170 of currency unit, as to which was made through solar installation for twenty years in Ibadan location while given room for the many economic figures of merit. **Table 5** is another input parameter table but know for Kano location. **Table 6** shows the same thing as **Table 2** but with 0.88 as the fraction of solar energy supplied know for Kano location. However **Table 7** does the same as **Table 3** with solar worth of savings of 17,764,928.49 of currency unit, for Kano. **Table 8** does the same as **Table 1** and **Table 9** has 0.558 as the fraction of energy supplied by solar and **Table 10** presents the solar savings over 20 years as 9,269,740.77 of currency unit, for Port Harcourt location. Both tables and figures shows that the overall cost of solar equipment and installation for locations with higher sunshine hours are lesser while the solar savings increases with time but later remain stable after some years. Most agricultural produce exhibit about 25% - 30% initial moisture content (wet-basis) before they are solar dried [8]. When decreasing initial moisture content, *i.e.* 50%, 40%, 30%, 20% (wet basis) were used in the simulation, the fraction of energy supplied by solar increased gradually. The required collector area in, m^2, consequently reduced, and hence a reduction in the overall solar dryer cost [8]. The lower the initial moisture content, of the produce, the higher the energy supplied by solar, and also the higher the fraction of energy

Table 1. Data/parameters input for Ibadan location.

	Data Input	
1	Location Name:	Ibadan
2	longitude:	3.90 East
3	Latitude:	7.43 North
4	Design Type:	Industrial
5	Ground reflectance:	0.2
6	Cover Material:	Glass
7	Number of Covers:	1
8	Plate Material:	"Copper Black" on copper
9	Insulating Material:	Blanket, mineral fiber
10	Collector Fluid:	Air
11	Glazing thickness:	20 mm
12	Plate Thickness	50 mm
13	Plate to cover spacing:	70 mm
14	Air channel depth:	30 mm
15	Air Mass flow rate:	0.3 kg/s
16	Surface azimuth:	0°
17	Collector Slope	15°
18	Insulation back thickness:	70 mm
19	Insulation edge thickness:	30 mm
20	Collector unit width:	1.258 m
21	Collector unit length:	1.258 mm
22	Mass of stock:	50,000 kg per month
23	Initial Moisture Content:	30% (% wet basis)
24	Final Moisture Content:	15% (% wet basis)
25	Crop Safe Drying Temperature:	58°C
26	Air Temperature After Dying:	32°C
27	Equilibrum Relative Humidity:	80

Table 2. Analysis of solar load output for Ibadan location.

	Analysis Output		
Month	**Monthly Heat Load**	**Energy Supply by Solar (MJ)**	**Fraction by Solar**
January	21,405.882	16,541.886	0.733
February	21,405.882	16,025.498	0.749
March	21,405.882	16,912.701	0.79
April	21,405.882	14,973.431	0.7
May	21,405.882	14,323.407	0.669
June	21,405.882	12,066.937	0.564
July	21,405.882	10,294.018	0.481
August	21,405.882	9835.477	0.459
September	21,405.882	11,461.054	0.535
October	21,405.882	15,391.858	0.719
November	21,405.882	16,532.451	0.772
December	21,405.882	16,864.141	0.788
Total	**256,870.588**	**171,222.859**	**0.667**

Table 3. Cost analysis input parameters for Ibadan, Kano and Port Harcourt location.

1	**Annual mortgage interest rate (%100):**	**0.14**
2	Term of mortgage (Years)	20
3	Down payment (as fraction of investment %100):	0.1
4	Collector area dependent costs (Monetary unit per m^2.):	15,000
5	Area Independent costs (Monetary unit):	20,000
6	Present cost of solar backup system fuel (Monetary unit per Giga Joule):	1280
7	Present cost of conventional system fuel (Monetary unit per Giga Joule)	1280
8	Efficiency of solar backup furnace (%100):	0.7
9	Efficiency of conventional system furnace (%100):	0.7
10	Property tax rate as fraction of investment (%100):	0
11	Effective income tax bracket (%100):	0
12	Extra ins., maint. & parasitic costs (as fraction of investment %/100):	0.1
13	General inflation rate per year (%/100):	0.165
14	Solar backup fuel inflation rate per year (%/100):	0.2
15	Conventional system fuel inflation rate per year (%/100)	0.2
16	Discount rate (after tax return on best alternative investment %/100):	0.8
17	Term of Economic analysis (Years):	20
18	Depreciation lifetime (Years)	20
19	Salvage value (as fraction of investment %/100)	0.2
20	Market Discount Rate (%/100):	0.08

supplied by solar. Consequently, the much lower the initial moisture content the smaller the initial cost of investment, the shorter the pay-back time (years), and the lower the collector area required in m^2 with all other parameters kept constant.

For Ibadan location as shown in **Tables 1-4**, with 50,000 kg/month of agricultural produce the average sunshine hours per month is about 159.8 hours. Therefore the drying rate will be 313 kg per hour. For Kano location as shown in **Tables 5-7**, with the same 50,000 kg per month of agricultural produce, the average sunshine hours

Table 4. Solar savings over 20 years for Ibadan location.

Collector Area (m^2): 85.46			Pay-back Time (yrs):8		Initial Cost of Investment: 1,325,615.30		
Year	**Fuel Savings**	**Extra Mortgage Payment**	**Extra Insurance, Maintenance, Energy**	**Extra Property Tax**	**Income Tax Savings**	**Solar Savings**	**Present Worth of Solar Saving**
0						−132,561.53	−132,561.53
1	154,287.98	−180,134.42	−132,561.53	0	0	−158,407.97	−146,674.05
2	378,501.89	−180,134.42	−154,434.18	0	0	43,933.29	37,665.71
3	454,202.27	−180,134.42	−179,915.82	0	0	94,152.03	74,740.91
4	545,042.72	−180,134.42	−209,601.93	0	0	155,306.37	114,154.82
5	654,051.27	−180,134.42	−244,186.25	0	0	229,730.59	156,350.78
6	784,861.52	−180,134.42	−284,476.98	0	0	320,250.12	201,811.9
7	941,833.82	−180,134.42	−331,415.69	0	0	430,283.72	251,066.42
8	1,130,200'59	−180,134.42	−386,099.27	0	0	563,966.89	304,693.76
9	1,356,240.7	−180,134.42	−449,805.65	0	0	726,300.63	363,331.14
10	1,627,488.8	−180,134.42	−524,023.59	0	0	923,330.84	427,680.83
11	1,952,986.6	−180,134.42	−610,487.48	0	0	1,162,364.71	498,518.3
12	2,343,583.9	−180,134.42	−711,217.91	0	0	1,452,231.6	576,701.15
13	2,812,300.7	−180,134.42	−828,568.87	0	0	1,803,597.43	663,179.03
14	3,374,760.9	−180,134.42	−965,282.73	0	0	2,229,343.71	759,004.68
15	4,049,713	−180,134.42	−1,124,554.38	0	0	2,745,024.24	865,346.12
16	4,859,655.7	−180,134.42	1,310,105.86	0	0	3,369,415.37	983,500.23
17	5,831,586.8	−180,134.42	−15,261,273.32	0	0	4,125,179.03	1,114,907.81
18	6,997,904.1	−180,134.42	−1,778,108.42	0	0	5,039,661.29	1,261,170.35
19	8,397,485	−180,134.42	−2,071,496.31	0	0	6,145,854.23	1,424,068.57
20	10,076,982	−180,134.42	−2,413,293.2	0	0	7,483,554.33	1,605,583.17
					Salvage Value	265,123.06	56,881.68
					Total Present Worth of Savings = 11,461,121.7		

Table 5. Data input for Kano location.

Data input	
Location Name:	Kano
Longitude:	8.53 East
Latitude:	12.05 North
Design Type:	Industrial
Ground reflectance:	0.2
Cover Material:	Glass
Number of covers:	1
Plate Material:	"Copper Black" on Copper
Insulating Material:	Blanket, mineral fiber
Collector Fluid:	Air
Glazing thickness:	20 mm
Plate thickness:	50 mm
Plate to cover spacing:	70 mm
Air channel depth:	30 mm
Air Mass flow rate:	0.3 kg/s

Continued

Surface azimuth:	0°
Collector slope:	15°
Insulating back thickness:	70 mm
Insulating edge thickness:	30 mm
Collector unit width:	1.258 m
Collector unit length:	1.258 mm
Mass of stock:	50,000 kg per month
Initial Moisture Content:	30% (% wet basis)
Final Moisture Content:	15% (% wet basis)
Crop Safe Drying Temperature:	58°C
Air Temperature After Drying:	32°C
Equilibrium Relative Humidity:	80

Table 6. Solar load analysis output for Kano location.

Analysis output			
Month	**Monthly Heat Load (MJ)**	**Energy Supply by Solar (MJ)**	**Fraction by Solar**
January	21,405.882	19,035.857	0.889
February	21,405.882	18,485.381	0.864
March	21,405.882	20,734.907	0.869
April	21,405.882	19,074.077	0.891
May	21,405.882	19,237.052	0.899
June	21,405.882	17,697.574	0.827
July	21,405.882	17,159.141	0.802
August	21,405.882	16,010.117	0.748
September	21,405.882	18,760.925	0.876
October	21,405.882	21,319.315	0.996
November	21,405.882	19,688.957	0.92
December	21,405.882	18,798.1	0.878
Total	**256,870.588**	**226,001.404**	**0.88**

Table 7. Solar savings over 20 years for Kano location.

Collector Area (m^2): 80.71			Pay-back Time (yrs):6		Initial Cost of Investment: 1,254,399.92		
Year	**Fuel Savings**	**Extra Mortgage Payment**	**Extra Insurance, Maintenance, Energy**	**Extra Property Tax**	**Income Tax**	**Solar Savings**	**Present Worth of Solar Saving**
0						−125,440	−125,439.99
1	53,064.45	−170,457.15	−125,439.99	0	0	−242,833	−224,845.08
2	499,970.12	−170,457.15	−146,137.59	0	0	183,375.4	157,214.84
3	599,964.15	−170,457.15	−170,250.29	0	0	259,256.7	205,806.33
4	719,956.98	−170,457.15	−198,341.59	0	0	351,158.2	258,111.79
5	863,948.37	−170,457.15	−231,067.95	0	0	462,423.3	314,717.51
6	1,036,738.05	−170,457.15	−269,194.17	0	0	597,086.7	376,265.93
7	1,244,085.66	−170,457.15	−313,611.2	0	0	760,017.3	443,462.8
8	1,492,902.79	−170,457.15	−365,357.05	0	0	957,088.6	517,085.18
9	1,791,483.35	−170,457.15	−425,640.97	0	0	1,195,385	597,990.23

Continued

10	**2,149,780.01**	**−170,457.15**	**−495,871.73**	**0**	**0**	**1,483,451**	**687,124.91**
11	2,579,736.02	−170,457.15	−577,690.56	0	0	1,831,588	785,536.83
12	3,095,683.22	−170,457.15	−673,009.5	0	0	2,252,217	894,386.19
13	3,714,819.87	−170,457.15	−784,056.07	0	0	2,760,307	1,014,959.03
14	4,457,783.84	−170,457.15	−913,425.32	0	0	3,373,901	1,148,681.97
15	5,349,340.61	−170,457.15	−1,064,140.5	0	0	4,114,743	1,297,138.59
16	6,419,208.73	−170,457.15	−1,239,723.68	0	0	5,009,028	1,462,087.5
17	7,703,050.47	−170,457.15	−1,444,278.09	0	0	6,088,315	1,645,482.58
18	9,243,660.57	−170,457.15	−1,682,583.98	0	0	7,390,619	1,849,495.34
19	11,092,392.68	−170,457.15	−1,960,210.33	0	0	8,961,725	2,076,539.84
20	13,310,871.22	−170,457.15	−2,283,645.04	0	0	10,856,769	2,329,300.33
					Salvage Value	250,880	53,825.85
						Total Present Worth of Savings = 17,764,928.49	

Table 8. Data Input for Port Harcourt location.

	Data Input	
1	Location Name:	Port Harcourt
2	longitude:	7.02 East
3	Latitude:	4.86 North
4	Design Type:	Industrial
5	Ground reflectance:	0.2
6	Cover Material:	Glass
7	Number of Covers:	1
8	Plate Material:	"Copper Black" on copper
9	Insulating Material:	Blanket, mineral fiber
10	Collector Fluid:	Air
11	Glazing thickness:	20 mm
12	Plate Thickness	50 mm
13	Plate to cover spacing:	70 mm
14	Air channel depth:	30 mm
15	Air Mass flow rate:	0.3 kg/s
16	Surface azimuth:	0°
17	Collector Slope	15°
18	Insulation back thickness:	70 mm
19	Insulation edge thickness:	30 mm
20	Collector unit width:	1.258 m
21	Collector unit length:	1.258 mm
22	Mass of stock:	50,000 kg per month
23	Initial Moisture Content:	30% (% wet basis)
24	Final Moisture Content:	15% (% wet basis)
25	Crop Safe Drying Temperature:	58°C
26	Air Temperature After Dying:	32°C
27	Equilibrum Relative Humidity:	80

Table 9. Analysis of solar load output for Port Harcourt location.

Analysis Output			
Month	**Monthly Heat Load**	**Energy Supply by Solar (MJ)**	**Fraction by Solar**
January	21,405.882	14,080.138	0.658
February	21,405.882	13,458.763	0.629
March	21,405.882	13,510.334	0.631
April	21,405.882	12,634.204	0.59
May	21,405.882	11,910.251	0.556
June	21,405.882	9519.288	0.445
July	21,405.882	9149.952	0.427
August	21,405.882	9567.973	0.447
September	21,405.882	10,126.62	0.473
October	21,405.882	12,250.308	0.572
November	21,405.882	13,074.296	0.611
December	21,405.882	13,996.183	0.654
Total	**256,870.588**	**143,278.311**	**0.558**

Table 10. Solar savings over 20 years for Port Harcourt location.

Collector Area (m^2): 75.96			Pay-back Time (yrs):8		Initial Cost of Investment: 1,183,184.54		
Year	**Fuel Savings**	**Extra Mortgage Payment**	**Extra Insurance, Maintenance, Energy**	**Extra Property Tax**	**Income Tax Savings**	**Solar Savings**	**Present Worth of Solar Saving**
0						−118,318.45	−118,318.45
1	205,427.8	−160,779.87	−118,318.45	0	0	−73,670.53	−68,213.45
2	317,134.1	−160,779.87	−137,841	0	0	18,513.23	15,872.11
3	380,560.92	−160,779.87	−160,584.76	0	0	59,196.29	46,991.92
4	456,673.11	−160,779.87	−187,081.25	0	0	108,811.99	79,980.06
5	548,007.73	−160,779.87	−217,949.66	0	0	169,278.2	115,207.9
6	657,609.28	−160,779.87	−253,911.35	0	0	242,918.06	153,079.58
7	789,131.13	−160,779.87	−295,806.72	0	0	332,544.54	194,036.54
8	946,957.36	−160,779.87	−344,614.83	0	0	441,562.66	238,562.56
9	1,136,348.83	−160,779.87	−401,476.28	0	0	574,092.68	287,189.27
10	1,363,618.9	−160,779.87	−467,719.86	0	0	735,118.86	340,502.27
11	1,636,342.32	−160,779.87	−544,893.64	0	0	930,668.8	399,147.9
12	1,963,610.78	−160,779.87	−634,801.09	0	0	1,168,029.81	463,840.71
13	2,356,332.94	−160,779.87	−739,543.27	0	0	1,456,009.79	535,371.78
14	2,827,599.52	−160,779.87	−861,567.91	0	0	1,805,251.74	614,617.89
	3,393,119.43	−160,779.87	−1,003,726.6	0	0	2,228,612.94	702,551.74
16	4,071,743.31	−160,779.87	−1,169,341.5	0	0	2,741,621.93	800,253.31
17	4,886,091.98	−160,779.87	−1,362,282.9	0	0	3,363,029.24	908,922.39
18	5,863,310.37	−160,779.87	−1,587,059.5	0	0	4,115,470.97	1,029,892.61
19	7,035,972.44	−160,779.87	−1,848,924.4	0	0	5,026,268.22	1,164,646.98
20	8,443,166.93	−160,779.87	−2,153,996.9	0	0	6,128,390.19	1,314,835.13
					Salvage Value	236,636.91	50,770.02
					Total Present Worth of Savings = 9,269,740.77		

per month is about 261.7 hours therefore the drying rate will be 191 kg per hour. For Port Harcourt location as shown in **Tables 8-10** with the same 50,000 kg produce the average sunshine hours per month is about 118 hours. Therefore the drying rate will be 424.3 kg per hour. A close observation of the drying rate of Ibadan, Kano and Port Harcourt of the produce revealed close and neighbouring values. Hence, a better way is to fix the drying rate in kg/hr. or kg/s., and calculate the amount that can be dried in each location. However, the large difference between any industrial solar dryer located at Ibadan, Kano and Port Harcourt will be the initial investment cost of equipment (to include solar collectors unit area cost), the solar savings for a given period of time and the pay-back period. The total cost which include, the installation cost, the equipment cost and the operating costs and maintenance cost at Ibadan and Port Harcourt will surely be on the high side when compared to that in Kano, hence relative advantage at locations with high average sunshine hours per day to those with lesser or lower average sunshine hours per day. However, a better solar saving could be achieved with lower inflation and interest rates. **Figure 1**, **Figure 2** and **Figure 3** show that Ibadan location has moderate solar fraction and

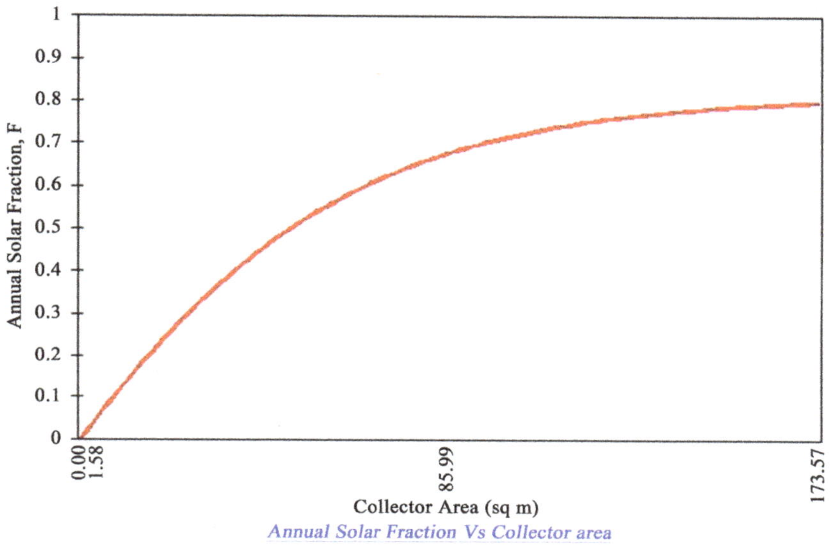

Figure 1. Annual solar fraction against collector area for Ibadan location.

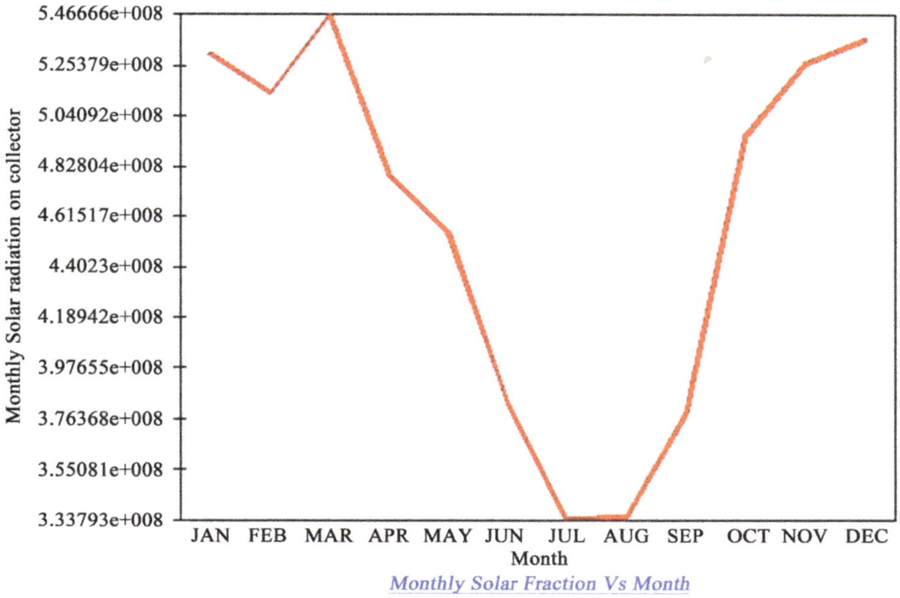

Figure 2. Monthly solar radiation on collector against month of the year for Ibadan location.

distribution when compared to Kano location in **Figures 4-6**, but for Port Harcourt from **Figures 7-9** with lower solar fraction both annually and monthly. Also affecting is the air flow rate required for drying, as it was revealed during simulation, that higher air flow rates, keeping other things constant, means increase in the annual solar fraction but for larger fan/blower and consequently cost of purchase [1]. The simulation iteration stops when 1) The number of solar modules obtained can fully supply the energy needed for drying 2) when the additional energy needed is less than 1/100th of the original energy supplied by the lead solar module. The simulation was done with input that represents industrial applications.

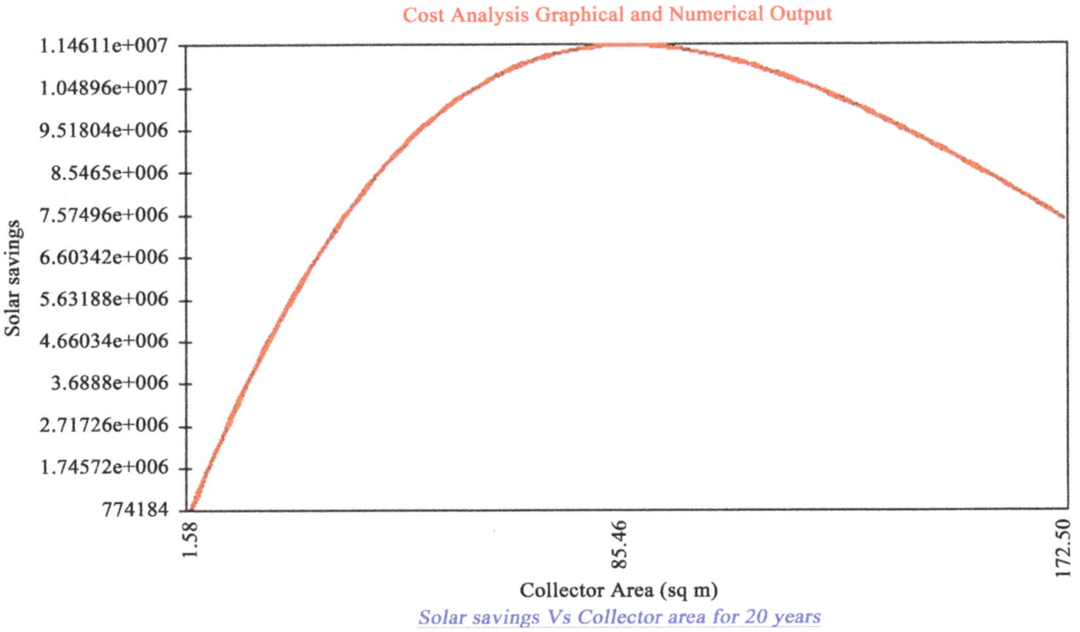

Figure 3. Solar savings against collector area for Ibadan location.

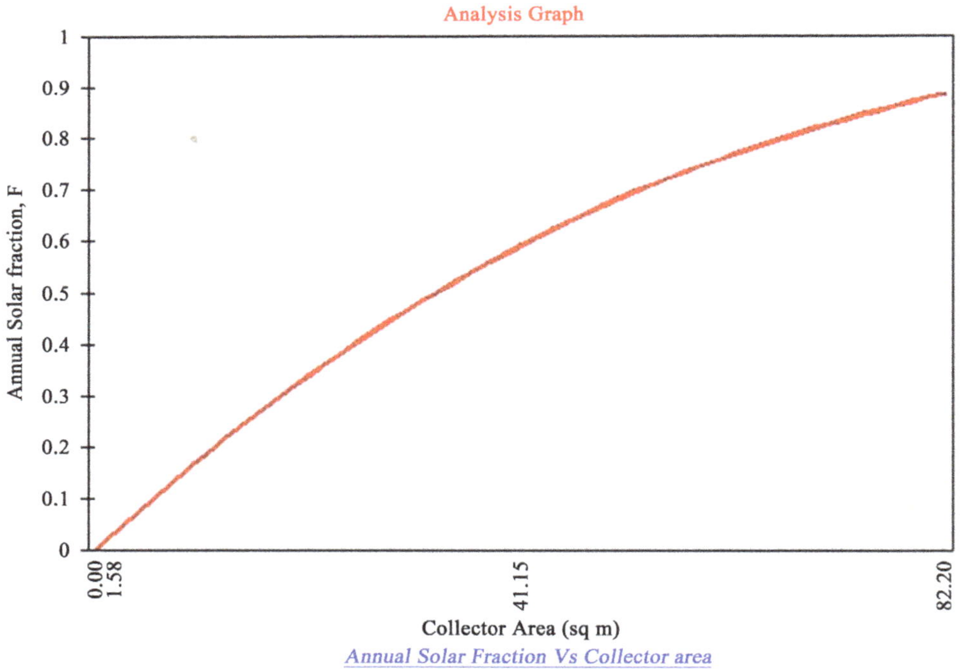

Figure 4. Annual solar fraction against collector area for Kano location.

Cost Analysis

In developing countries like Nigeria, items like mortgage interest rate, down payment and the likes, sounds unrealistic, hence the cost analysis is recommended to be adapted. The C_A and C_E are area dependent cost and area

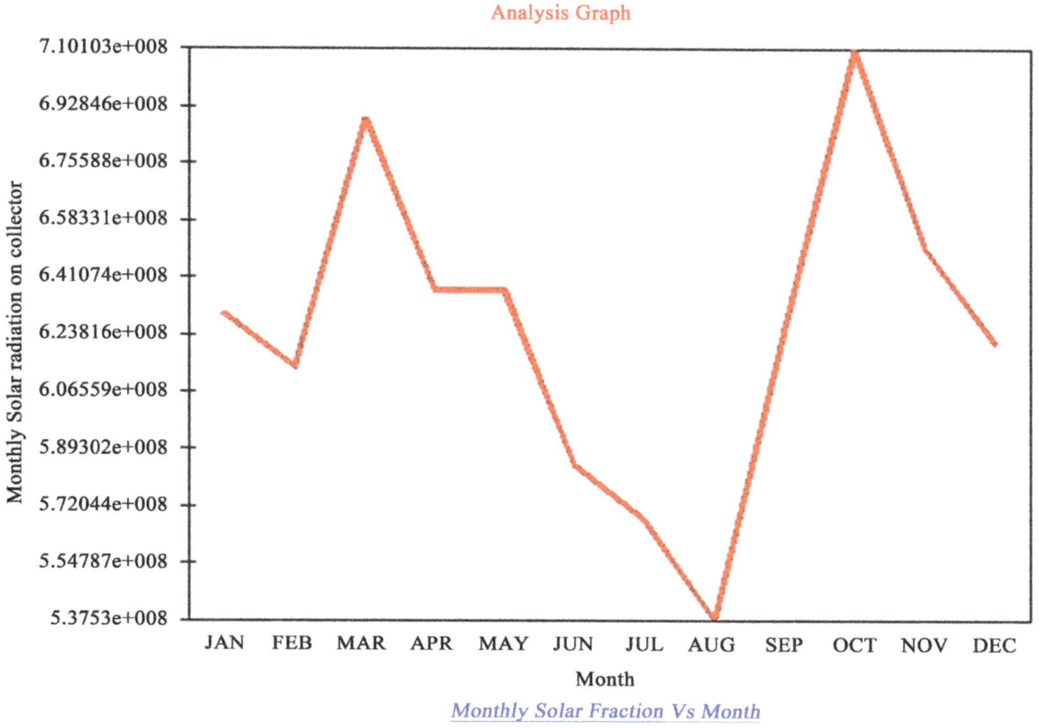

Figure 5. Monthly solar radiation on collector against month of the year for Kano location.

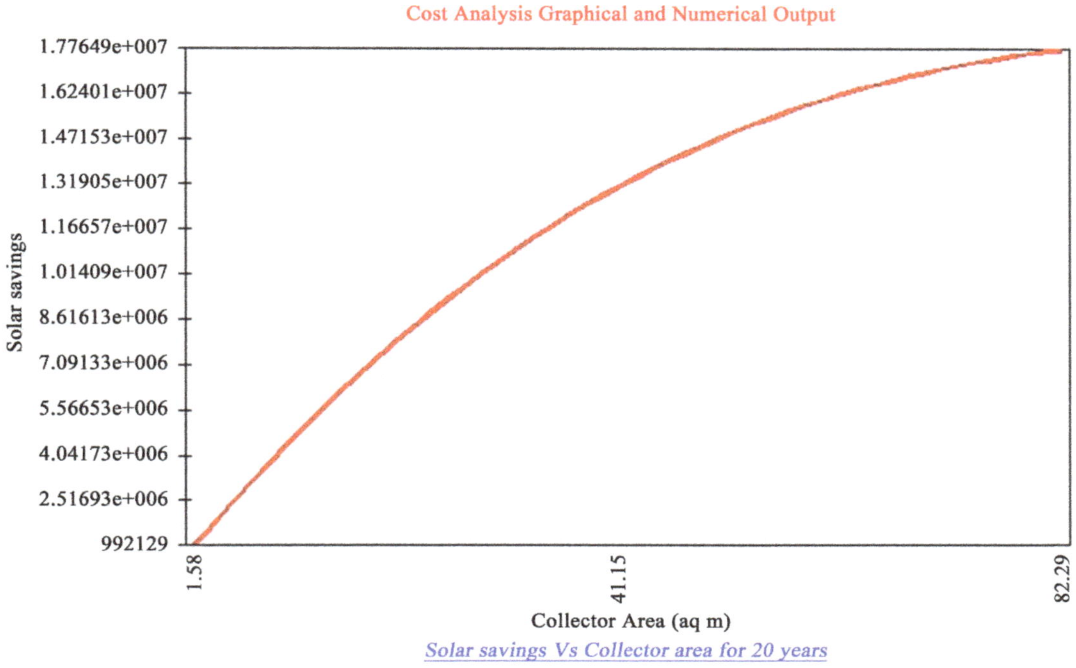

Figure 6. Solar savings against collector area for Kano location.

independent cost respectively and they contribute to the overall cost of the installation. Based on simulation experience, $C_A \approx nC_E$ and for industrial set up it was assumed that, $n = 3$. If a smaller initial moisture content is used, the collector area required for a solar fraction supply will be smaller and hence larger C_A/C_E. Smaller C_E will surely reduce Cost, C_s. Therefore the larger the ratio of C_A/C_E, the more reduced "C_s" and hence, higher solar savings.

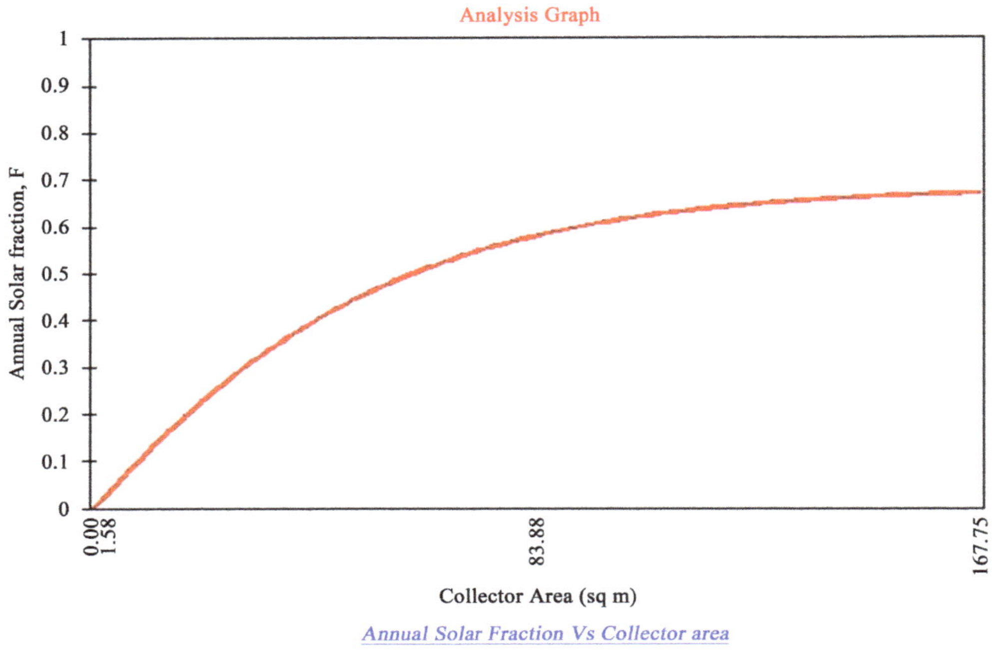

Figure 7. Annual solar fraction against collector area for Port Harcourt location.

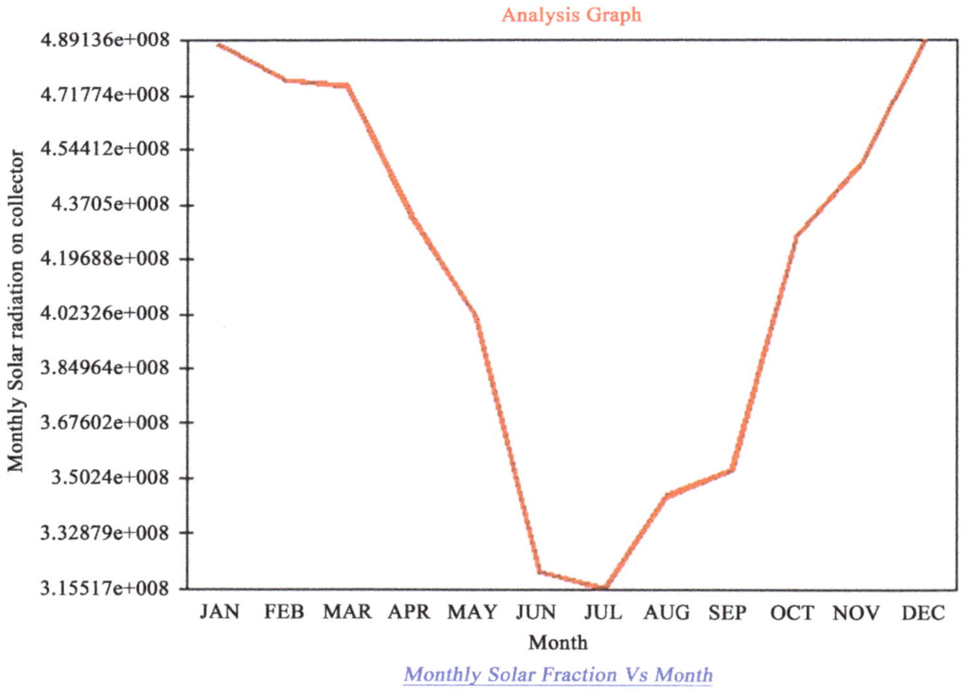

Figure 8. Monthly solar radiation on collector against month of the year for Port Harcourt location.

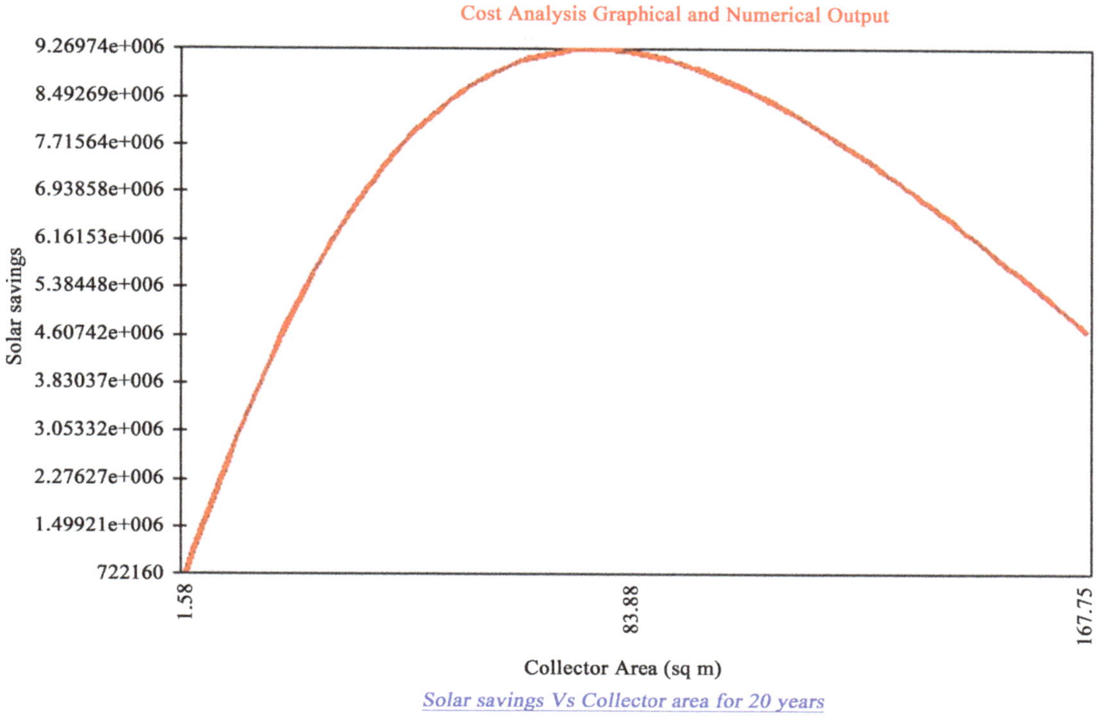

Figure 9. Solar savings against collector area for Port Harcourt location.

4. Recommendations

The selection of other materials for use in the design should also be based on their availability and affordability because nobody will want to buy any equipment that is too expensive and for which spare parts are not locally available. To help in performing economic analysis using this software, prices of solar air heating systems of different configurations can be obtained on the internet sites of international vendors and manufacturers. Though varieties of economic figures of merit like payback times, cash flow etc., have been proposed and applied but the life cycle costing method is sufficiently the most inclusive since it take into account any level of detail the user wishes to include even, the dynamic nature of time value of money and hence recommended. The kinetics of moisture within the agricultural produce had not been dealt with in this paper.

5. Conclusion

Pre-investment, investment and operating cost that may be attached with industrial solar drying processes are actually functions of different meteorological data. These data are the essentials of optimum profit when the application of Solar drying equipment becomes pragmatically imperative. Although there are many values of solar collector that could easily support profitability but the optimal collector area needs to be examined and appropriately applied. However, the study revealed that solar collector area of 85.46 m^2, 80.71 m^2 and 75.96 m^2 supplied about 67%, 88% and 55.8% of the annual energy needed which are the simulated optimum solar energy value with payback period of 8 years, 6 years and 8 years and also salvage value of 265,123.06, 250,880 and 236,636.91 for Ibadan, Kano and Port-Harcourt locations respectively. Though many solar economic figures of merits are available but larger number of those methods are designed to evaluate fuel payment for an alternative or conventional process and energy supply.

References

[1] Soha, M.S. and Chandra, R. (1994) Solar Drying Systems and Their Testing Procedures: A Review. *Energy Conversion and Management*, **35**, 219-267.

[2] Duffie, J.A. and Beckman, W.A. (1980) Solar Thermal Engineering. John Willey & Sons Inc., Hoboken.

[3] Fagbenle, R.O. (1993) Estimation of Diffuse Solar Radiation in Ibadan, Nigeria. *International Journal of Solar Energy*, **13**, 145-153. http://dx.doi.org/10.1080/01425919208909781

[4] Lasode, J.A. (1994) *Journal of Applied Science, Engineering and Technology*, **4**, 32-43.

[5] Fagbenle, R.O. (1991) On Monthly Average Daily Extraterrestrial Solar Radiation for Nigeria Latitudes. *Nigeria Journal of Renewable Energy*, **2**, 1-8.

[6] Fagbenle, R.O. (2004-2005) Heat Transfer Lecture Note. University of Ibadan, Ibadan, 12-19.

[7] Fagbenle, R.O. (1990) Estimation of Total Solar Radiation in Nigeria Using Meteorological Data. *Nigeria Journal of Renewable Energy*, **1**, 1-8.

[8] Otherno, H. (1993) Design Factors of Small Scale Thermo-Syphon Solar Crop Dryers. *Workshop on Solar Drying*, University of Accra, Accra, 2-10.

17

A Review of Price Forecasting Problem and Techniques in Deregulated Electricity Markets

Nitin Singh, S. R. Mohanty

Department of Electrical Engineering, Motilal Nehru National Institute of Technology Allahabad, Allahabad, India

Email: nitins@mnnit.ac.in, soumya@mnnit.ac.in

Abstract

In deregulated electricity markets, price forecasting is gaining importance between various market players in the power in order to adjust their bids in the day-ahead electricity markets and maximize their profits. Electricity price is volatile but non random in nature making it possible to identify the patterns based on the historical data and forecast. An accurate price forecasting method is an important factor for the market players as it enables them to decide their bidding strategy to maximize profits. Various models have been developed over a period of time which can be broadly classified into two types of models that are mainly used for Electricity Price forecasting are: 1) Time series models; and 2) Simulation based models; time series models are widely used among the two, for day ahead forecasting. The presented work summarizes the influencing factors that affect the price behavior and various established forecasting models based on time series analysis, such as Linear regression based models, nonlinear heuristics based models and other simulation based models.

Keywords

Electricity Price Forecasting, Time Series Models, ARIMA, GARCH, ANN, Fuzzy ARTMAP

1. Introduction

With the Introduction of deregulation of power industry, new challenges have been encountered by the participants of the electricity market due to which forecasting of wind power, electric loads and energy price have become a major issue globally [1]. Deregulation however, has been associated with the expectation of greater consumer participation and efficiency gains for both consumers and share-holders. Globally energy price forecasting has come up as important area of research due to deregulation of whole sale market. Major market participants such as generators, power suppliers, investors and trades wish to maximize the profitability [2]-[4]. Unlike load forecasting, electricity price forecasting is much more complex because of the unique characteristics and uncertainties in operation as well as bidding strategies [5]. In other commodity markets like stock market, agricultural market price forecasting is always being at the center of studies because of its importance [6]-[9].

Electricity is also a commodity, and its price should also be forecasted along with time but if the same methods were used for forecasting electricity prices as other commodity prices, the forecasted price will exhibit lower accuracy without any surprise due to volatile nature of electricity price among all commodities. Many techniques and models have been developed for forecasting whole sale electricity prices, especially for short term price forecasting [3]. The state of art techniques for electricity price forecasting are categorized into equilibrium analysis [5], simulation methods [10], econometric methods [11], time series [12]-[14], intelligent systems [15]-[17] and volatility analysis [18]. Time series and intelligent systems are commonly used for day-ahead price forecasting. This paper reviews established approaches and mainly focusing on soft computing models.

2. Factors Influencing Price Forecasting

In deregulated power markets, fluctuation is a common behavior of price which is because of many different economic as well as technical factors. Some researchers have only used historical data of prices [19] or both prices and demand to forecast spot price excluding other factors such as weather, fuel cost and generation reserve. The various factors that affect the spot price are shown in **Figure 1** [20].

2.1. Electric Power Demand

One of the important factors in spot price is system's total demand. Studies show that if system demand increases, spot price also increases.

2.2. Whether Conditions

Electricity demand certainly depends on environmental condition and especially daily temperature. Weather fluctuation will affect demand and hence spot price will also be affected.

2.3. Fuel Cost

Fuel cost is one of the main parts of generation cost that its variation has a major impact on electricity spot price.

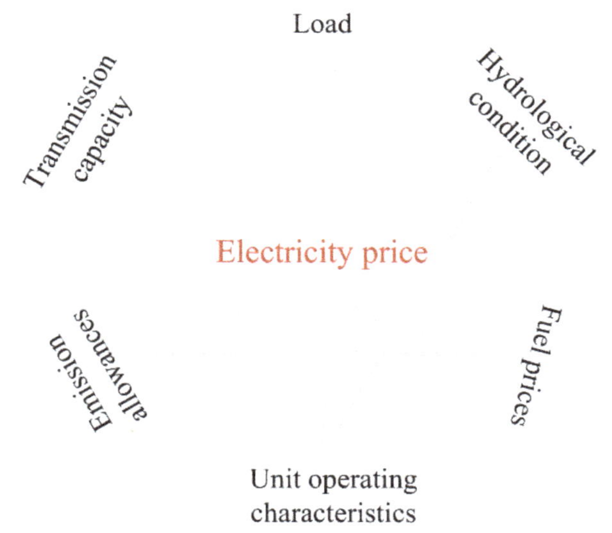

Figure 1. Factors affecting electricity prices.

2.4. Available Transmission Capacity

Electric power is provided by generator that may be located far from location of consumers. It should be transmitted to consumers via transmission network facilities. There is some physical constraint in transmission networks that is an obstruction for market participants to buy or sell energy. This issue can affect important changes on spot price and may increase it.

2.5. Generation Reserves

Having enough generation reserve is an important factor for electricity spot price, *i.e.* when demand increase suddenly if there is enough generation reserve capacity available as well as deliverable, consumers will be served. But if there is not sufficient generation reserve available, consumer would face with lack of received energy and therefore to make the balance between supply and demand electricity spot price increases.

3. Electricity Load and Price Forecasting Problems and Methods

3.1. Load Forecasting

Forecasts, in particular have become important after restructuring of the power systems as many countries have deregulated their power system and turned electricity into commodity from necessity. Many countries are still in the process and soon electricity will be a commodity with players in all across global market. Load series is not only complex nut also exhibits several levels of seasonality: the prediction is not only depends on the previous hour load but also on the load of the same hour on previous day, and same denominations in the previous week [21]-[23].

Various techniques and models have been developed for the forecasting the electrical load with varying degrees of success, but the still the models based on the linear regression scores over the other reported models. These models allow the system operators and engineer, physically interpretation of the components so that their behavior can be understood. Models based on Artificial Intelligence (AI) were also developed for forecasting of electrical load, such as expert systems, fuzzy inference, fuzzy neural models and neural network (NN) based models. Neural networks due to their intrinsic capability to learn complex and non linear relationships that are otherwise difficult by other conventional methods, have gained popularity among all artificial intelligence based models [24] [25].

3.2. Price Forecasting

With introduction of the deregulated electricity markets major emphasis is on maximizing the profits of the various market players. As far as forecasting is concerned electricity prices and load are mutually interlinked, due their dependability on each other and error in one will propagate to other. Non-storability, Seasonal behavior and Transportability are the major issues which makes electricity price so specific. These issues make it impossible to treat the electricity at par with any other commodity and forbid the application of forecasting methods common in other commodity markets [26].

Electricity price forecasting can be categorized into three different categories based on time horizons: Short-term forecasting, medium-term forecasting and long-term forecasting as shown in **Figure 2**. Short term price forecasting (mainly one day ahead) will be mainly used by the market players to maximize profits in the spot markets. Knowledge of medium term forecasting will allow the successful negotiations of bilateral contracts between suppliers and consumer while long term forecasting will influence the decisions on transmission expansion and enhancement, generation augmentation and distribution planning.

4. Prices Forecasting Methods

Survey reveals that various methods have been developed for forecasting. A rough tree of classification is shown in the **Figure 3**, this classification is not comprehensive and other approaches or methods are possible, these methods can be used for load forecasting as well as price forecasting [1] [3].

Mainly for price forecasting the approaches can be classified into two categories [6] [27]-[35] 1) time series and 2) simulation approach, time series mainly relies on the historical data of market prices. In simulation approach requires precise modeling of power system equipments and their cost information, because of large

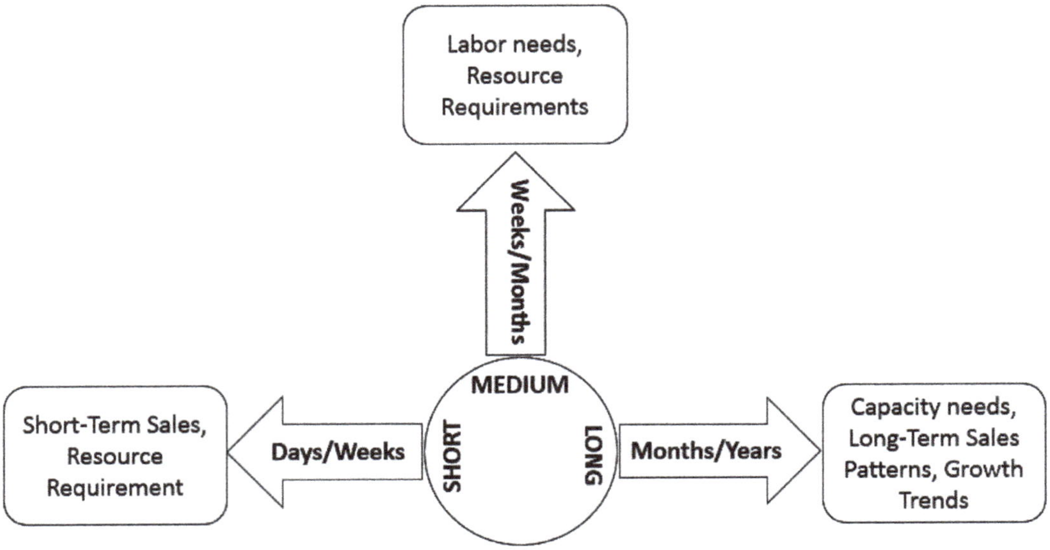

Figure 2. Types of forecasting.

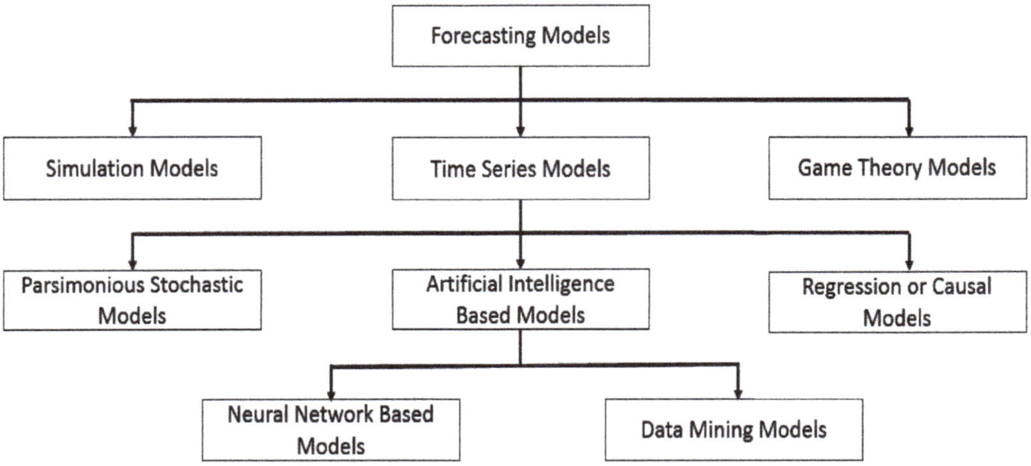

Figure 3. Classification of forecasting techniques.

amount of data involved simulation method can be computationally intensive.

Time series approach can be further classified into the following, linear regression based models and non linear heuristic models. Regression-based models include auto-regressive moving average (ARMA) models, and its extension, auto regressive integrated moving average (ARIMA) models, and their variants. While these models are aimed at modeling and forecasting the changing price itself, generalized autoregressive conditional heterokedasticity (GARCH) is aimed at modeling the volatility of electricity prices [20].

Nonlinear heuristic based models uses artificial neural network and other artificial intelligent techniques for modeling the input-output data relation without complete information of the connections. Other soft computing methods are also used to extend the data representation capability of the regression based or ANN models.

5. Price Forecasting Methodology

A typical procedure of price forecasting is shown in the **Figure 4** [20]. The flow chart is depicting the process of time series based forecasting. The process of forecasting usually starts with the input data, the major input data for the price forecasting are the past market prices, record of a few weeks to several months is taken as input.

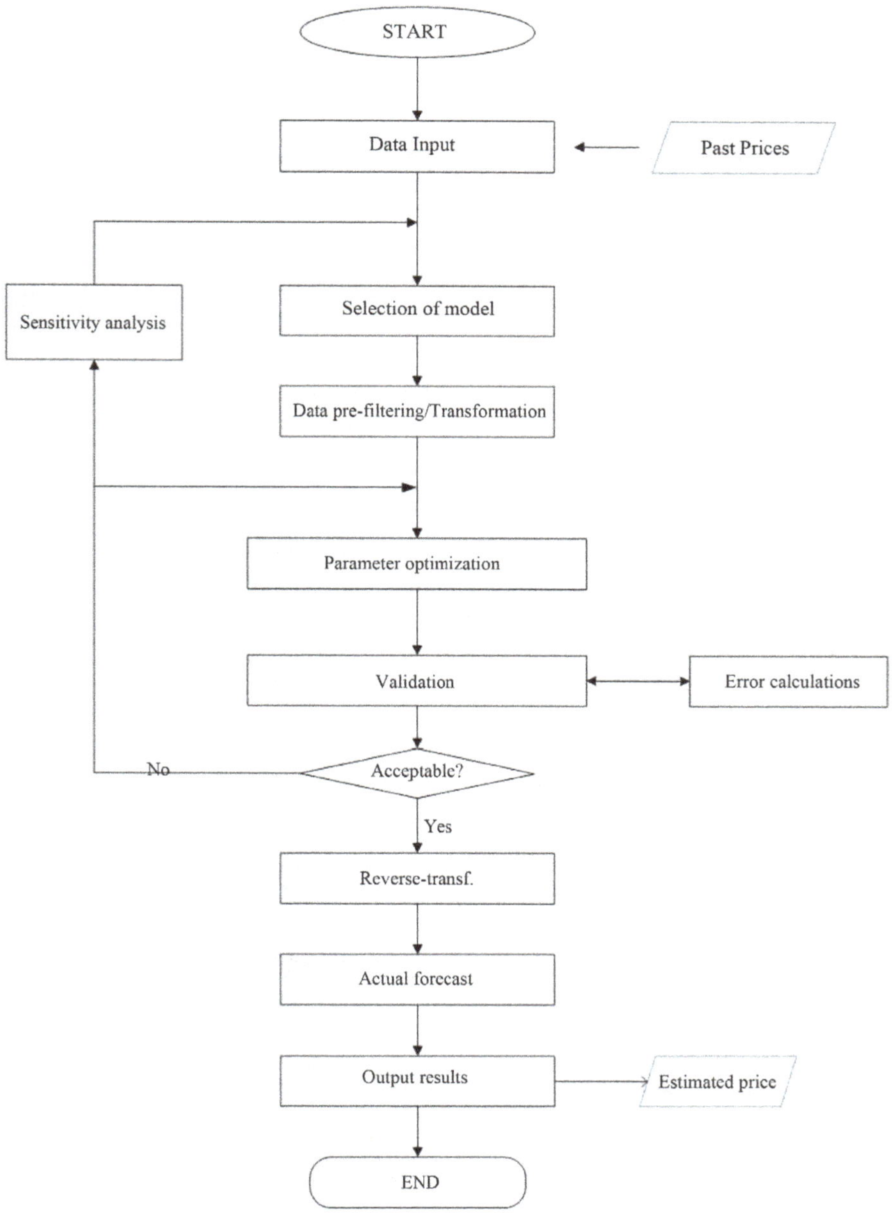

Figure 4. Flow chart showing forecasting procedure.

Some complex forecasting models require additional input as demand and/or temperature data.

Simple statistical analysis on the input data set (e.g. mean and volatility) will give some hint of model selection and later model validation. The scope of forecast (e.g. price profile of its volatility, etc.) will be an important factor for the selection and design of forecasting models/techniques. The accuracy of results needed will be an important factor of the selection and design of models/techniques. The model validation is carried out after optimizing the parameters of models to check the performance of the model. The process of validation is repeated if the results are not satisfactory with different starting parameters. If the validation is successful the model is applied to do the actual forecast.

6. Pre-Processing of Data for Time Series Models Using Wavelet Transforms

Wavelet Transform is a mathematical model which analyses data and provides time and frequency representation simultaneously (time-scale analysis) [36]. It is used for analyzing non-stationary signals in power systems such as price time series [37]-[40], voltage and current waveforms [41]. The wavelet transforms decomposes the original time domain signal into several other scales with different levels of resolution in what is called multi-resolution decomposition [42].

Wavelet Transform is most suited for the non-stationary data (mean and autocorrelation of series are not constant), the price series data is non-stationary and volatile in nature, that is why use of wavelet transform gives accurate forecasting results [43]. FT (Fourier Transform) decompose the original price series into a linear combinations as sine and cosine functions whereas by using Wavelet Transform (WT) the series is decomposed into a sum of more flexible functions *i.e.* localized in both time and frequency. Wavelet Transform can be classified into two: Continuous Wavelet Transform (CWT) and Discrete Wavelet Transform (DWT) [40] [44]-[46].

The CWT of a continuous time signal $x(t)$ is defined as (1):

$$\psi_{a,b}(t) = \int_{-\infty}^{\infty} x(t)\psi^{*}_{a,b}(t)\,\mathrm{d}t \tag{1}$$

where $\psi(t)$ is a mother wavelet, a is a scaling parameter, b is a translating parameter and

$$\psi^{*}_{a,b}(t) = \frac{1}{\sqrt{a}}\psi^{*}\left(\frac{t-b}{a}\right) \tag{2}$$

Each wavelet is created by scaling and translating operations in a mother wavelet. The mother wavelet is an oscillate function with finite energy and zero average.

The DWT of a sampled signal $x(n)$ is defined as (3):

$$\psi_{c,d} = \sum_{n} x(n)\psi^{*}_{c,d}(n) \tag{3}$$

where

$$\psi^{*}_{c,d}(n) = \frac{1}{\sqrt{a^{c}_{o}}}\psi^{*}\left(\frac{n - db_{o}a^{c}_{o}}{a^{c}_{o}}\right) \tag{4}$$

where, c and d are scaling and sampling numbers respectively. General block diagram for level 3 decomposition is shown in **Figure 5**.

Technically, the price data is transformed into low and high coefficients. The low coefficients are an approximated version that is associated with low pass filtering and possess the similar characteristics as of original price series, while latter with high pass filtering which contains information regarding peaks that occur in the original price signal. Results are significantly affected by the selection of mother wavelet. For price forecasting application generally Daubenchies wavelet transforms are suitable because they have compact or narrow window function which is suitable for local analysis of non-stationary price series.

7. Forecasting Models Based on Linear Regression

7.1. ARIMA Model

ARMA stands for Auto-Regressive Moving Average, ARMA is suitable model for stationary time series but most

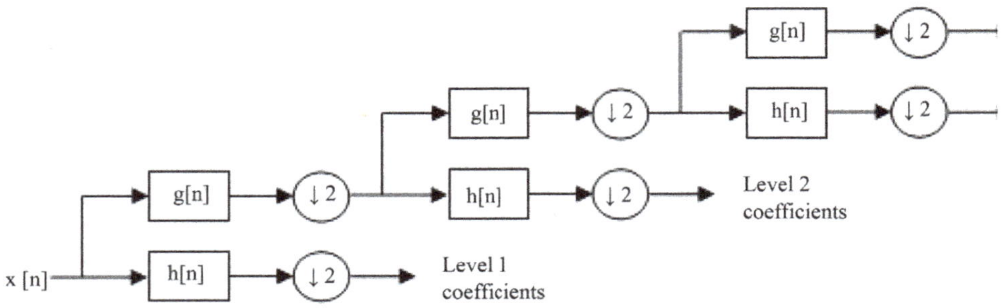

Figure 5. Level 3 wavelet decomposition.

of the price series are non-stationary. To overcome this problem and to allow ARMA model to handle non-stationary data, the new model is introduced for non-stationary data, the model is called Auto-Regressive Integrated Moving Average (ARIMA), it has been successfully applied to forecast the commodity prices [47]-[49]. The application of ARIMA methodology for the study of time series analysis is due to box and Jenkins [50].

There are many ARIMA models; generally ARIMA model is defined as ARIMA (p, q, d) where: p is the number of autoregressive terms, q is the number of lagged forecast error in the prediction equation, d is the number of non-seasonal differences. If there is no differencing (*i.e.* $d = 0$), then ARIMA model can be called an ARMA model [47].

Consider a time series x_t, then the first order differencing is defined as:

$$x_t' = x_t - x_{t-1} \tag{5}$$

where, L can be used to express differencing

$$x_t' = x_t - x_{t-1} = x_t - Lx_t = (1-L)x_t \tag{6}$$

Thus, ARIMA (p, d, q) is defined as:

$$\underbrace{\left(1-\varphi_1 L-\varphi_2 L^2-\cdots-\varphi_p L^p\right)}_{AR(p)}\underbrace{(1-L)^d}_{I(d)} x_t = c+\underbrace{\left(1-\psi_1 L-\psi_2 L^2-\cdots-\psi_q L^q\right)}_{MA(q)}\varepsilon_t \tag{7}$$

$$\varphi(L)(1-L)^d x_t = c+\psi(L)\varepsilon_t \tag{8}$$

ARIMA models are derived from autoregressive (AR), moving average (MA) and auto-regressive moving average (ARMA). In AR, MA and ARMA models conditions of stationary are satisfied; therefore they are applicable only to stationary series. ARIMA model captures the incremental evolution in the price instead of price value.

7.2. GARCH Model

GARCH stands for Generalized Autoregressive Conditional Heteroskedasticity while the (ARIMA) models are aimed at modeling and forecasting the changing price itself, (GARCH) model is aimed at modeling the volatility of prices [51] [52]. GARCH models consider the moments of a time series as variant (*i.e.* the error term: real value minus forecasted value does not have zero mean and constant variance as with an ARIMA process). The error term is now assumed to be serially correlated and can be modeled by an Auto Regressive (AR) process. Thus, a GARCH process can measure the implied volatility of a time series due to price spikes [53]-[55]. The model GARCH (p, q) is defined as follows:

Consider a time series x_t with a constant mean offset, then

$$x_t = \mu + \varepsilon_t \tag{9}$$

where μ is offset and $\varepsilon_t = \sigma_t z_t$.

$$\sigma_t^2 = c + \sum_{i=1}^{q}\varphi_i \varepsilon_{t-i}^2 + \sum_{i=1}^{p}\psi_i \sigma_{t-i}^2 \tag{10}$$

where p is the order of GARCH terms σ^2 and q is the order of ARCH terms ε^2.

As we can easily see in Equation (10), in GARCH (p, q) model is $p = 0$, *i.e.* a GARCH (0, q) model becomes an ARCH (q) model. GARCH model can only specified for stationary time series so below equation must be satisfied for stationary time series.

$$\sum_{i=1}^{q}\varphi_i + \sum_{i=1}^{p}\psi_i < 1 \tag{11}$$

GARCH process can measure the implied volatility due to price spikes.

8. Forecasting Models Based on Nonlinear Heuristics

8.1. Artificial Neural Network Based Model

Most of the time series models are linear predictors, while electricity price is a non-linear function of its input features, making it difficult for the time series techniques to completely capture the behavior of price signal. Therefore the researchers have come up with the idea of using Neural Network (NNs) for electricity price forecasting [56]-[59].

Neural networks are highly interconnected simple processing units designed to model how the human brain performs a particular task [60]. Basic structure of the neural network is shown in the **Figure 6**. The network generally consists of three to four layers and during training process, the neurons in the input layer pass the raw information onto the rest of the neurons in the other layers. The connection weights between different layers keep on updating with the ongoing learning process [60].

A neural network uses a learning function to modify the variable connection weights at the input of each processing element *i.e.* neuron. The ANN models could be differentiated based on type of learning function, learning algorithm and no. of hidden layers etc. Generally a three layered neural networks are chosen for forecasting the electricity price.

ANN based models have gained popularity due to their property to solve undefined relationship between input and output variables, approximate complex nonlinear function and implement multiple training algorithms. However, neural network also suffers from the disadvantage that the network will not be flexible enough to model the data well with too few units, and on the contrary, it will be over-fitting with too many units [20].

In order to overcome such weakness, different evolutionary techniques have been combined with ANNs recently [58] [61]-[65]. ANN model with feature selection technique and relief algorithm [59] and particle swarm optimization is used for ANN training [66].

8.2. Radial Basis Function Neural Network Model Based Model

Radial basis function Neural Network (RBFNN) has comparatively less chance to trap in local minima and has faster learning rate [67]. RBFNN uses radial basis function as the activation for the hidden layer neurons as compared to the artificial neural network (ANN). Similar to the ANN architecture the RBFNN also contains three layers *i.e.* input layer, output layer and only one hidden layer. The difference arises in terms of center neurons activation function and training method. The training of RBFNN consists of three steps: 1) centre selection; 2) width selection of basis function and 3) weight calculation for output layer.

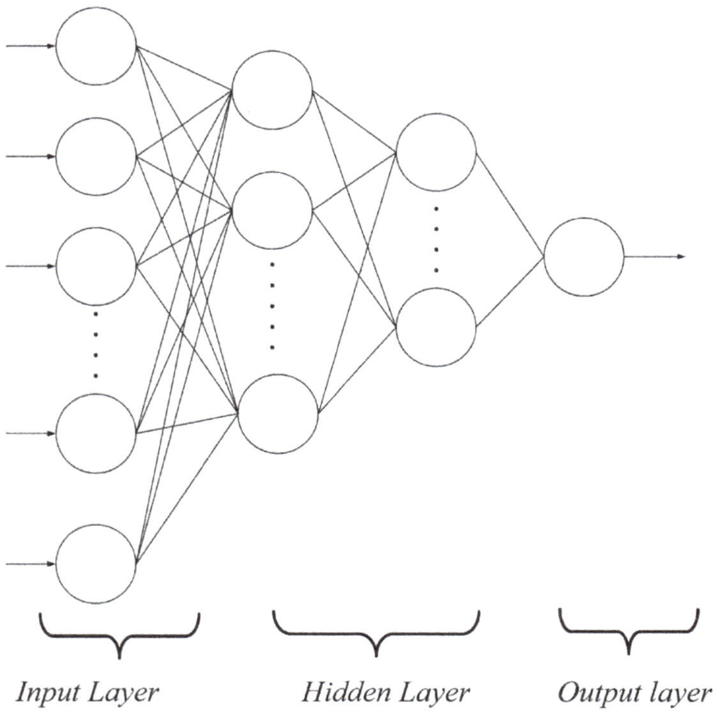

Figure 6. Architecture of artificial neural network.

The model of RBFNN can be described as follows:

$$f(I) = \psi\left(\frac{\|I - c_i\|}{r_i^2}\right)^2 \tag{12}$$

$f(I)$ is the output of ith neuron of hidden layer and I is an input training vector as described in III. A; $\Psi(.)$ is radial basis function used in non-linear mapping, C_i is center for i^{th} hidden layer neuron and r_i is radius for i^{th} hidden layer neuron.

$$\|I - c_i\| = \sqrt{(I_1 - c_{i1})^2 + (I_2 - c_{i2})^2 + \cdots + (I_q - c_{iN_i})^2} \tag{13}$$

In (13), $\|I - c_i\|$ is Euclidean distance and it can be calculated using (12), where, q is number of inputs in one training pattern. The width (σ) of the basis function is decided by the singular values of G_{tr}, which are generated using (14). Here, $d_{\max}$ is maximum euclidean distance between final centre points C_i and all training input points. Weights of the output layer can be measured using Equation (15). G_{tr}^{+} and Y_{tr} are pseudo inverse of G_{tr} and output training patterns matrix respectively.

$$G_{tr}\left(\|I_q - c_i\|^2\right) = \exp\left(-\frac{N_h}{d_{\max}}\|I_q - c_i\|^2\right) \tag{14}$$

$$W = G_{tr}^{+} \times Y_{tr} \tag{15}$$

The basic structure of RBFNN is shown in **Figure 7**, the numbers of neurons in input layer (N_i) and output layer neuron (N_o) are selected on the basis of training patterns developed. The nonlinearity of the system decides

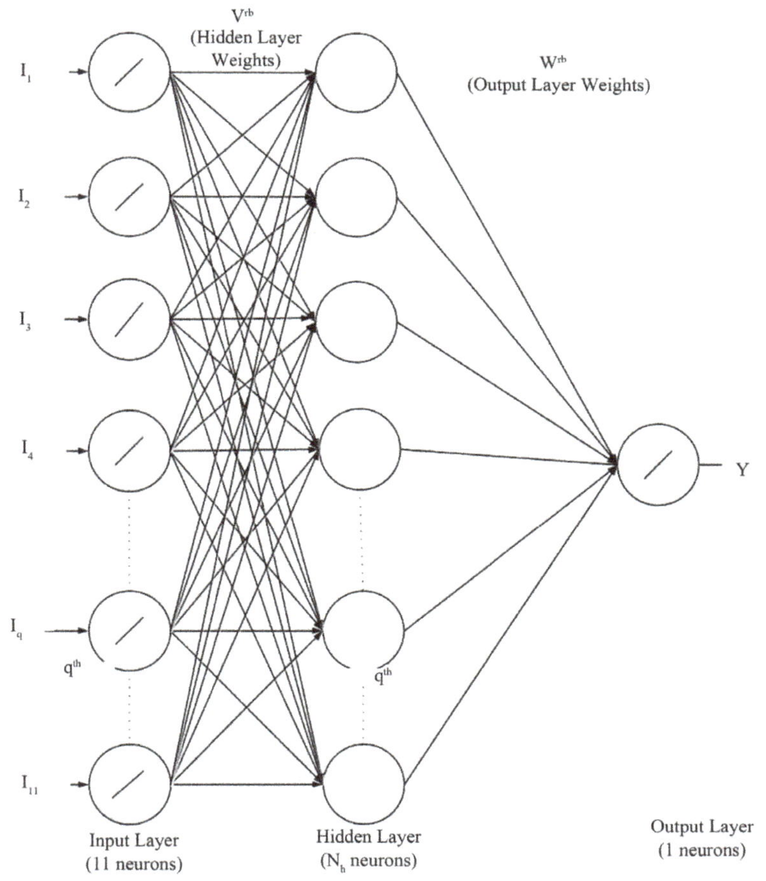

Figure 7. Architecture of radial basis function neural network (RBFNN).

the number of neurons in the hidden layer (N_h). The data flow start from the input layer and traverse through the hidden layer and arrives at the output layer. Input as well as output layers of RBFNN have linear activation functions, however the hidden layer neurons has a radial basis function (Gaussian) activation function.

$$Y = W^{\mathrm{T}} \times G_{tst}^{\mathrm{T}} \tag{16}$$

where, G_{tst} defined as (17) and suffix *tst* denote any testing or real time input pattern for which output is desired from a trained RBFNN. Output Y can be calculated using (16).

$$G_{tst}\left(\left\|I_{tst} - c_i\right\|^2\right) = \exp\left(-\frac{\left\|I_{tst} - c_i\right\|^2}{2\sigma^2}\right) \tag{17}$$

Input layer weight matrix has value 1 for all its elements, because input is direct and linearly mapped to hidden layer. For training of RBFNN K-mean clustering is applied. In RBFNN weight matrices, V_{rb} and W_{rb} contain weights of hidden layer and output layer, respectively.

8.3. Fuzzy Inference System Based Model

An FIS performs input-output mapping based on fuzzy logic. Fuzzy evaluates the intermediate states between discrete crisp states and is able to handle the concept of partial truth instead of absolute truth. Traditional adaptive fuzzy system include ANFIS and neuro-fuzzy methods are intended to combine the advantages of ANN and fuzzy logic with the difference that ANFIS architecture has linear output function [68], whereas neuro-fuzzy systems are essentially a subset of ANN applied to controls and classification problem [69].

Wang-mendel suggested an algorithm for implementing FIS for time series prediction [70] and the same approach was extended to forecast the electricity price. The approach is model free and heuristic in nature. A common framework called the fuzzy rule base is constructed to combine both numerical and linguistic information. The numerical information is sampled from measurements, and the linguistic information interprets the numerical information [71]. The FIS is able to bridge the gap between interpretability and accuracy by providing a verbally interpretable rule base and numerical accuracy through training. The FIS using wang-mendel learning algorithm does not require iterative training making it more efficient than ARMA or GARCH time series techniques and ANN or neuro-fuzzy intelligent systems [70].

Compared to the black box nature of Artificial Neural Network (ANN) the Fuzzy Inference System (FIS) provides a transparent linguistic rule base instead of a black box. The rules may be modified manually to include expert knowledge. The rule base provides FIS the advantage of interpretability and transparency. FIS also provides flexibility in choosing predefined membership function. The FIS algorithm can be modified for higher accuracy and efficiency.

8.4. Fuzzy ARTMAP Based Model

The Fuzzy ARTMAP is a new concept for electricity price forecasting [72], it has already been applied for wind speed forecasting [73] and load forecasting [74]. Mostly conventional neural networks suffers from plasticity-stability dilemma, *i.e.* the information related to the plasticity or adaptivity to the new inputs or change in inputs at the same time stable in response [75] [76]. The fuzzy ARTMAP structure shown in **Figure 8** addresses this dilemma by incorporating a feedback mechanism between the competitive and input layers to allow new information to be learned without eliminating previously obtained knowledge, in this it becomes more stable and shows a faster convergence capability [77]. ARTMAP is a class of neural architectures that perform incremental supervised learning of recognition categories and multidimensional maps in response to input vectors presented in arbitrary order. An ARTMAP system embodies twin art modules (ARTa and ARTb) to fabricate stable recognition categories corresponding to the arbitrary input patterns. ARTa uses ART-1 while ARTb uses FUZZY ART. This set up enables to switch the binary modules set theory notations to transform into a corresponding feature in the fuzzy ART module.

Example; the intersection operator ($\cap$) of ART_1 is replaced by the operator (^) in the FUZZY ART. The architecture, called fuzzy ARTMAP, achieves by synthesis of fuzzy logic and adaptive resonance theory (ART) neural network by employing a close formal similarity between two computations of fuzzy subsets and ART category, resonance, and learning. Fuzzy ARTMAP also actualize a new min-max learning rule that collectively

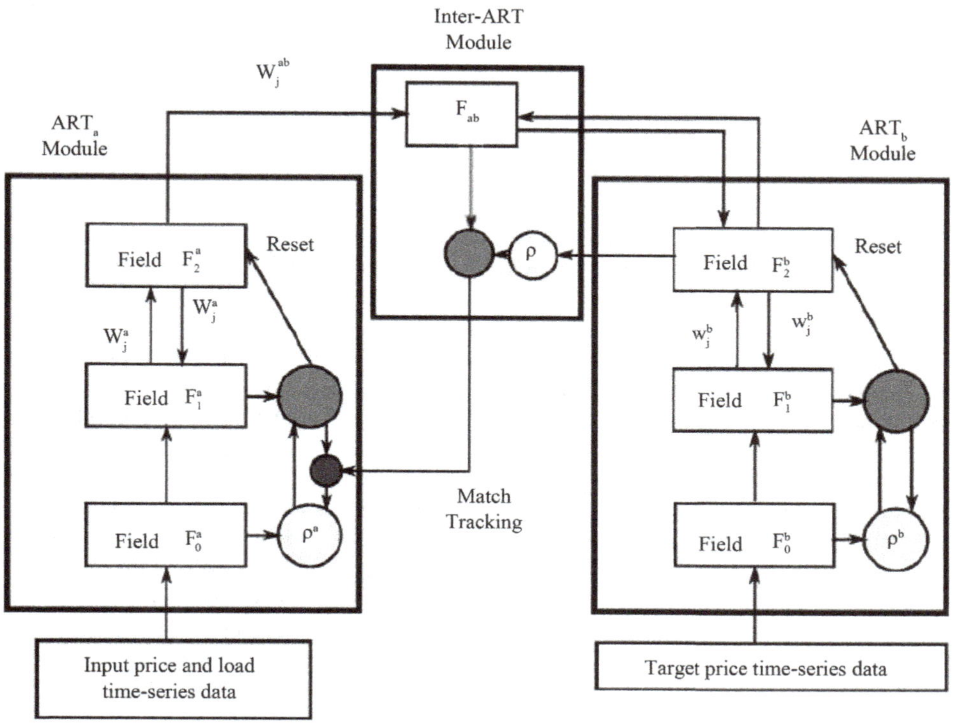

Figure 8. Architecture of fuzzy ARTMAP [72].

minimizes predictive error and maximizes generalization, or code compression. This is achieved by a match tracking process that increases the ART vigilance parameter by the minimum amount needed to correct a predictive error.

So as a result, the system automatically learns a minimal number of recognition categories, or "hidden units" to meet the criteria of accuracy. Category proliferation is prevented by normalizing input vectors at a preprocessing stage. A normalization procedure called complement coding leads to a symmetric theory in which the AND operator (^) and the OR operator (v) of fuzzy logic plays complementary roles.

In training, the best matching category is [75];

$$J = \arg\max_{0 \le j \le N} T_j\left(I_{tr}\right) \tag{18}$$

where,

$$T_j\left(I_{tr}\right) = \begin{cases} \dfrac{\left|I_{tr} \wedge w_j\right|}{\alpha + \left|w_j\right|}, & \text{if } \dfrac{\left|I_{tr} \wedge w_j\right|}{\left|I_{tr}\right|} \\ 0, & \text{Otherwise} \end{cases} \tag{19}$$

where, T_j = choice function, α = choice parameter, ^ = Fuzzy MIN operator, ρ = vigilance parameter and $\dfrac{\left|I_{tr} \wedge w_j\right|}{\left|I_{tr}\right|} \ge \rho$ is the vigilance criteria. If a vigilance criterion satisfies then resonance occurs. During training, the vigilance criterion varies from baseline vigilance which is initial value. If vigilance criteria pass then category J becomes representative membership function for time series, and the weight vector of the winning category w_j is updated by following Equation (20):

$$w_j^{(\text{new})} = \beta\left(I_{tr} \wedge w_j^{(\text{old})}\right) + \left(1-\beta\right) w_j^{(\text{old})} \tag{20}$$

where β is the learning rate. If a vigilance criterion fails then category J is deactivated for the current price series by setting choice function equals to zero. If ART_b does not predict the correct output for ART_a, then vigilance parameter is increased. This is called match tracking, in match tracking vigilance parameter is slightly increased to a new value:

$$\rho = \frac{|I_{tr} \wedge w_j|}{|I_{tr}|} + \varepsilon \tag{21}$$

where ε is a learning precision.

The scheme resizes a category on predictive success by amplifying the vigilance parameter ρ by a minimal amount essential to verify the predictive error in the ART_b. The parameter ρ holds an inverse relationship with the category size. A lower value leads to a broadly generalized category with higher compressed code. This parameter rates the minimum faith that ART_a should have while accepting a category during hypothesis testing which focuses ART_a on a new cluster. The failures at ART_a increase ρ to that threshold value which in turn triggers ART_a under a process called match tracking.

This technique reduces generalization essential to correct a predictive error. The combination of these techniques *i.e.* ARTMAP and Match tracking leads to a faster learning and erudition from a rare event. The fuzzy ART reduces to ART_1 for a binary input and works as self for a binary input and works as self for an analog vector. Thus the crisp logics of ART-1 with their fuzzy counterparts form a potent module.

Once the training stage is completed, the Fuzzy ARTMAP network is used as a classifier of the input price series which is given to the ART_a. ART_b is not used during classifying process and the learning capability of the network is deactivated during classifying process (*i.e.* $\beta = 0$). In this stage we get predicted classified labels in the output. These output labels are defuzzified for getting the forecasted price series. To find the best training parameters for the neural network some models use optimization algorithms for good results and comparatively low processing time.

9. Forecasting Models Based on Simulation Methods

Simulation methods usually simulate generator dispatch patterns over an extended period of time. These methods mimic the actual dispatch with system operating requirements and constraints. Despite of the high data requirement by these models, they can provide detailed insights into the price curve.

The simulation methods which are currently being used by the electric power industry range from the bubble-diagram type contract path models to production simulation models with full electrical representation, such as GE-MAPS software [10]. The production simulation models by nature of their chronological simulation patterns, will consider time varying systems limits and characteristics. Some important issues that must be addressed in any market simulation program that forecast the LMPs for the electricity market are [10]:

- Detailed transmission model
- Unit commitment
- Economic dispatch with transmission constraints
- Secure dispatch
- Chronological simulation
- Large-scale study capability
- Data resources
- Benchmark and application

Simulation model known as MAPS, has been develop which stands for market assessment and portfolio strategies, this model incorporates a full representation of the electrical transmission model. The detailed power flow data and secure dispatch of generators, tracking transmission line flows, loss determination, and transaction evaluation are well integrated, providing an accurate through time simulation of system operation.

The MAPS model is able to simulate large power system for one or multiple year within optimum period. The MAPS model can be applied to solve the following issues:

- Analyze market power issues
- Evaluate alternative market structures
- Estimate stranded generation investments
- Assess economics of building new generation
- Assessing transmission costs

- Understanding market behavior

The general input output structure of MAPS is shown in the **Figure 9**.

The data requirement of MAPS is similar to any free-standing production cost program or load flow models. Through its integration of generation and transmission models, it captures hour by hour market dynamics while simulating the transmission constraints of the system. Market simulation programs minimize the system cost to serve loads subject to transmission constraints, unit commitment and economic dispatch with transmission constraints are the core functions of typical market simulation programs.

The program automatically provides the location market clearing prices for any bus, identifies the bottlenecks of transmission networks, and produces the generation schedules and power flows on the transmission grid, which are important in deregulated markets. Simulation methods are intended to provide detailed insights into system prices. However, these methods suffer from two drawbacks. First they require detailed system operation data and second simulation methods are complicated to implement and their computational cost is very high.

10. Forecasting Models Based on Game Theory Models

There has been great deal of research to understand electric power markets, and various methods for modeling, analyzing and selecting bidding strategies for power suppliers. Gaming theory is a natural platform for market competition [78]. It is of great interest to model the strategies of the market participants and identify solution to those games. Since participants in oligopolistic electricity markets shift their bidding curves in order to maximize their profits, these model provides the solution to these games and profit can be considered as the outcome of the power transaction game. In this group of models, equilibrium models [5], take the analysis of strategic market equilibrium as the key point. The gaming models are generally used by the market operators for deciding the market strategies. The detailed discussion on game theory can be found in [79]-[83].

11. Forecasting Models Accuracy

Mainly following types of accuracy parameters are defined in the literature by authors' for validating the accuracy of the proposed model. For the maximum accuracy of models values of these measures must be in permissible limits. Error is defined as the difference between the actual value and the forecasted value for the corresponding period.

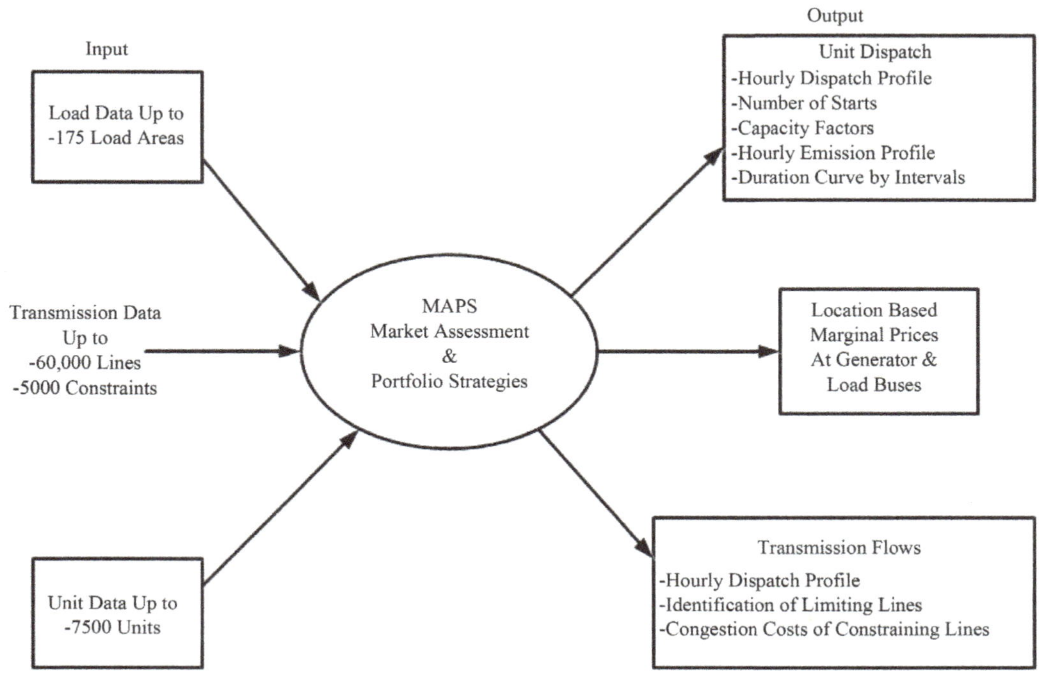

Figure 9. Architecture of MAPS [10].

$$\varepsilon_t = A_t - F_t \tag{22}$$

where, ε_t is the error for the period t, A_t is the actual value for the period t, F_t is the forecasted value for the period t, then measures of aggregate error:

11.1. Mean Absolute Error

$$MAE = \frac{\sum_{t=1}^{N} |\varepsilon_t|}{N} \tag{23}$$

11.2. Mean Absolute Percentage Error

$$MAPE = \frac{\sum_{t=1}^{N} \left|\frac{\varepsilon_t}{A_t}\right|}{N} \tag{24}$$

11.3. Mean Absolute Deviation

$$MAD = \frac{\sum_{t=1}^{N} |\varepsilon_t - \bar{\varepsilon}_t|}{N} \tag{25}$$

11.4. Percentage Mean Absolute Deviation

$$PMAD = \frac{\sum_{t=1}^{N} |\varepsilon_t - \bar{\varepsilon}_t|}{\sum_{t=1}^{N} |A_t|} \tag{26}$$

11.5. Mean Square Error

$$MSE = \frac{\sum_{t=1}^{N} \varepsilon_t^2}{N} \tag{27}$$

11.6. Root Mean Square Error

$$RMSE = \sqrt{\frac{\sum_{t=1}^{N} \varepsilon_t^2}{N}} \tag{28}$$

where N represents the number of observations used for analysis.

12. Conclusions

In the presented work a study of different price forecasting methodologies is done in the deregulated environment. The restructuring of power markets has created an increasing need to forecast accurate future prices among the market participants with the purpose of profit maximization. Price forecasting is a difficult task due to special characteristic of price series such as non-constant mean and variance, outliers, seasonal and calendar effects.

Electricity price forecasting models includes statistical and non statistical models. Time series models, econometric models and intelligent systems methods are three main statistical models. Non-statistical methods include equilibrium analysis and simulation methods. Methods based on time series are more commonly used for electricity price forecasting due to their flexibility and ease of implementation. The main drawback of time series models is that they are usually based on the hypothesis of stationarity, whereas the price series violates this assumption.

The scope of forecast (e.g. price profile, or its volatility etc.) is an important factor for the selection and design of forecasting models/techniques. The complexity of model(s) also largely determines the number of required input data. Depending on the target of forecast, the procedure may apply data filtering and transformation before the model is optimized for the given price data. Wavelet transform is generally used for smoothening the price data, and removing seasonal effect, outliers and other irregularity effects, the result of the approximated series under wavelet transform is better than the original price data and more stable mean and variance with no outliers.

It can be concluded that there is no universal tool for price forecasting which can be used for every market and operator. For specific applications it becomes essential to select the specific tool/techniques, and following points should be kept in mind:

1) Type of forecast (*i.e.* long term, medium term, short term).
2) Available resources for processing, storing the historical data of the price.
3) Importance of accuracy in forecasting.

By combining wavelet transform with ARIMA, GARCH, Neural Network and other models, the performance characteristics of these models can be increased by reducing forecasting errors.

References

[1] Hu, L., Taylor, G., Wan, H.-B. and Irving, M. (2009) A Review of Short-Term Electricity Price Forecasting Techniques in Deregulated Electricity Markets. *Universities Power Engineering Conference* (*UPEC*), 2009 *Proceedings of the* 44*th International*, Glasgow, 1-4 September 2009, 1-5.

[2] Aggarwal, S.K., Saini, L.M. and Kumar, A. (2009) Electricity Price Forecasting in Deregulated Markets: A Review and Evaluation. *International Journal of Electrical Power & Energy Systems*, **31**, 13-22. http://dx.doi.org/10.1016/j.ijepes.2008.09.003

[3] Taylor, G., Irving, M. and Hu, L.L. (2008) A Fuzzy-Logic Based Bidding Strategy for Participants in the UK Electricity Market. Padova, 1-5.

[4] Hu, L. (2006) Optimal Bidding Strategy for Power Producers in the UK Electricity Market. MPhil Thesis, School of Engineering and Design, Brunel University, London.

[5] Bunn, D.W. (2000) Forecasting Loads and Prices in Competitive Power Markets. *Proceedings of the IEEE*, **88**, 163-169. http://dx.doi.org/10.1109/5.823996

[6] González, A.M., Roque, A.M.S. and García-González, J. (2005) Modeling and Forecasting Electricity Prices with Input/Output Hidden Markov Models. *IEEE Transactions on Power Systems*, **20**, 13-24. http://dx.doi.org/10.1109/TPWRS.2004.840412

[7] Georgilakis, P.S. (2006) Market Clearing Price Forecasting in Deregulated Electricity Markets Using Adaptively Trained Neural Networks. *Advances in Artificial Intelligence*, *Lecture Notes in Computer Science*, **3955**, 56-66.

[8] Mandal, P., Senjyu, T. and Funabashi, T. (2006) Neural Networks Approach to Forecast Several Hour Ahead Electricity Prices and Loads in Deregulated Market. *Energy Conversion and Management*, **47**, 2128-2142. http://dx.doi.org/10.1016/j.enconman.2005.12.008

[9] Vucetic, S., Tomsovic, K. and Obradovic, Z. (2001) Discovering Price-Load Relationships in California's Electricity Market. *IEEE Transactions on Power Systems*, **16**, 280-286. http://dx.doi.org/10.1109/59.918299

[10] Bastian, J., Zhu, J., Banunarayanan, V. and Mukerji, R. (1999) Forecasting Energy Prices in a Competitive Market. *IEEE Computer Applications in Power*, **12**, 40-45. http://dx.doi.org/10.1109/67.773811

[11] Kian, A. and Keyhani, A. (2001) Stochastic Price Modeling of Electricity in Deregulated Energy Markets. *Proceedings of the* 34*th Annual Hawaii International Conference on System Sciences*, Maui, 6 January 2001, 7.

[12] Nogales, F.J., Contreras, J., Conejo, A.J. and Espinola, R. (2002) Forecasting Next-Day Electricity Prices by Time Series Models. *IEEE Transactions on Power Systems*, **17**, 342-348. http://dx.doi.org/10.1109/TPWRS.2002.1007902

[13] Szkuta, B.R., Sanabria, L.A. and Dillon, T.S. (1999) Electricity Price Short-Term Forecasting Using Artificial Neural Networks. *IEEE Transactions on Power Systems*, **14**, 851-857. http://dx.doi.org/10.1109/59.780895

[14] Zhang, L. and Luh, P.B. (2005) Neural Network-Based Market Clearing Price Prediction and Confidence Interval Estimation with an Improved Extended Kalman Filter Method. *IEEE Transactions on Power Systems*, **20**, 59-66. http://dx.doi.org/10.1109/TPWRS.2004.840416

[15] Lucarella, D., Venturini, A., Canazza, V., Li, G. and Liu, C.C. (2005) An Intelligent System for Price Forecasting Accuracy Assessment. *Proceedings of the* 13*th International Conference on Intelligent Systems Application to Power Systems*, Arlington, 6-10 November 2005, 92-99.

[16] Amjady, N. and Hemmati, M. (2006) Energy Price Forecasting—Problems and Proposals for Such Predictions. *IEEE Power and Energy Magazine*, **4**, 20-29. http://dx.doi.org/10.1109/MPAE.2006.1597990

[17] Zhou, M., Yan, Z., Ni, Y.X., Li, G. and Nie, Y. (2006) Electricity Price Forecasting with Confidence-Interval Estimation through an Extended ARIMA Approach. *IEE Proceedings—Generation, Transmission and Distribution*, **153**, 187. http://dx.doi.org/10.1049/ip-gtd:20045131

[18] Nicolaisen, J.D., Richter, C.W. and Sheble, G.B. (2000) Price Signal Analysis for Competitive Electric Generation Companies. *International Conference on Electric Utility Deregulation and Restructuring and Power Technologies*, London, 4-7 April 2000, 66-71.

[19] Antunes, J.F., de Souza Araujo, N.V. and Minussi, C.R. (2013) Multinodal Load Forecasting Using an ART-ARTMAP-Fuzzy Neural Network and PSO Strategy. 2013 *IEEE Grenoble PowerTech*, Grenoble, 16-20 June 2013, 1-6. http://dx.doi.org/10.1109/ptc.2013.6652373

[20] Niimura, T. (2006) Forecasting Techniques for Deregulated Electricity Market Prices—Extended Survey. 2006 *IEEE PES Power Systems Conference and Exposition*, Atlanta, 29 October-1 November 2006, 51-56. http://dx.doi.org/10.1109/psce.2006.296248

[21] Ramos, J.L.M., Exposito, A.G., Santos, J.M.R., Lora, A.T. and Guerra, A.R.M. (2002) Influence of ANN-Based Market Price Forecasting Uncertainty on Optimal Bidding. *PSCC Power Systems Computation Conference*, Seville, 24-28 June 2002, 24-28.

[22] Rashidi-Nejad, M., Gharaveisi, A., Khajehzadeh, A. and Salehizadeh, M. (2006) Eelctricity Price Forecasting Using WaveNet. 2006 *Large Engineering Systems Conference on Power Engineering*, Halifax, 26-28 July 2006, 131-137. http://dx.doi.org/10.1109/lescpe.2006.280375

[23] Osorio, G.J., Pousinho, H.M.I., Matias, J.C.O. and Catalao, J.P.S. (2012) Intelligent Approach for Forecasting in Power Engineering Systems. 2012 *IEEE 16th International Conference on Intelligent Engineering Systems*, Lisbon, 13-15 June 2012, 297-302. http://dx.doi.org/10.1109/ines.2012.6249848

[24] Negnevitsky, M., Mandal, P. and Srivastava, A.K. (2009) An Overview of Forecasting Problems and Techniques in Power Systems. *IEEE Power & Energy Society General Meeting*, Calgary, 26-30 July 2009, 1-4. http://dx.doi.org/10.1109/pes.2009.5275480

[25] Mori, H. and Kosemura, N. (2002) A Data Mining Method for Short-Term Load Forecasting in Power Systems. *Electrical Engineering in Japan*, **139**, 12-22. http://dx.doi.org/10.1002/eej.1150

[26] Mandal, P., Senjyu, T., Urasaki, N., Funabashi, T. and Srivastava, A.K. (2007) A Novel Approach to Forecast Electricity Price for PJM Using Neural Network and Similar Days Method. *IEEE Transactions on Power Systems*, **22**, 2058-2065. http://dx.doi.org/10.1109/TPWRS.2007.907386

[27] Varadan, V., Leung, H. and Bosse, E. (2006) Dynamical Model Reconstruction and Accurate Prediction of Power-Pool Time Series. *IEEE Transactions on Instrumentation and Measurement*, **55**, 327-336. http://dx.doi.org/10.1109/TIM.2005.861492

[28] Mount, T.D., Ning, Y. and Cai, X. (2006) Predicting Price Spikes in Electricity Markets Using a Regime-Switching Model with Time-Varying Parameters. *Energy Economics*, **28**, 62-80. http://dx.doi.org/10.1016/j.eneco.2005.09.008

[29] Sueyoshi, T. and Tadiparthi, G.R. (2005) A Wholesale Power Trading Simulator with Learning Capabilities. *IEEE Transactions on Power Systems*, **20**, 1330-1340. http://dx.doi.org/10.1109/TPWRS.2005.851948

[30] Batlle, C. and Barquin, J. (2005) A Strategic Production Costing Model for Electricity Market Price Analysis. *IEEE Transactions on Power Systems*, **20**, 67-74. http://dx.doi.org/10.1109/TPWRS.2004.831266

[31] Lu, X., Dong, Z.Y. and Li, X. (2005) Electricity Market Price Spike Forecast with Data Mining Techniques. *Electric Power Systems Research*, **73**, 19-29. http://dx.doi.org/10.1016/S0378-7796(04)00125-7

[32] Figueiredo, M., Ballini, R., Soares, S., Andrade, M. and Gomide, F. (2004) Learning Algorithms for a Class of Neurofuzzy Network and Application. *IEEE Transactions on Systems, Man, and Cybernetics—Part C: Applications and Reviews*, **34**, 293-301.

[33] Luh, P.B. and Guo, J.-J. (2003) Selecting Input Factors for Clusters of Gaussian Radial Basis Function Networks to Improve Market Clearing Price Prediction. *IEEE Transactions on Power Systems*, **18**, 665-672. http://dx.doi.org/10.1109/TPWRS.2003.811012

[34] Hong, Y.-Y. and Hsiao, C.-Y. (2002) Locational Marginal Price Forecasting in Deregulated Electricity Markets Using Artificial Intelligence. *IEE Proceedings—Generation, Transmission and Distribution*, **149**, 621. http://dx.doi.org/10.1049/ip-gtd:20020371

[35] Davison, M., Anderson, C.L., Marcus, B. and Anderson, K. (2002) Development of a Hybrid Model for Electrical Power Spot Prices. *IEEE Transactions on Power Systems*, **17**, 257-264. http://dx.doi.org/10.1109/TPWRS.2002.1007890

[36] Martinez, R., Srivastava, A.K., Mandal, P., Haque, A.U. and Meng, J. (2012) A Hybrid Intelligent Algorithm for Short-Term Energy Price Forecasting in the Ontario Market. 2012 *IEEE Power and Energy Society General Meeting*, San Diego, 22-26 July 2012, 1-7.

[37] Zhang, B.-L. and Dong, Z.-Y. (2001) An Adaptive Neural-Wavelet Model for Short Term Load Forecasting. *Electric Power Systems Research*, **59**, 121-129. http://dx.doi.org/10.1016/S0378-7796(01)00138-9

[38] Huang, C.-M. and Yang, H.-T. (2001) Evolving Wavelet-Based Networks for Short-Term Load Forecasting. *IEE Proceedings—Generation, Transmission and Distribution*, **148**, 222. http://dx.doi.org/10.1049/ip-gtd:20010286

[39] Conejo, A.J., Contreras, J., Espínola, R. and Plazas, M.A. (2005) Forecasting Electricity Prices for a Day-Ahead Pool-Based Electric Energy Market. *International Journal of Forecasting*, **21**, 435-462. http://dx.doi.org/10.1016/j.ijforecast.2004.12.005

[40] Conejo, A.J., Plazas, M.A., Espinola, R. and Molina, A.B. (2005) Day-Ahead Electricity Price Forecasting Using the Wavelet Transform and ARIMA Models. *IEEE Transactions on Power Systems*, **20**, 1035-1042. http://dx.doi.org/10.1109/TPWRS.2005.846054

[41] Zheng, G., Yan, X.-M., Li, H.-W. and Liu, D. (2004) Classification of Voltage Sag Based on Wavelet Transform and Wavelet Network. *Proceedings of* 2004 *International Conference on Machine Learning and Cybernetics*, **1**, 466-470. http://dx.doi.org/10.1109/icmlc.2004.1380734

[42] Mallat, S.G. (1989) A Theory for Multiresolution Signal Decomposition: The Wavelet Representation. *IEEE Transactions on Pattern Analysis and Machine Intelligence*, **11**, 674-693. http://dx.doi.org/10.1109/34.192463

[43] Xu, H.T. and Niimura, T. (2004) Short-Term Electricity Price Modeling and Forecasting Using Wavelets and Multivariate Time Series. 858-862.

[44] Atiya, A., Talaat, N. and Shaheen, S. (1997) An Efficient Stock Market Forecasting Model Using Neural Networks. *Proceedings of International Conference on Neural Networks*, Vol. 4, Houston, 9-12 June 1997, 2112-2115. http://dx.doi.org/10.1109/icnn.1997.614231

[45] Mishra, S., Sharma, A. and Panda, G. (2011) Wind Power Forecasting Model Using Complex Wavelet Theory. 2011 *International Conference on Energy, Automation and Signal*, Bhubaneswar, 28-30 December 2011, 1-4. http://dx.doi.org/10.1109/iceas.2011.6147151

[46] Al Wadia, M.T.I.S. and Tahir Ismail, M. (2011) Selecting Wavelet Transforms Model in Forecasting Financial Time Series Data Based on ARIMA Model. *Applied Mathematical Sciences*, **5**, 315-326.

[47] Contreras, J., Espinola, R., Nogales, F.J. and Conejo, A.J. (2003) ARIMA Models to Predict Next-Day Electricity Prices. *IEEE Transactions on Power Systems*, **18**, 1014-1020. http://dx.doi.org/10.1109/TPWRS.2002.804943

[48] Weiss, E. (2000) Forecasting Commodity Prices Using ARIMA. *Technical Analysis of Stocks & Commodities*, **18**, 18-19.

[49] Chinn, M.D., LeBlanc, M. and Coibion, O. (2001) The Predictive Characteristics of Energy Futures: Recent Evidence for Crude Oil, Natural Gas, Gasoline and Heating Oil. UCSC Dept. of Economics Working Paper No. 490.

[50] Hagan, M.T. and Behr, S.M. (1987) The Time Series Approach to Short Term Load Forecasting. *IEEE Transactions on Power Systems*, **2**, 785-791. http://dx.doi.org/10.1109/TPWRS.1987.4335210

[51] Hamilton, J.D. (1994) Time Series Analysis, 2. Princeton University Press, Princeton.

[52] Enders, W. (2008) Applied Econometric Time Series. John Wiley & Sons, Hoboken.

[53] Garcia, R.C., Contreras, J., van Akkeren, M. and Garcia, J.B.C. (2005) A GARCH Forecasting Model to Predict Day-Ahead Electricity Prices. *IEEE Transactions on Power Systems*, **20**, 867-874. http://dx.doi.org/10.1109/TPWRS.2005.846044

[54] Bollerslev, T. and Ole Mikkelsen, H. (1996) Modeling and Pricing Long Memory in Stock Market Volatility. *Journal of Econometrics*, **73**, 151-184. http://dx.doi.org/10.1016/0304-4076(95)01736-4

[55] Tan, Z., Zhang, J., Wang, J. and Xu, J. (2010) Day-Ahead Electricity Price Forecasting Using Wavelet Transform Combined with ARIMA and GARCH Models. *Applied Energy*, **87**, 3606-3610. http://dx.doi.org/10.1016/j.apenergy.2010.05.012

[56] Baba, N. and Kozaki, M. (1992) An Intelligent Forecasting System of Stock Price Using Neural Networks. *International Joint Conference on Neural Networks*, Vol. 1, Baltimore, 7-11 June 1992, 371-377. http://dx.doi.org/10.1109/ijcnn.1992.287183

[57] Snyder, J., Sweat, J., Richardson, M. and Pattie, D. (1992) Developing Neural Networks to Forecast Agricultural Commodity Prices. *Proceedings of the 25th Hawaii International Conference on System Sciences*, Vol. 4, Kauai, 7-10 Jan 1992, 516-522. http://dx.doi.org/10.1109/hicss.1992.183442

[58] Mandal, P., Senjyu, T., Uezato, K. and Funabashi, T. (2005) Several-Hours-Ahead Electricity Price and Load Forecasting Using Neural Networks. *IEEE Power Engineering Society General Meeting*, San Francisco, 12-16 June 2005,

2205-2212. http://dx.doi.org/10.1109/pes.2005.1489530

[59] Amjady, N. and Daraeepour, A. (2008) Day-Ahead Electricity Price Forecasting Using the Relief Algorithm and Neural Networks. *5th International Conference on European Electricity Market*, Lisboa, 28-30 May 2008, 1-7. http://dx.doi.org/10.1109/eem.2008.4579109

[60] Haykin, S. (2004) Neural Networks: A Comprehensive Foundation. 2nd Edition.

[61] Srinivasan, D., Yong, F.C. and Liew, A.C. (2007) Electricity Price Forecasting Using Evolved Neural Networks. *International Conference on Intelligent Systems Applications to Power Systems*, Toki Messe, 5-8 November 2007, 1-7. http://dx.doi.org/10.1109/isap.2007.4441660

[62] Mori, H. and Awata, A. (2006) A Hybrid Method of Clipping and Artificial Neural Network for Electricity Price Zone Forecasting. *International Conference on Probabilistic Methods Applied to Power Systems*, Stockholm, 11-15 June 2006, 1-6. http://dx.doi.org/10.1109/pmaps.2006.360234

[63] Azevedo, F. and Vale, Z. (2006) Forecasting Electricity Prices with Historical Statistical Information Using Neural Networks and Clustering Techniques. 2006 *IEEE PES Power Systems Conference and Exposition*, Atlanta, 29 October-1 November 2006, 44-50. http://dx.doi.org/10.1109/psce.2006.296247

[64] Zhang, X., Wang, X.-F., Chen, F.-H., Ye, B. and Chen, H.-Y. (2005) Short-Term Electricity Price Forecasting Based on Period-Decoupled Price Sequence. *Proceedings of CSEE*, **25**, 1-6.

[65] Luh, P.B., Kasiviswanathan, K. and Zhang, L. (2003) Energy Clearing Price Prediction and Confidence Interval Estimation with Cascaded Neural Networks. *IEEE Transactions on Power Systems*, **18**, 99-105. http://dx.doi.org/10.1109/TPWRS.2002.807062

[66] Bashir, Z.A. and El-Hawary, M.E. (2009) Applying Wavelets to Short-Term Load Forecasting Using PSO-Based Neural Networks. *IEEE Transactions on Power Systems*, **24**, 20-27. http://dx.doi.org/10.1109/TPWRS.2008.2008606

[67] Singh, N.K., Tripathy, M. and Singh, A.K. (2011) A Radial Basis Function Neural Network Approach for Multi-Hour Short Term Load-Price Forecasting with Type of Day Parameter. 2011 *6th IEEE International Conference on Industrial and Information Systems*, Kandy, 16-19 August 2011, 316-321. http://dx.doi.org/10.1109/iciinfs.2011.6038087

[68] Jang, J.-S.R. (1993) ANFIS: Adaptive-Network-Based Fuzzy Inference System. *IEEE Transactions on Systems, Man, and Cybernetics*, **23**, 665-685. http://dx.doi.org/10.1109/21.256541

[69] Hong, Y.-Y. and Lee, C.-F. (2005) A Neuro-Fuzzy Price Forecasting Approach in Deregulated Electricity Markets. *Electric Power Systems Research*, **73**, 151-157. http://dx.doi.org/10.1016/j.epsr.2004.07.002

[70] Wang, L.-X. and Mendel, J.M. (1992) Generating Fuzzy Rules by Learning from Examples. *IEEE Transactions on Systems, Man, and Cybernetics*, **22**, 1414-1427. http://dx.doi.org/10.1109/21.199466

[71] Nauck, D.D. (2000) Data Analysis with Neuro-Fuzzy Methods.

[72] Mandal, P., Haque, A.U., Meng, J., Srivastava, A.K. and Martinez, R. (2013) A Novel Hybrid Approach Using Wavelet, Firefly Algorithm, and Fuzzy ARTMAP for Day-Ahead Electricity Price Forecasting. *IEEE Transactions on Power Systems*, **28**, 1041-1051. http://dx.doi.org/10.1109/TPWRS.2012.2222452

[73] Ul Haque, A. and Meng, J. (2011) Short-Term Wind Speed Forecasting Based on Fuzzy Artmap. *International Journal of Green Energy*, **8**, 65-80. http://dx.doi.org/10.1080/15435075.2010.529784

[74] Lopes, M.L.M., Minussi, C.R. and Lotufo, A.D.P. (2005) Electric Load Forecasting Using a Fuzzy ART&ARTMAP Neural Network. *Applied Soft Computing*, **5**, 235-244. http://dx.doi.org/10.1016/j.asoc.2004.07.003

[75] Christodoulou, C. and Georgiopoulos, M. (2000) Applications of Neural Networks in Electromagnetics. Artech House, Inc., Norwood.

[76] Dagher, I., Georgiopoulos, M., Heileman, G.L. and Bebis, G. (1999) An Ordering Algorithm for Pattern Presentation in Fuzzy ARTMAP That Tends to Improve Generalization Performance. *IEEE Transactions on Neural Networks*, **10**, 768-778. http://dx.doi.org/10.1109/72.774217

[77] Zornetzer, S.F., Davis, J.L. and Lau, C. (1990) An Introduction to Neural and Electronic Networks. Academic Press Professional, Inc., Waltham.

[78] Osborne, M.J. and Rubinstein, A. (1994) A Course in Game Theory. MIT Press, Cambridge, MA.

[79] Bajpai, P. and Singh, S.N. (2004) Bidding and Gaming in Electricity Market: An Overview and Key Issues. *Proceedings of National Power System Conference* (*NPSC*), Chennai, 27-30 December 2004, 338-346.

[80] Kleindorfer, P.R., Wu, D.-J. and Fernando, C.S. (2001) Strategic Gaming in Electric Power Markets. *European Journal of Operational Research*, **130**, 156-168. http://dx.doi.org/10.1016/S0377-2217(00)00048-5

[81] David, A.K. and Wen, F.S. (2000) Strategic Bidding in Competitive Electricity Markets: A Literature Survey. 2000 *Power Engineering Society Summer Meeting*, Vol. 4, Seattle, 16-20 July 2000, 2168-2173. http://dx.doi.org/10.1109/pess.2000.866982

[82] Hobbs, B.F., Metzler, C.B. and Pang, J.-S. (2000) Strategic Gaming Analysis for Electric Power Systems: An MPEC Approach. *IEEE Transactions on Power Systems*, **15**, 638-645. http://dx.doi.org/10.1109/59.867153

[83] Pepyne, D.L., Guan, X.H. and Ho, Y.-C. (2001) Gaming and Price Spikes in Electric Power Markets. *IEEE Transactions on Power Systems*, **16**, 402-408. http://dx.doi.org/10.1109/59.932275

18

Study and Mitigation of Subsynchronous Oscillations with SSC Based SSSC

Mohan P. Thakre*, Vijay S. Kale, Koteswara Raju Dhenuvakonda*, Bhimrao S. Umre, Anjali S. Junghare

Department of Electrical Engineering, Visvesvaraya National Institute of Technology, Nagpur, India
Email: thakre_mohan@yahoo.com

Abstract

This paper proposes a powerful subsynchronous component based (SSC) controller to mitigate the subsynchronous resonance (SSR) with statics synchronous series compensator (SSSC). The mitigation of SSR is achieved by increasing the network damping at those frequencies which are close to the torsional frequency of the turbine-generator shaft. The increase of network damping is done by the extraction of subsynchronous component of voltage and current from the measured signal of the system. From the knowledge of subsynchronous components, a series voltage is injected by SSSC into the transmission line to make the subsynchronous current to zero which is the main cause of turbine oscillations. To analyze the effectiveness of the proposed control scheme, IEEE first benchmark model has taken. The results show the accuracy of the proposed control scheme to mitigate the Torque amplification of SSR.

Keywords

Current Control, SSC, SSSC, SSR, Torsional Oscillation, Voltage Source Converter

1. Introduction

Series capacitor compensation has been extensively employed in power system to increase the power transfer capability of long HV and EHV lines, load sharing among parallel lines and enhance the steady state and transient stability limits [1]. This is achieved by the partial compensation of transmission line reactance. However, the use of series compensation may lead to some new problems to power system operation viz the possibility of subsynchronous resonance (SSR), turbine-generator shaft oscillations with bellow the system frequency. Series capacitors may excite subsynchronous oscillations with any fault or disturbance, when the natural frequency of system aligns with the complement of one of the torsional modes of the turbine-generator shaft [2].

*Both of the two authors are the first authors.

SSR is an electric power system condition where the electrical network exchanges energy with a turbine generator at one or more of the natural frequencies of the combined system below the synchronous frequency of the system [3]. Under such condition, a small voltage induced by rotor oscillation can result in large subsynchronous current; this current will produce an oscillatory component of rotor torque whose phase is such that it enhances the rotor oscillations with large magnitude that will damages the turbine shaft [4] [5].

The fast development of modern power electronic devices led to the development of FACTs devices like TCSC, STATCOM and SSSC. A large number of methods and solutions have been addressed by the different authors to avoid the problem of SSR with the concern of FACTS devices [6]-[16]. Irrespective of solution, the main problem is, how fast and accurate estimation of subsynchronous components from the measured signal of system. The damage of turbine shaft due to SSR is avoidable by the design of appropriate protection scheme with the knowledge of subsynchronous component of current and voltage [6].

The voltage sourced converter-based SSSC is essentially an ac voltage source which, with a constant dc voltage and fixed control inputs, would operate only at the selected output frequency, and its output impedance at other frequencies would theoretically be zero. The SSSC considered is a voltage source inverter, and is equipped with a proportional integrator controller (PI) that regulates the generator terminal voltage. In a practical SSSC, the voltage-sourced converter on the dc side is terminated by a finite energy storage capacitor to maintain the desired dc operating voltage. Thus the dc capacitor in effect interacts with the ac system via the operating switch array of the converter. This interaction may conceivably influence the subsynchronous behaviour of a practical SSSC [7].

This paper is organized as follows: a study system model *i.e.*, IEEE first benchmark model with SSSC is introduced in Section 2. In Section 3, mathematical analysis is presented for extraction of subsynchronous component of voltage. This analysis will be helpful in determining the value of voltage injected in series to the line with SSSC. Consequently, in Section 4, the design of subsynchronous controller is depicted. Section 5 shows the parameters of the study system and the specifications of SSSC. In Section 6, simulation results obtained for IEEE first benchmark model with SSSC controller with three-phase fault. The Section 7 concludes the total work of the paper.

2. IEEE First Benchmark Model with SSSC Controller

Figure 1 shows the single-line diagram of IEEE first benchmark model with an SSSC installed downstream the step-up transformer located at the output of the power station. The generated voltage is denoted by v_s and the grid current is denoted by i respectively. The SSSC is modelled as a controlled ideal voltage source. The injected voltage is denoted by v_{SSSC}. In classical control scheme the principle of the SSR mitigation is to replace the fundamental frequency voltage created by (at least a portion of) the inserted fixed capacitor banks by injecting an equal voltage that has been produced by the SSSC. As the capacitive reactance from the capacitor bank is eliminated (or reduced), the electrical resonance of the system becomes shifted, thus avoiding the risk of SSR. The effectiveness of this control strategy has been described in several publications and has been proved both analytically and through real time simulations [12] [13].

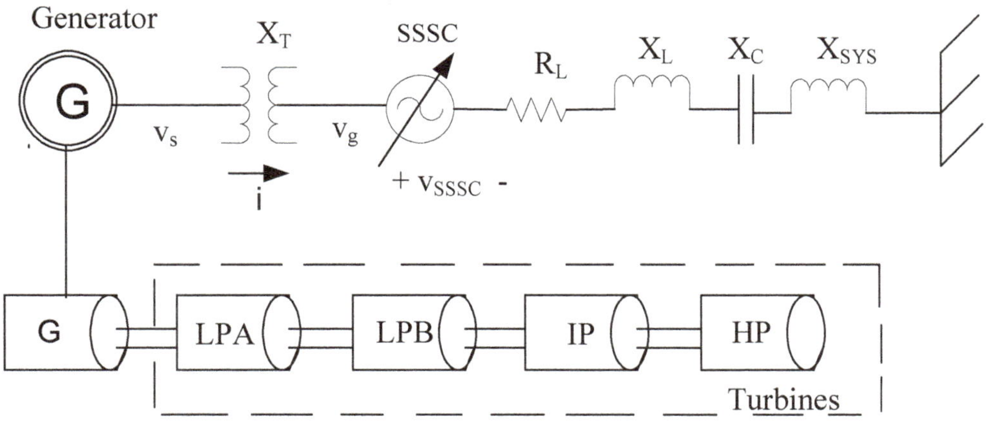

Figure 1. Single-line diagram of power plant with generation unit and SSSC.

3. Sub-Synchronous Component of Voltage

To derive the subsynchronous component of the voltage at the generator terminals, consider the generic case of a synchronous generator connected to a transmission line.

The per-unit voltage at the generator terminals in $\alpha\beta$-plane as

$$v_s^{(\alpha\beta)} = v_{s,\alpha}(t) + jv_{s,\beta}(t) = \omega(t)V_s e^{j(\omega_0 t+\delta(t))} \tag{1}$$

where V_s is the amplitude of voltage at the generator terminals at rated speed, δ is phase displacement $\omega(t)$ is the per-unit rotor Speed and ω_0 is the fundamental frequency expressed in radian per second. The generator rotor oscillates around its fundamental frequency ω_0, its speed can be represented by

$$\omega(t) = \omega_0 + A\sin(\omega_m t) \tag{2}$$

where A is the amplitude of the oscillation and ω_m is the oscillation frequency of the rotor. Substituting (2) in (1), the α-component of the output voltage can be represented as

$$\begin{aligned} v_{s,\alpha}(t) &= \left[\omega_0 + A\sin(\omega_m t)\right]V_s\cos\left[\omega_0 t + \delta(t)\right] \\ &= \omega_0 V_s \cos\left[t+\delta(t)\right] + \frac{AV_s}{2}\left\{-\sin\left[(\omega_0-\omega_m)t+\delta(t)\right] + \sin\left[(\omega_0+\omega_m)t+\delta(t)\right]\right\} \end{aligned} \tag{3}$$

The derivative of the rotor angle is given by

$$\frac{\mathrm{d}}{\mathrm{d}t}\delta(t) = \left[\omega(t)-\omega_0\right]\omega_B = A\sin(\omega_m t)\omega_B \tag{4}$$

where, ω_B is the base frequency in radians per second. Therefore by integrating both sides of Equation (4), we get δ_0 the rotor angle in steady-state condition is derived. The rotor angle can be written as

$$\delta(t) = \delta_0 - A\frac{\omega_B}{\omega_m}\cos(\omega_m t) \tag{5}$$

The α and β components of the output voltage are obtained by Substituting Equation (5) in (3). The α component is obtained as:

$$v_{s,\alpha}(t) = \omega_0 V_s \cos(\omega_0 t+\delta_0) + \frac{AV_s}{2\omega_m}\left\{(\omega_0-\omega_m)\sin\left[(\omega_0-\omega_m)t+\delta_0\right] + (\omega_0+\omega_m)\sin\left[(\omega_0+\omega_m)t+\delta_0\right]\right\} \tag{6}$$

The β component of the voltage is expressed as

$$v_{s,\beta}(t) = \omega_0 V_s \cos(\omega_0 t+\delta_0) + \frac{AV_s}{2\omega_m}\left\{-(\omega_0-\omega_m)\cos\left[(\omega_0-\omega_m)t+\delta_0\right] - (\omega_0+\omega_m)\cos\left[(\omega_0+\omega_m)t+\delta_0\right]\right\} \tag{7}$$

When a small distribution is applied to the generator rotor, the resultant voltage will be constituted by the sum of three terms; they are fundamental frequency component, sub-synchronous component of frequency and super-synchronous component of frequency. The component of super-synchronous frequency is higher than the sub-synchronous frequency component. For super-synchronous frequency the network presents a small positive damping, thus the super synchronous voltage does not represent a risk for the power plant. Therefore super synchronous component of the voltage will not be taken in to account [6].

From Equations (6) and (7), the subsynchronous component of the measured voltage is given by

$$v_{s,sub}^{\alpha\beta}(t) = -\frac{AV_s}{2\omega_m}(\omega_0-\omega_m)e^{j\left[(\omega_0-\omega_m)t+\delta_0+\frac{\pi}{2}\right]} \tag{8}$$

The grid voltage vector can be transformed in the synchronous reference plane as

$$v_s^{dq}(t) = v_s^{(\alpha\beta)}(t)e^{-j\omega_0 t} = v_{s,f}^{(dq)}(t) + v_{s,sub}^{(dq)}(t) \tag{9}$$

where $v_{s,f}^{(dq)}$ is the d-q voltage vector at the fundamental frequency and the sub-synchronous frequency component is written as

$$v_{s,sub}^{(dq)}(t) = v_{s,sub}^{(\alpha\beta)}(t)\mathrm{e}^{-j\omega_0 t} = -\frac{AV_s}{2\omega_m}(\omega_0 - \omega_m)\mathrm{e}^{-j\left[\omega_m t + \delta_0 + \frac{\pi}{2}\right]} \tag{10}$$

When the generator rotor oscillates around its rated speed, the voltage at the terminals can be expressed in the synchronous dq co-ordinate system as

$$v_s^{dq}(t) = v_{s,f}^{dq}(t) + v_{s,sub}^{(dq)}(t) + v_{s,sub}^{(dq)}(t) \tag{11}$$

The frequency f_{sub} and f_{sup} denotes the sub-synchronous and the super-synchronous component of the measured grid voltage respectively. The network presents a small positive damping for frequencies above the fundamental. Therefore, the super-synchronous component is not taken in this paper. The subsynchronous voltage rotates clockwise in the synchronous reference frame. Consider the generator rotor oscillates with angular frequency ω_m. The dq_m-plane denotes a new set of co-ordinate systems that rotates synchronously with synchronous voltage vector, Equation (10) can be rewritten as

$$v_s^{dq}(t) = v_{s,f}^{dq}(t) + v_{s,sub}^{(dq_m)}(t)\mathrm{e}^{-j\omega_m t} \tag{12}$$

In order to extract the sub-synchronous component from the measured signal, (11) can be rearranged so that $v_{s,f}^{dq}$ and $v_{s,sub}^{(dq)}$ become isolated and then applying low-pass filtering on the resulting expression, the estimation control system (ECS) can be expressed as follows.

$$v_{s,f}^{dq}(t) = H_f(p)\left[v_s^{dq}(t) - v_{s,sub}^{dq}(t)\mathrm{e}^{j(\omega_m t)}\right] \tag{13}$$

$$v_{s,sub}^{dq_m\ dq_m}(t) = H_{sub}(p)\left[v_s^{dq}(t)\mathrm{e}^{j(\omega_m t)} - v_{s,f}^{dq}(t)\mathrm{e}^{j(\omega_m t)}\right] \tag{14}$$

where indicated with p, the operator $\left(\frac{\mathrm{d}}{\mathrm{d}t}\right)$, $H_f(p)$, and $H_{sub}(p)$ represents the transfer function of a low pass filter (LPF) for the fundamental and for the subsynchronous component, respectively. Equation (14) can be written in the synchronous dq-frame as

$$v_{s,sub}^{dq}(t) = H_{sub}(p + j\omega_m)\left[v_s^{dq}(t) - v_{s,f}^{dq}(t)\right] \tag{15}$$

Equations (13) and (15) can thus be combined together in order to extract the fundamental and the subsynchronous components of the measured voltage. **Figure 2** shows the block diagram of the LPF-based estimation of subsynchronous components.

4. Subsynchronous Controller

Consider the generator is modeled as an ideal voltage source behind the sub-transient inductance of the generator. To make the subsynchronous current to zero, the objective of the common control system is to produce and inject the subsynchronous component of the internal bus current/voltage by STATCOM/SSSC [9] [12]. Assume that the voltage downstream of the SSSC is equal to zero, *i.e.*, the voltage drop over the impedance downstream the compensator is treated as a disturbance. With the signal references given in **Figure 1**, the law governing the sub-synchronous current controller (SSCC) can be written in the Laplace domain as

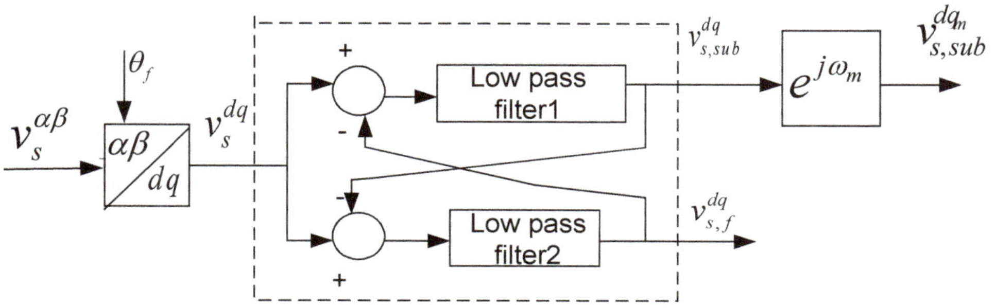

Figure 2. Block diagram of the LPF based estimation of subsynchronous components.

$$v_{SCCC_{sub}}^{(dq_m)}(s) = v_{g,sub}^{dq_m}(s) + \left[R + j(\omega - \omega_m)(L_T + L^{"})\right] i_{sub}^{(dq_m)}(s) + \left(K_p + \frac{K_i}{s}\right)\left[i_{sub}^{(dq_m)}(s) - i_{sub}^{(dq_m)*}(s)\right] \quad (16)$$

where R, L_T and $L^{"}$ are the resistance of the system upstream the SSSC, the leakage inductance of the transformer and the sub transient inductance of the generator, respectively. The current reference is $i_{sub}^{(dq_m)*}$, while K_p and K_i are the proportional and the integral gains of the PI-regulator, respectively.

Figure 3 shows the Blok diagram of Subsynchronous controller in which the measured three-phase voltages are transformed to the $\alpha\beta$-plane and then to the synchronous dq-coordinate hi system by using the transformation angle θ_f. The output of the estimation block is the fundamental and the subsynchronous voltage components, both in the synchronous dq-frame. The subsynchronous component is further transformed in the subsynchronous dq_m-coordinate systems using the transform angle θ_m, obtaining by integrating oscillating frequency ω_m. The resulting quantities are then sent to the subsynchronous component controller (SSCC). The result of SSCC is again converted into $\alpha\beta$ and then in to. These are given to the PWM generator. From PWM generator gate pulses are taken and given to the three single phase VSC based bridges *i.e.* SSSC.

5. Study System Parameters

The system investigated for the study is the well-known IEEE first benchmark model. The system consists of 892.4 MVA turbine-generator connected to an infinite bus through radial series compensated line. The voltage and frequency are 539 KV and 60 Hz respectively. Program has been written to figure out turbine natural frequencies [2] [4]. Here five mass systems has been taken in to study, the obtained natural frequencies are 1.8002 Hz, 16.1335 Hz, 24.4785 Hz, 32.237 Hz, 47.4563 Hz. For 55% series compensation the resonant frequency is 28.14 Hz [5]. **Tables 1-3** give the complete parameters of the IEEE first benchmark model with transmission line.

Rating of SSSC

In this paper three-phase VSC based bridge is used for SSSC. The amount of power needed for mitigation of SSR is related to several factors (such as series-compensation level, fault duration and its location) that cannot be predicted accurately. As the sub-synchronous frequency component of voltage and current is low, the rating of SSSC is low (0.1%). The voltage rating is 8 KV (either DC source or capacitor). The power rating is 12 MVA. The results are obtained for active power 0.1 pu [12]. The results are shown with and without SSSC.

6. Mat Lab Simulation Circuit and Results

To know the effectiveness of the proposed control strategy to mitigate SSR due to Torque Amplification, the IEEE FBM with SSSC has been simulated using the Matlab-Simulink. **Figure 4** shows the MAT Lab simulation circuit with the control circuit along with SSSC. The SSSC injects the voltage in series with the line according to the firing angle given by the firing angle generator. The firing angle control is obtained from the subsynchronous component estimation control circuit. For the value of 55% compensation, a three-phase fault is applied at 1sec to the grid. The fault clearing time has been set to 0.05 s. Due to the unstable mode, when the fault is cleared, large oscillations will be experienced between the different sections of the turbine-generator shaft

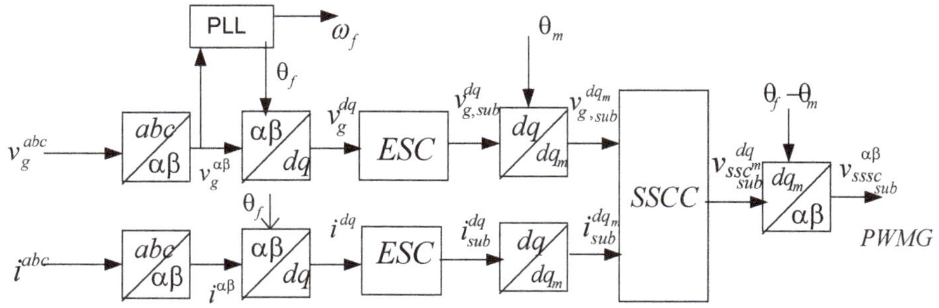

Figure 3. Block diagram of subsynchronous controller.

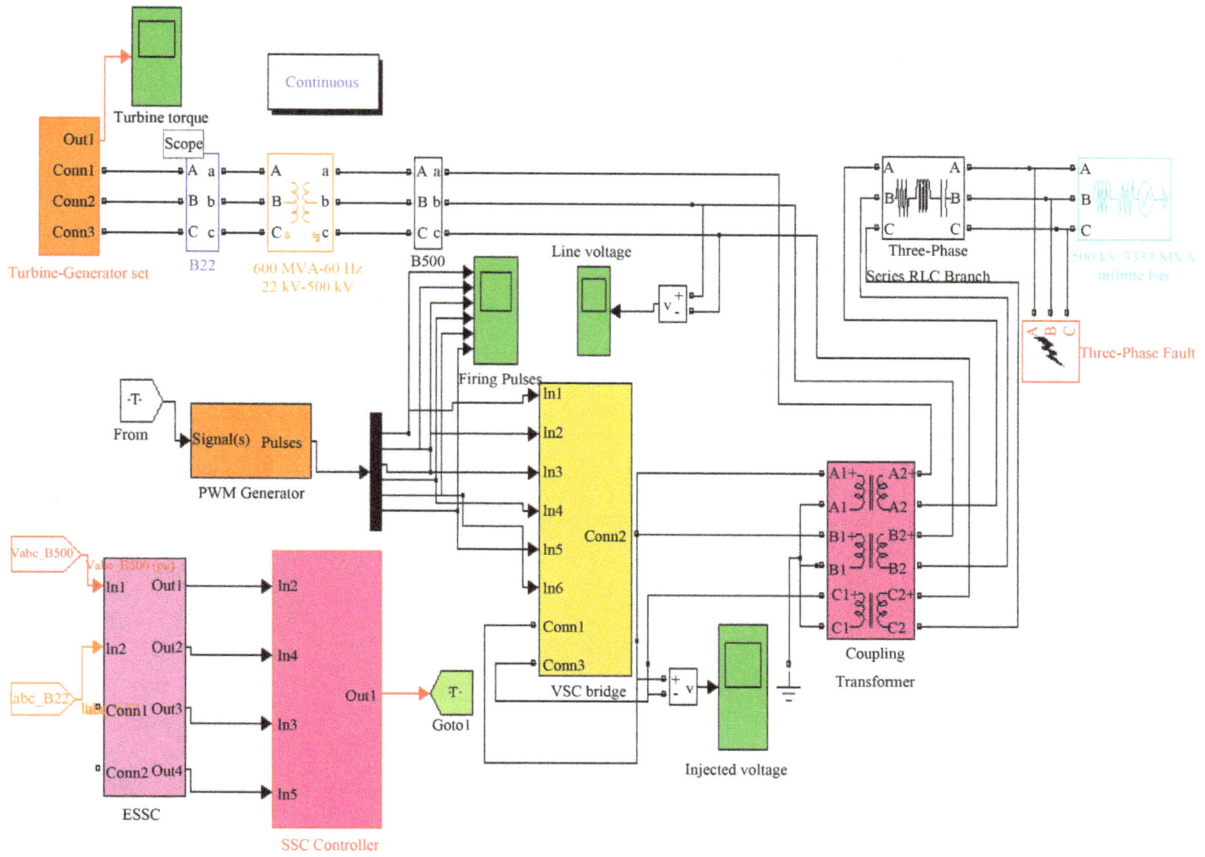

Figure 4. Simulation circuit of IEEE first benchmark model with SSSC.

Table 1. IEEE first benchmark network parameters.

Network resistance	R_L	0.0113 pu
Transformer reactance	X_T	0.142 pu
Transformation ratio		22/539 KV
Line reactance	X_L	0.50 pu
Transmission line reactance	X_{sys}	0.08 pu

Table 2. Synchronous machine parameters.

Reactance	Value [per unit]	Time constant	Value [sec]
X_a	0.130	T'_{d0}	4.3
X_d	1.79	T''_{d0}	0.032
X'_d	0.169	T'_{q0}	0.85
X''_d	0135	T''_{q0}	0.05
X_q	1.71		
X'_q	0.228		
X''_q	0.200		

shown in **Figure 5**. **Figure 6** shows the electromagnetic torque and rotor speed both are increasing drastically will lead to shaft damage of turbine-generator shaft.

Table 3. IEEE first benchmark shaft parameters.

Inertia	$H\ [s^{-1}]$	Shaft section	Spring constant [pu·T/rad]
HP turbine	0.092897	HP-IP	19.303
IP turbine	0.155589	IP-LPA	34.929
LPA turbine	0.858670	LPA-LPB	52.038
LPB turbine	0.884215	LPB-GEN	70.858
Generator	0.868495		
HP turbine	0.092897	HP-IP	19.303

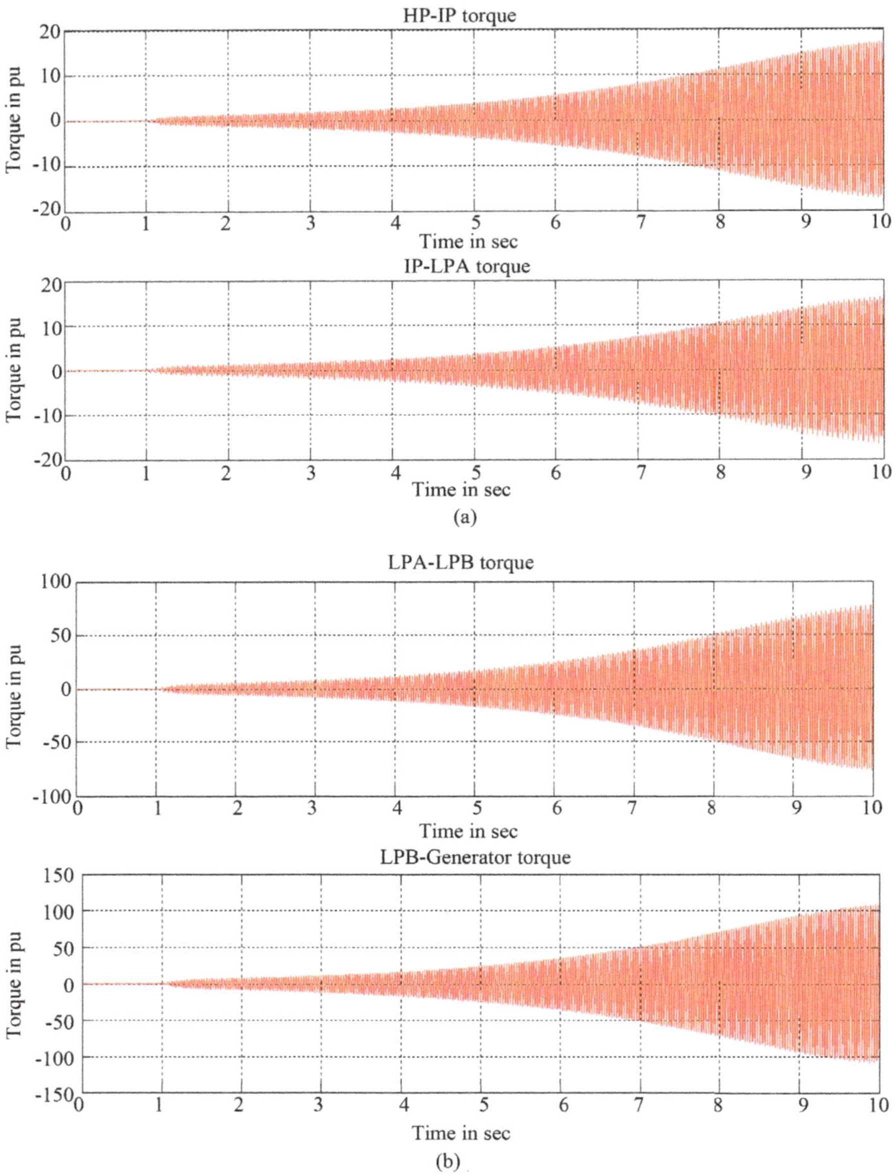

Figure 5. Simulated turbine-generator shaft torques for IEEE first benchmark model without SSSC. (a) HP-IP and IP-LPA Torques without SSSC; (b) LPA-LPB and LPB-generator torques without SSSC.

To avoid the shaft damage of torque amplification effect due to SSR, SSSC is connected. **Figure 7** and **Figure 8** shows the mitigation of SSR with series injected voltage supplied by SSSC. The voltage injected is very small shown in **Figure 9**. Approximately 250 Volts this is the main achievement of the proposed control system. Because of series injected voltage, the turbine oscillations are reduced to such a value which will not damage the turbine-generator shaft, thus avoiding the risk of SSR.

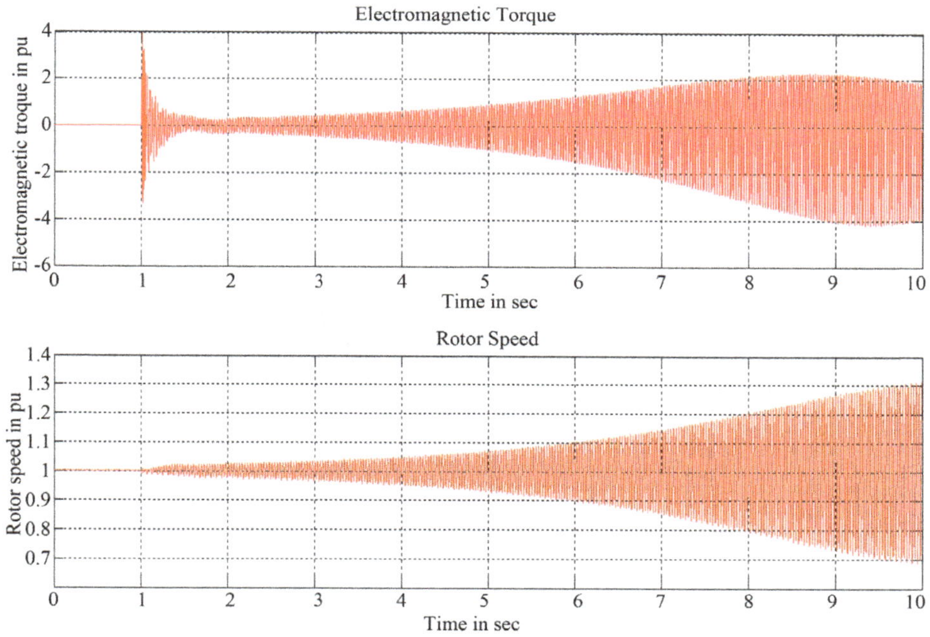

Figure 6. Simulated electromagnetic torque and speed for IEEE first benchmark model without SSSC.

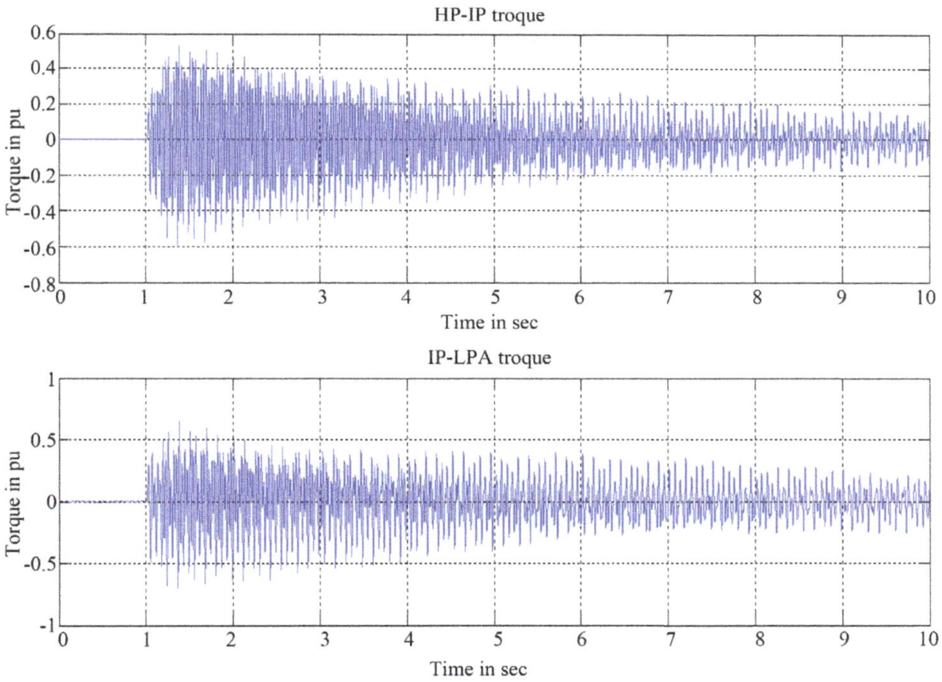

(a)

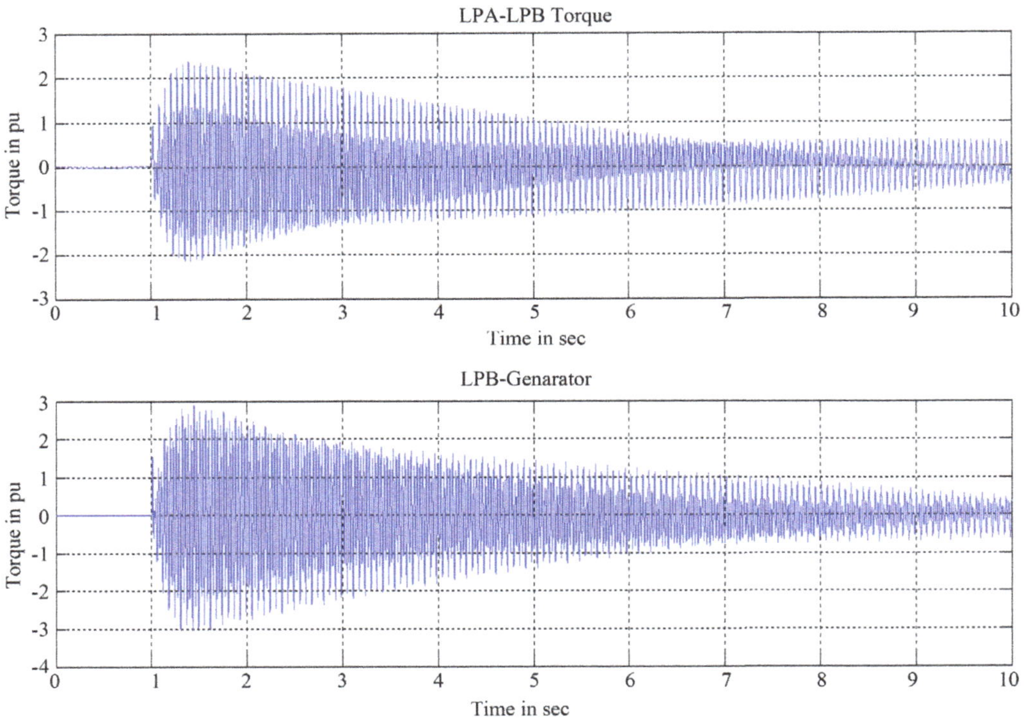

(b)

Figure 7. Simulated turbine-generator shaft torques for IEEE first benchmark model with SSSC. (a) HP-IP and IP-LPA torques with SSSC; (b) LPA-LPB and LPB-generator torques with SSSC.

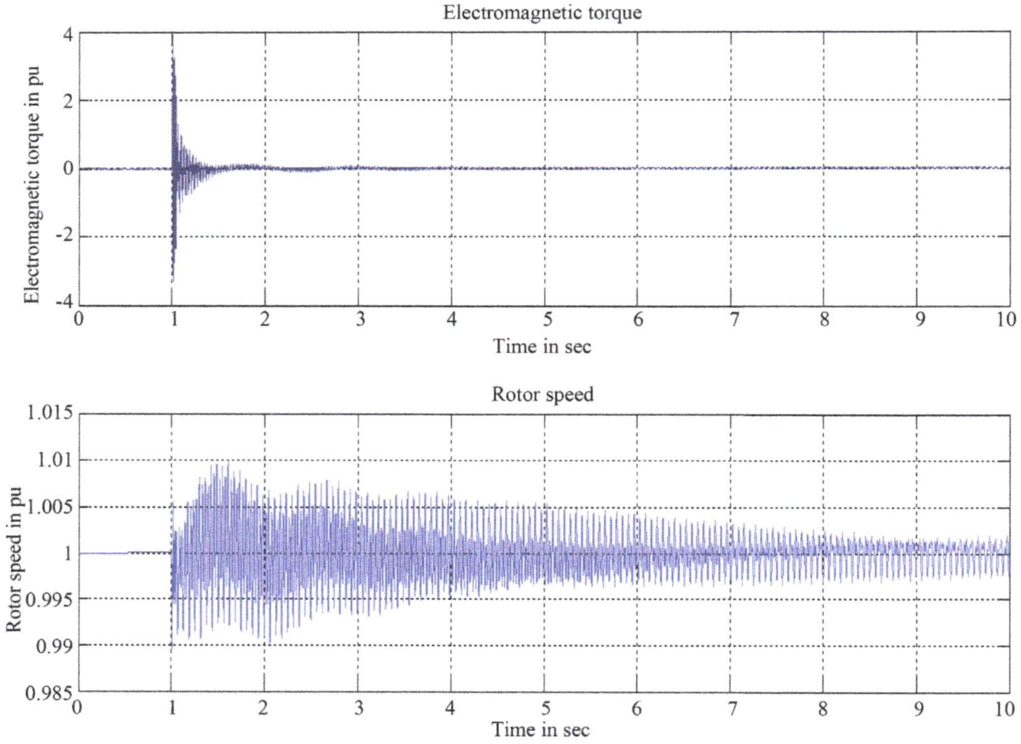

Figure 8. Simulated electromagnetic torque and speed for IEEE first benchmark model with SSSC.

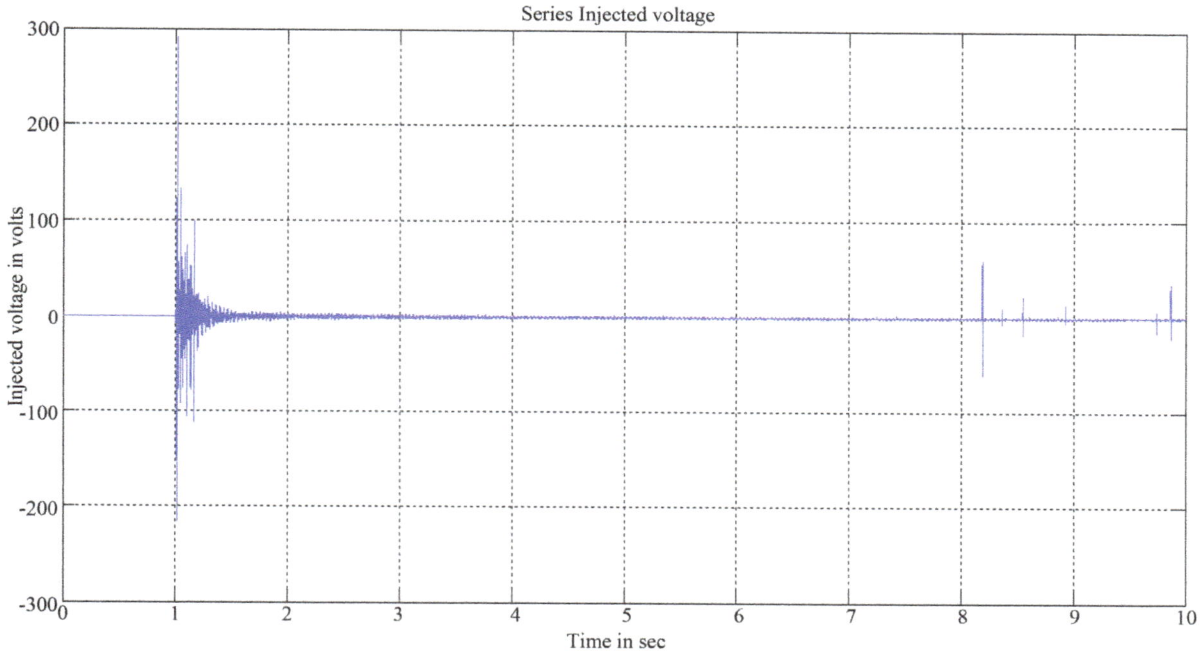

Figure 9. Series injected voltage of SSSC during three-phase fault.

7. Conclusion

In this Research work, an accurate SSC based control scheme is proposed to mitigate the oscillations due to SSR with SSSC. The SSSC is constituted by three-phase VSC connected in series with the power line. Based on the control scheme the SSR mitigation is obtained by increasing the network damping at those frequencies which are close to the natural mode frequencies of the turbine-generator shaft. In the control scheme the estimation of subsynchronous components are proposed and are used for SSR mitigation by injecting the voltage in series with the line by SSSC. It has been shown that SSR mitigation is achieved by injecting a low amount of voltage in the grid, leading to reduced power rating for the SSSC. Finally, simulation results have shown the effectiveness of the proposed control scheme.

References

[1] Anderson, P. and Farmer, R. (1996) Series Compensation of Power Systems. PBLSH, Encinita.

[2] Anderson, P.M., Agrawal, B.L. and Ness, J.V. (1989) Subsynchronous Resonance in Power Systems. IEEE Press, New York.

[3] IEEE SSR Working Group (1985) Terms, Definitions and Symbols for Sub-Synchronous Oscillations. *IEEE Transactions on Power Apparatus and Systems*, **PAS-104**, 1326-1334. http://dx.doi.org/10.1109/TPAS.1985.319152

[4] Kundur, P. (1994) Power System Stability and Control. McGraw-Hill, Inc.

[5] IEEE SSR Task Force (1977) First Benchmark Model for Computer Simulation of Subsynchronous Resonance. *IEEE Transactions on Power Apparatus and Systems*, **PAS-96**, 1565-1571.

[6] Bongiorno, M., Svensson, J. and Ängquist, L. (2008) Online Estimation of Sub-Synchronous Voltage Components in Power Systems. *IEEE Transactions on Power Delivery*, **23**, 410-418. http://dx.doi.org/10.1109/TPWRD.2007.905557

[7] Hingorani, N.G. and Gyugyi, L. (2000) Understanding FACTS. IEEE Press, Piscataway.

[8] Padiyar, K.R. and Swayam Prakash, V. (2003) Tuning and Performance Evaluation of Damping Controller for a STATCOM. *International Journal of Electrical Power & Energy Systems*, **25**, 155-166. http://dx.doi.org/10.1016/S0142-0615(02)00029-7

[9] Padiyar, K.R. and Prabhu, N. (2006) Design and Performance Evaluation of Sub-Synchronous Damping Controller with STATCOM. *IEEE Transactions on Power Delivery*, **21**, 1398-1405.

[10] Umre, B.S., Khedkar, M.K., Trupti Hande, M.S. and Modak, J.P. (2007) Application of STATCOM for Reducing

Stresses Due to Torsional Oscillations in Turbine-Generator Shaft. *IEEE Power Electronics Specialists Conference*, 17-21 June 2007, 865-869.

[11] Perkins, B.K. and Iravani, M.R. (1997) Dynamic Modeling of a TCSC with Application to SSR Analysis. *IEEE Transactions on Power Systems*, **12**, 1619-1625. http://dx.doi.org/10.1109/59.627867

[12] Bongiorno, M., Ängquist, L. and Svensson, J. (2008) A Novel Control Strategy for Subsynchronous Resonance Mitigation. *IEEE Transactions on Power Electronics*, **23**, 735-743. http://dx.doi.org/10.1109/TPEL.2007.915178

[13] Prabhu, N. Thirumalaivasan, R. and Janaki, M. (2013) Damping of SSR Using Subsyanchronous Current Suppressor with SSSC. *IEEE Transactions on Power Systems*, **28**, 64-74. http://dx.doi.org/10.1109/TPWRS.2012.2193905

[14] J.A. Castillo J., D. Olguín S., A.R. Messina and C.A. Rivera S. (2007) Analysis and Study of Subsynchronous Torsional Interaction with FACTS Devices. *Electric Power Components and Systems*, **35**, 1233-1253.

[15] Kumar, L.S. and Ghosh, A. (1999) Modeling and Control Design of a Static Synchronous Series Compensator. *IEEE Transactions on Power Delivery*, **14**, 1448-1453. http://dx.doi.org/10.1109/61.796239

[16] Pillai, G., Gosh, A. and Joshi, A. (2001) Robust Control of SSSC to Improve Torsional Damping. *Proc.* 38*th IEEE Power Engineering Society Winter Meeting*, **3**, 1115-1120.

19

Cymbal Structural Optimization for Improving Piezoelectric Harvesting Efficiency with Taguchi's Orthogonal Experiment

Guangqing Shang[1], Wei Ning[2], Chunhua Sun[1*]

[1]Department of Mechanic and Electronic Engineering, Suzhou Vocational University, Suzhou, China
[2]College of Mechanical Engineering, Shanxi University of Technology, Hanzhong, China
Email: [*]chh_sunny@163.com

Abstract

To improve piezoelectric harvesting efficiency of Cymbal, optimization design of Cymbal parameters was studied with the method of Taguchi's orthogonal experiment. The effective factors of piezoelectric harvesting property were firstly analyzed. The orthogonal experiment schedule was then designed. The finite element model of Cymbal was built via ASPL tool in ANSYS software and static analysis was done. The experimental results were gotten with developed program. The optimization level of each factor was gained. Under the synthetical optimization level of each design factor, the piezoelectric analysis was tested and the open voltage of 236.476 V was revealed with improving 35.73% than the maximum voltage of 174.228 V in the orthogonal experiment. The average voltage of 229.98 V was measured with the manufactured optimized Cymbal structure design. The relative error was 2.54% between simulation and measured data. It indicated that the optimization design schedule was reasonable. Cymbal harvester with the optimized parameters could scavenge larger voltage.

Keywords

Energy Harvest, Piezoelectric Effects, Cymbal Harvester, Taguchi's Orthogonal Experiment

1. Introduction

The special structural design of piezoelectric Cymbal harvester makes it possess some special characteristics,

[*]Corresponding author.

such as the ease of fabrication and the ability to tailor performance. The comprehensive piezoelectric effects of d_{33} and d_{31} are easily excited for Cymbal harvester under the axial external force [1]. Besides being used as the actuator and sensor, the piezoelectric Cymbal harvester has recently been focused to be an energy transducer. The harvesting energy properties of Cymbal have been researched via the theoretical analysis method, the finite element method and experiment [2]-[8]. The harvesting energy basic rule and effective factors of Cymbal have been mastered. So basic foundation for applying piezoelectric Cymbal harvester has laid. However, structural optimization and synthetic effect of various factors and the dominance degree of each factor have not been given a clean answer. In this study, the Taguchi experimental design is adopted for studying the synthetic effect of various factors on the output electrical characteristic of piezoelectric Cymbal harvester. The dominance degree of each factor is analyzed according to the result of Taguchi experiment. A optimized structure of Cymbal is determined.

2. Effective Factors of Cymbal Harvester

Figure 1 shows the structure of piezoelectric ceramics Cymbal harvester. The factors of a Cymbal harvester on the piezoelectric harvesting characteristics deal with two kinds: physical and geometrical ones. The physical factors include the materials of endcap, piezoelectric ceramic and binder, while the geometrical factors involve the endcap diameter D, endcap thickness t_1, PZT thickness t_2, cavity depth h, the dimple diameter d_1 and the cavity diameter d_2. In this paper, the above five geometric factors except D are mainly discussed.

Aluminum is used as the endcap material, PZT-5A is to be piezoelectric ceramics. The diameter of the endcap D is the same as that of the PZT piezoelectric ceramics disk and equal to 29 mm. Based on Taguchi's orthogonal experiment design, the piezoelectric harvesting characteristic of Cymbal harvester is discussed to find out the above five optimized structural parameters.

3. The Taguchi's Orthogonal Experiment Design

The open voltage generated by Cymbal harvester is used as optimization objective, the five above-discussed geometric parameters as factors. The aim of the orthogonal experiment is to optimize the geometric parameters of a Cymbal harvester and to obtain the highest voltage. Meanwhile, an alternative aim is to find out the dominance degrees and the optimal values of these factors.

According to Taguchi parameter design methodology, a $L_{16}(4^5)$ standard orthogonal array with five factors in four levels each in 16 runs is employed [9]. The $L_{16}(4^5)$ considering the five above-discussed parameters are shown in **Table 1**. Each row of the orthogonal array represents a specified set of factor levels to be tested.

4. The Finite Element Analysis

As shown in **Table 1**, 16 experiments need to run for obtaining the open voltage according to the orthogonal experiment design. For saving time, manpower, material and financial resources, the finite element method is adopted. The technologies of APDL parametric modeling and command stream are used to develop the application program for Cymbal piezoelectric analysis based on two development platform of ANSYS 12.0. The developed primary menu, dialogue of inputting parameters and the finite element model are shown in **Figures 2-4**, respectively.

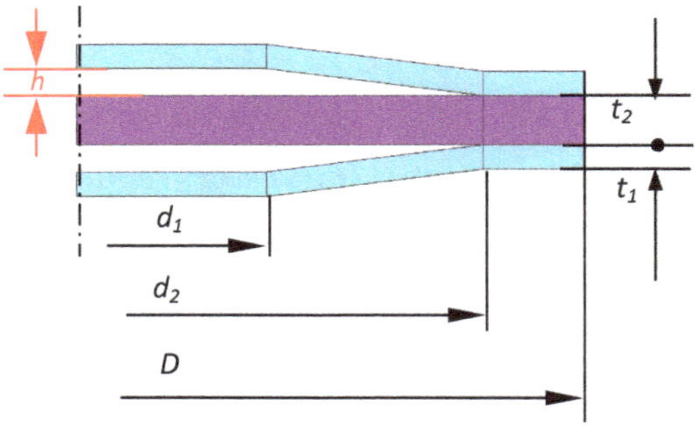

Figure 1. Cymbal structural parameters.

Table 1. Schedule and results of Taguchi's orthogonal experiments.

Run No.	Factors					Results
	1 (t_1/mm)	2 (t_2/mm)	3 (h/mm)	4 (d_1/mm)	5 (d_2/mm)	Open voltage/V
1	1 (0.3)	1 (1)	1 (1.5)	1 (4)	1 (16)	30.652
2	1 (0.3)	2 (2)	2 (1.6)	2 (5)	2 (18)	58.626
3	1 (0.3)	3 (3)	3 (1.8)	3 (6)	3 (20)	91.838
4	1 (0.3)	4 (4)	4 (2.0)	4 (8)	4 (22)	169.639
5	2 (0.4)	1 (1)	2 (1.6)	3 (6)	4 (22)	117.835
6	2 (0.4)	2 (2)	1 (1.5)	4 (8)	3 (20)	174.228
7	2 (0.4)	3 (3)	4 (2.0)	1 (4)	2 (18)	32.557
8	2 (0.4)	4 (4)	3 (1.8)	2 (5)	1 (16)	41.766
9	3 (0.5)	1 (1)	3 (1.8)	4 (8)	2 (18)	94.535
10	3 (0.5)	2 (2)	4 (2.0)	3 (6)	1 (16)	45.925
11	3 (0.5)	3 (3)	1 (1.5)	2 (5)	4 (22)	97.034
12	3 (0.5)	4 (4)	2 (1.6)	1 (4)	3 (20)	50.936
13	4 (0.6)	1 (1)	4 (2.0)	2 (5)	3 (20)	49.544
14	4 (0.6)	2 (2)	3 (1.8)	1 (4)	4 (22)	50.795
15	4 (0.6)	3 (3)	2 (1.6)	4 (8)	1 (16)	87.227
16	4 (0.6)	4 (4)	1 (1.5)	3 (6)	2 (18)	81.143
Sum of level 1: $\sum_{i1}$(V)	350.755	292.566	383.057	164.94	205.57	
Sum of level 2: $\sum_{i2}$(V)	366.386	329.574	314.624	246.97	266.861	
Sum of level 3: $\sum_{i3}$(V)	288.43	308.656	278.934	336.741	366.546	
Sum of level 4: $\sum_{i4}$(V)	268.709	343.484	297.665	525.629	435.303	
Average of level 1: X_{i1} (V)	87.689	73.142	95.764	41.235	51.393	
Average of level 2: X_{i2} (V)	91.597	82.394	78.656	61.743	66.715	
Average of level 3: X_{i3} (V)	72.108	77.164	69.734	84.185	91.637	
Average of level 4: X_{i4} (V)	67.177	85.871	74.416	131.407	108.826	
Range R_i (V)	24.42	12.729	26.03	90.172	57.433	

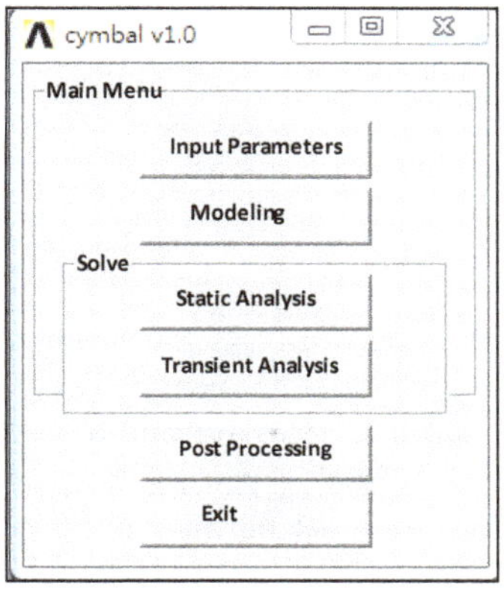

Figure 2. The developed main menu.

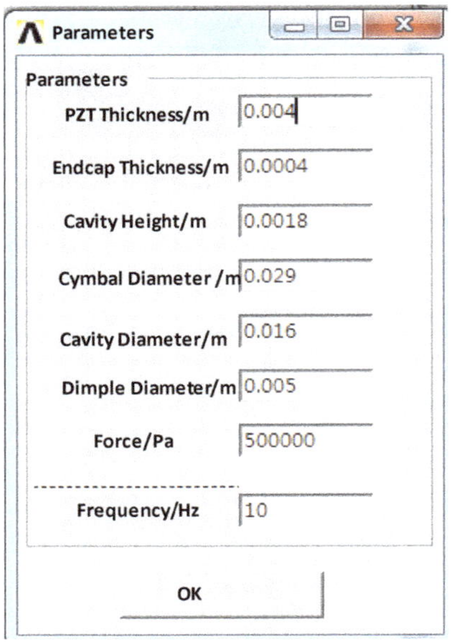

Figure 3. Dialogue of input parameters.

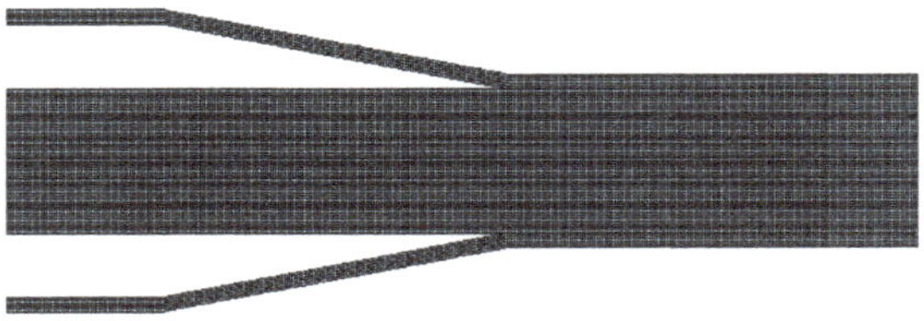

Figure 4. Cymbal finite element model.

Table 1 lists the simulation results calculated by using the developed program. Meanwhile, **Table 1** is also shown the average and range of each level of factors.

5. Analysis of Simulation Results

From the results shown in **Table 1**, it can be seen that the maximum voltage of 174.228 V is obtained under the parametric conditions of No. 6. However, from the average of each factor in four levels, the optimized parameters are: level 2 of factor 1 (endcap thickness of $t_1 = 0.4$ mm), level 4 of factor 2 (PZT thickness of $t_2 = 4$ mm), level 1 of factor 3 (cavity depth of $h = 1.5$ mm), level 4 of factor 4 (dimple diameter of $d_1 = 8$ mm) and level 4 of factor 5 (cavity diameter of $d_2 = 22$ mm). Using the above optimized parameters, the open voltage is up to 236.476 V. The optimized result increase 35.73% compared with the maximum value of 174.228 V using the orthogonal experiment in **Table 1**. **Figure 5** shows the potential distribution under the optimized conditions.

According to the range value of R, factors in order of dominance degree from big to small list as follows: dimple diameter d_1, cavity diameter d_2, cavity depth h, endcap thickness t_1 and PZT thickness t_2.

6. Test Measurement

Figure 6 shows the manufactured samples according to the optimized parameters of Cymbal. During measuring, the material test system is used as the loading equipment, Tektronix oscilloscope is adopted to collect electric signal. The test results are shown in **Table 2**. The average of test results is 229.98 V while data processing. Compared to the simulation result of 236.476 V, the absolute error is 2.54%. The error mainly comes from the following: 1) without considering effect of binder during simulation; 2) material difference between simulation and test. It shows that the simulation and test fits well and Taguchi's orthogonal experiment design is available for optimizing Cymbal structure.

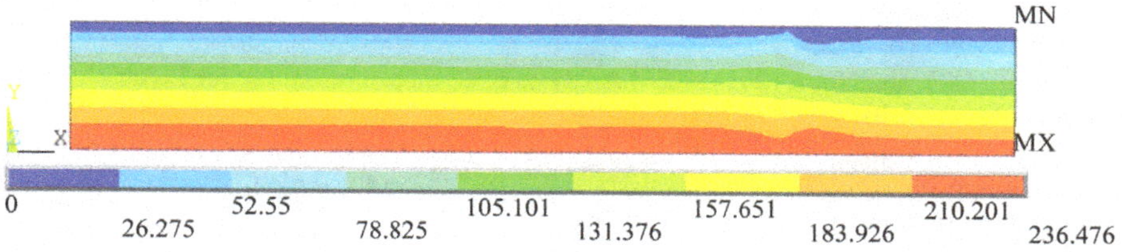

Figure 5. Voltage distribution with optimized parameters.

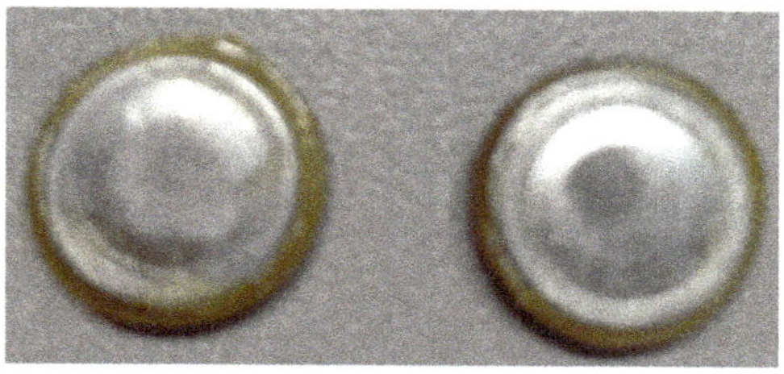

Figure 6. Cymbal samples with optimized parameters.

Table 2. The measured voltage data.

Sample No.	Open voltage(V)	Sample No.	Open voltage(V)
1	230.5	6	228.8
2	230.1	7	229.4
3	229.6	8	230.2
4	230.4	9	230.4
5	229.8	10	230.6

7. Conclusions

To improve the efficiency of Cymbal harvesting energy, Taguchi's orthogonal experiment design is adopted. The finite element program is developed to simulate the experimental conditions via ANSYS 12.0 two development platforms. Effects of synthetical factors on piezoelectric harvesting efficiency of Cymbal are analyzed. The results show that the dominance degree of each factor lists following: dimple diameter d_1, cavity diameter d_2, cavity depth h, endcap thickness t_1 and PZT thickness t_2. The excited open voltage of Cymbal with the optimized parameters is larger than the maximum of orthogonal experiment. It indicates that the method of the orthogonal experiment design for optimal Cymbal harvester is available and the optimized structure of Cymbal harvester can be used for savaging higher electrical potential.

The future of work will focus on the application of the optimal Cymbal and coupling with the environment.

Acknowledgements

This research was supported by the National Natural Science Foundation of China (No. 51175359) and the 4th "333 Engineering" Research Funding Project of Jiangsu Province (BRA2014086).

References

[1] Kim, H.W., Priya, S., Uchino, K. and Robert, E. (2005) Piezoelectric Energy Harvesting under High Pre-Stressed Cyclic Vibrations. *Journal of Electroceramics*, **15**, 27-34. http://dx.doi.org/10.1007/s10832-005-0897-z

[2] Pan, Z.M. and Liu, B. (2006) Mathematical Model of Cymbal-Type Piezoelectric Transducers. *China Mechanical Engineering*, **17**, 283-286.

[3] Guo, Z.Y., Ye, M., Cheng, B., Bai, Z.F. and Cao, B.G. (2007) Influence of Shape Parameters on Electricity Generation by Cymbal Transducer. *Mechanical Science and Technology*, **26**, 1454-1457.

[4] Lu, Y.G. and Yan, Z.F. (2013) Finite Element Analysis on Energy Harvesting with Cymbal Transducer. *Journal of Vibration and Shock*, **32**, 157-162.

[5] Sun, C.H., Shang, G.Q. and Yu, J. (2011) FEM Analysis of Cymbal Transducer for Electricity Generation. *Mechanical Science and Technology*, **30**, 138-141.

[6] Ochoa, P., Pons, J.L., Villegas, M. and Fernandez, J.F. (2005) Mechanical Stress and Electric Potential in Cymbal Piezoceramics by FEA. *Journal of the European Ceramic Society*, **25**, 2457-2461. http://dx.doi.org/10.1016/j.jeurceramsoc.2005.03.082

[7] Xing, Z.B., Sun, C.L., Liu, G.G. and Zhao, X.Z. (2007) An Experimental Study of Cymbal Transducers. *Piezoelectrics & Acoustooptics*, **29**, 273-275.

[8] Wu, L., Chure, M.-C., Wu, K.-K. and Tung, C.-C. (2014) Voltage Generated Characteristics of Piezoelectric Ceramics Cymbal Transducer. *Journal of Materials Science and Chemical Engineering*, **2**, 32-37. http://dx.doi.org/10.4236/msce.2014.210005

[9] Zhao, X.M. (2006) Experimental Design Methods. Science Press, Beijing.

20

Comparative Assessment of Combined-Heat-and-Power Performance of Small-Scale Aero-Derivative Gas Turbine Cycles

Barinyima Nkoi, Barinaadaa Thaddeus Lebele-Alawa

Mechanical Engineering Department, Faculty of Engineering, Rivers State University of Science and Technology, Port Harcourt, Nigeria
Email: nkoi.barinyima@ust.edu.ng, lebele-alawa.thaddeus@ust.edu.ng

Abstract

This paper considers comparative assessment of combined-heat-and-power (CHP) performance of three small-scale aero-derivative industrial gas turbine cycles in the petrochemical industry. The bulk of supposedly waste exhaust heat associated with gas turbine operation has necessitated the need for CHP application for greater fuel efficiency. This would render gas turbine cycles environmentally-friendly, and more economical. However, choosing a particular engine cycle option for small-scale CHP requires information about performances of CHP engine cycle options. The investigation encompasses comparative assessment of simple cycle (SC), recuperated (RC), and intercooled-recuperated (ICR) small-scale aero-derivative industrial gas turbines combined-heat-and-power (SS-ADIGT-CHP). Small-scale ADIGT engines of 1.567 MW derived from helicopter gas turbines are herein analysed in combined-heat-and-power (CHP) application. It was found that in this category of ADIGT engines, better CHP efficiency is exhibited by RC and ICR cycles than SC engine. The CHP efficiencies of RC, ICR, and SC small-scale ADIGT-CHP cycles were found to be 71%, 60%, and 56% respectively. Also, RC engine produces the highest heat recovery steam generator (HRSG) duty. The HRSG duties were found to be 3171.3 kW for RC, 2621.6 kW for ICR, and 3063.1 kW for SC. These outcomes would actually meet the objective of aiding informed preliminary choice of small-scale ADIGT engine cycle options for CHP application.

Keywords

Aero-Derivative Gas Turbines, Combined-Heat-and-Power, Heat Recovery Steam Generator, CHP Efficiency

1. Introduction

Gas turbine is a very satisfactory means of producing mechanical power. It is designed to be highly effective in producing aligned high thrust and power [1] [2]. In contemplating environmentally-friendly gas turbine cycles in the petrochemical industry identification is made of combined-heat-and-power (CHP) as one prominent application that would make gas turbine operation very pleasant to the environment in the aspects of reducing heat energy loss to the environment, and reducing global warming. It also enhances fuel efficiency. CHP simply defined is the simultaneous generation of mechanical power and heat energy in a single system from same fuel input [3].

Some processes in the petrochemical industry occur at relatively moderate temperatures (below 600˚C), and steam is generally the source of their heat energy supply. Such processes include the likes of refining and transformation of crude oil by separation, conversion, and purification carried out in refineries [4], and steam cracking of heavier feedstock, polymerisation, and processing of aromatics occurring in petrochemical plants [5]. Steam could be generated by conventional boilers or heat recovery steam generators in CHP application. It is worth stating that combined-heat-and-power (CHP) generation of steam and power is presently a key energy saving, as well as environmentally-friendly technology in the petrochemical industry [5].

The benefit of CHP is illustrated in **Figure 1** where a CHP plant with a single 100 units fuel source yields about 75% overall cycle efficiency as against separate power plant and boiler source of an aggregate of 147 units of fuel yielding about 51% overall cycle efficiency [6]. In light of this, the performance of small-scale aero-derivative industrial gas turbines (SS-ADIGT) derived from helicopter gas turbines discussed by the author in ref [7] is herein analysed in CHP application.

The decision to use aero-derivative gas turbines is mainly based on economical and operational advantages. Gas turbine manufacturers have found that to reduce cost of designing and developing new gas turbines, a more effective approach is to develop high performance industrial gas turbines by modifying aircraft gas turbine engines [8]. Also, by introducing aero-derivative's removable gas generator, better flexibility is provided which in

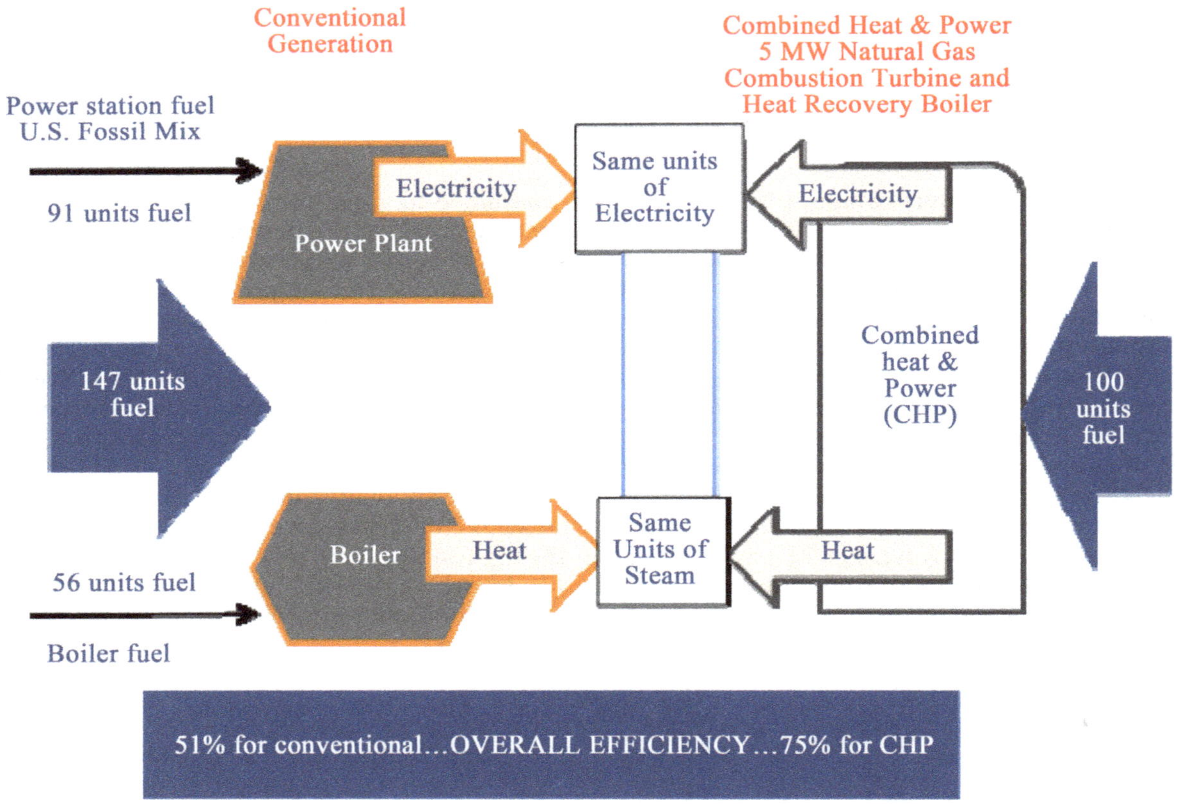

Figure 1. Energy saving benefit of CHP over traditional system [6].

turn lead to reducing maintenance operation and enhancing gas turbine availability in industrial applications [9]. More so, implementing aero-derivative technology for industrial gas turbine has resulted in low maintenance downtime, good part-load efficiencies, and higher rate of return [10]. Besides, aero-derivative gas turbines can meet stringent NO_x control requirements because they are suitable for power augmentation by steam injection. For instance, the GE LM series industrial aero-derivative gas turbines are meeting NO_x requirements as low as 25 parts per million (ppm) using steam injection. Other merits of aero-derivative gas turbines include low weight-to-power ratio, compactness, and hence, lesser erection and startup time [11] [12]. Moreover, aero-derivative gas turbines are most suitable for highly efficient cogeneration plants, more flexible combined-cycle plants, and in mechanical drive applications for production and distribution of oil and gas [13].

However, deciding on choice of small-scale ADIGT cycle option for CHP application poses some difficulty for engineers and decision-makers. Hence, the objective of this paper is to carry out performance comparison of simple cycle (SC), recuperated (RC), and intercooled-recuperated (ICR) small-scale ADIGT cycles in CHP, that would aid good and informed choice of turbine cycle option for the purpose of use in small-scale CHP application. The novelty of this research work is in the area of comparing the performances of SC, RC, and ICR aero-derivative gas turbine cycles in small-scale CHP. Previous works only considered CHP analysis of the simple engine cycle, and as such, considering CHP performance of advanced cycles (RC and ICR) actually presents a wider range of options for small-scale CHP engine cycle choices.

2. Materials and Methods

2.1. CHP Modelling

CHP systems are either developed as "topping cycles" or "bottoming cycles" as illustrated in **Figure 2** and **Figure 3** respectively [14]. Topping cycles describe systems where there occur primary power generation and subsequent heat utilization, whereas bottoming cycles pertain to systems where heat is primarily generated in a

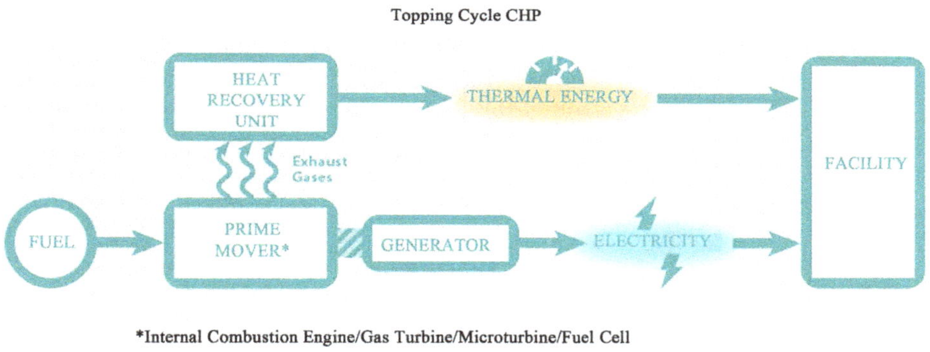

Figure 2. Topping cycle CHP (Source: [14]).

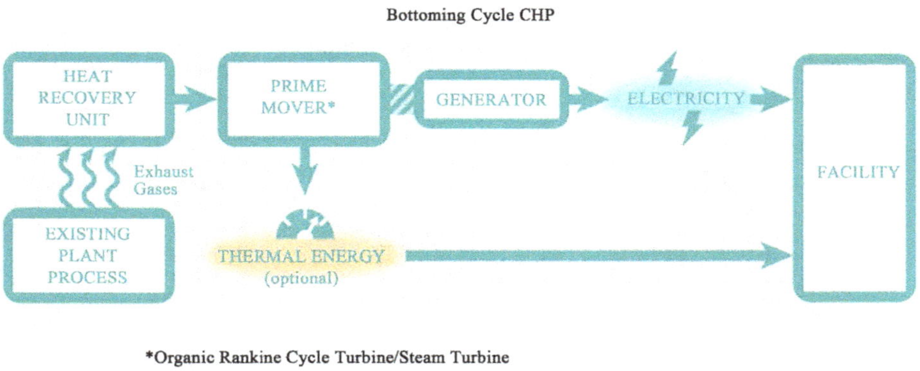

Figure 3. Bottoming cycle CHP (Source: [14]).

process with subsequent utilisation for power generation [15]. In this work, topping cycle arrangement is adopted where power is primarily generated from gas turbine and heat recovery steam generator (HRSG) is designed to match for the purpose of process steam production. Performance parameters of the aero-derivative gas turbines discussed by the author in ref [7] are employed to determine the parameters of the HRSG.

2.1.1. HRSG Performance

A set of heat exchangers that utilises the exhaust heat of a gas turbine to produce steam is referred to as heat recovery steam generator (HRSG). Three types of HRSG are identified, namely, unfired, supplementary fired, and exhaust fired. The most common and widely used HRSG is the unfired type because it is simple in design and cheap [16]. HRSG of the unfired type is considered in this research without considering the material dimension of the heat exchangers. It is pertinent to declare that only the thermodynamic performance in terms of temperature profile of exhaust gas, steam temperature and flow, and heat capacity, of the HRSG are being modelled in this research. Pinch and approach point technology is applied in modelling the HRSG performance, and with a single steam pressure mode of operation.

2.1.2. Pinch and Approach Points Technology

Approach point is the difference between the temperature of saturated steam and the temperature of water entering the evaporator, whereas pinch point is the difference between the gas temperature leaving the evaporator and the temperature of saturated steam [16]. Steam generation is directly affected by the pinch and approach points. Also affected is the exhaust gas and steam temperature profile. For the design case of an unfired HRSG, selection is usually made of the values of pinch and approach points; pinch point ranges from 10˚C to 30˚C whereas approach point ranges from 5˚C to 15˚C based on the sizes of evaporators that can be built and shipped economically, and to maximise heat transfer rate between exhaust gas and steam streams. **Figure 4** illustrates pinch point, approach point, exhaust gas and steam temperature profiles of HRSG.

Using the notations in **Figure 4** above the path 4-y-x-1 indicates gas turbine exhaust gas temperature profile whereas the path a-b-c-d-e indicates steam temperature profile. Pinch point $= T_x - T_c$; approach point $= T_c - T_b$; process a-b occurs in the economiser; c-d in the evaporator; and d-e in the super-heater.

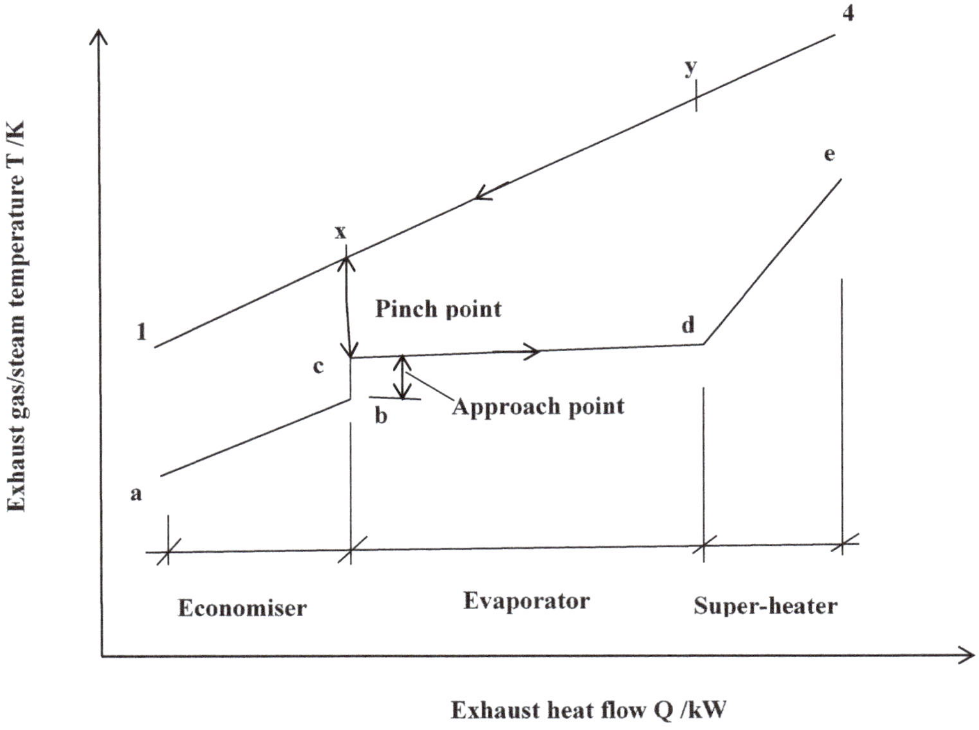

Figure 4. Single steam pressure HRSG exhaust gas/steam temperature profiles.

2.2. ADIGT-CHP Design Point Performance

To model the design point performance of a CHP plant is to match the parameters of HRSG with the design point of the gas turbine given particular consideration to desired steam flow or temperature and saturation pressure. In doing so, pinch and approach points are selected by the engineering judgement; and from gas turbine exhaust gas flow, the HRSG temperature profile, duty, and steam flow are established. Using pinch technology and thermodynamic properties of steam, the computation of CHP HRSG gas/steam temperature profile and steam flow is done as follows: Gas turbine exhaust gas temperature and mass flow are imported from gas turbine performance simulation while the HRSG pinch and steam saturation pressure (which fixes the steam saturation temperature—T_c) are selected by the engineering judgement. In this design the steam saturation pressure is 10bar. With the notations of **Figure 4** the temperature of exhaust gas at pinch point (T_x) is given by Equation (1)

$$T_x = T_c + \text{pinch} = T_c + 15 \tag{1}$$

where $\text{Pinch} = 15$.

The superheated steam temperature (T_e) is chosen as required by the industrial process heat demand. The steam flow (w_s) is computed from total heat transfer in super-heater and evaporator using heat balance above pinch as defined by Equation (2)

$$Q_{4x} = Q_{\text{evap}} + Q_{\text{super}};\ Q_{4x} = w_g c_{pa}(0.99)(T_4 - T_x) = w_s\left[(h_e - h_c) + 0.02(h_d - h_c)\right]$$

$$\therefore w_s = \frac{w_g c_{pa}(0.99)(T_4 - T_x)}{(h_e - h_c) + 0.02(h_d - h_c)} \tag{2}$$

where 0.99 = heat loss factor,

0.02 = blow down factor,

$\dot{W}_g$ = exhaust gas flow,

$\dot{c}_{pa}$ = specific heat at constant pressure of air,

h_e = specific enthalpy of super-heated steam,

h_c = specific enthalpy of saturated water,

h_d = specific enthalpy of saturated steam,

T_4 = gas turbine exhaust temperature,

Q_{evap} = evaporator duty,

Q_{super} = super-heater duty.

Equation (3) defines the super-heater duty (Q_{super})

$$Q_{\text{super}} = w_s(h_e - h_d) \tag{3}$$

Gas temperature drop in the super-heater (ΔT_{4y}) is given by Equation (4)

$$\Delta T_{4y} = \frac{Q_{\text{super}}}{w_g c_{pa}(0.99)} \tag{4}$$

This implies that exhaust gas temperature to evaporator (T_y) is calculated using Equation (5)

$$T_y = T_4 - \Delta T_{4y} \tag{5}$$

Evaporator duty (Q_{evap}) is determined with the aid of Equation (6)

$$\text{Evaporator duty } Q_{\text{evap}} = w_s(h_d - h_c) \tag{6}$$

Similarly Equation (7) defines Economiser duty (Q_{econ})

$$\text{Economiser duty } Q_{\text{econ}} = w_s(1.02)(h_c - h_a) \tag{7}$$

Gas temperature drop in the economiser (ΔT_{x1}) is given by Equation (8)

$$\Delta T_{x1} = \frac{Q_{econ}}{w_g c_{pa}(0.99)} \tag{8}$$

This implies that exhaust gas exit temperature from the economiser (T_1) is calculated using Equation (9)

$$T_1 = T_x - \Delta T_{x14y} \tag{9}$$

Total HRSG duty (Q_{HRSG}) is computed by Equation (10)

$$\text{HRSG duty}\left(Q_{\mathrm{HRSG}}\right) = Q_{\mathrm{evap}} + Q_{\mathrm{super}} + Q_{\mathrm{econ}} \tag{10}$$

$$\text{Electrical efficiency} = \frac{P_E}{P_T} = \eta_E$$

The electrical efficiency could be assumed, such that

$$\text{Useful electric power generated } P_E = \eta_E + P_T$$

where P_T = gas turbine power.

Heat to power ratio of the CHP is given by Equation (11)

$$\text{Heat to power ratio} = \frac{Q_{\mathrm{HRSG}}}{P_E} \tag{11}$$

Equations (1) to (11) are referred from [17]

Equation (12) is used to compute First Law CHP efficiency (η_1)

$$\text{First Law efficiency } \eta_1 = \frac{\dot{W}_E + \dot{W}_{ST}}{\dot{m}_f \cdot h_f} \tag{12}$$

where $\dot{W}_E$ = Electrical energy rate,

$\dot{W}_{ST}$ = Steam energy rate,

$\dot{m}_f$ = Fuel mass flow in combustor,

Δh_f = LHV = Low heating value of fuel.

Equation (13) is used to compute Second Law CHP efficiency (η_2)

$$\text{Second law efficiency } \eta_2 = \frac{\dot{W}_E + \dot{W}_{ST}}{\dot{m}_f\left[\Delta h_f - T_R \cdot \Delta S_f\right]} \tag{13}$$

The denominator of Equation (13) is the availability rate of the fuel consumed, where ΔS_f = Entropy released by fuel combustion. T_R = Temperature at exhaust. Equations (12) and (13) and referred from [18] [19].

3. Results and Discussions

3.1. Small-Scale-ADIGT-CHP Design Point Performance Analysis

The CHP design point simulation for the SS-ADIGT was done using TURBOMATCH (a gas turbine engine performance simulation code) [7]-[22] and the HRSG gas/steam temperature profiles are shown in **Figure 5** while the CHP DP performance results are indicated in **Table 1**.

The technical performance of the simple, recuperated, and intercooled-recuperated SS-ADIGT cycles derived from helicopter engines have been analysed by the author in [7]. The superheated steam temperature for the CHP was set within the range of steam temperatures obtainable in refinery and petrochemical plants which is about 100˚C - 500˚C.

3.2. Small-Scale ADIGT-CHP Off-Design Performance

The HRSG would normally not operate at the design point due to variations in the inlet gas conditions and steam parameters. The inlet gas conditions in turn would depend on gas turbine off-design variation in ambient conditions, firing temperature, altitude, power setting, etc. This makes the CHP plant exhibits varying outputs. The CHP off-design performance was simulated with TURBOMATCH engine off-design. The off-design performances of the SS-ADIGT-CHP with changing conditions of the engines are shown in **Figures 6-10**.

At design and off-design conditions the RC and ICR ADIGT engines exhibit better CHP efficiency than the

Table 1. SS-ADIGT-CHP design point performance results.

Parameter	Values for the SS-ADIGT engines		
	Simple cycle	Recuperated	ICR
Steam saturation temperature (K)	457	457	457
Pinch point	15	15	15
Approach point	8	8	8
Superheated steam temperature (K)	673	673	673
Steam mass flow (kg/s)	1.10	1.14	0.94
Economiser feed water temperature (K)	388	388	388
Super-heater duty (kW)	530.07	548.80	453.69
Evaporator duty (kW)	2198.23	2275.89	1881.45
Economiser duty (kW)	334.74	346.57	286.50
HRSG duty (kW)	3063.05	3171.26	2621.63
Gas turbine exhaust mass flow (kg/s)	5.65	5.64	5.64
Gas turbine exhaust temperature (K)	885	901	826
Gas temperature at evaporator exit (K)	806	819	758
Gas exit (stack) temperature (K)	422	420	429
Gas turbine power (kW)	1567	1567	1567
GT Thermal efficiency	**0.296**	**0.336**	**0.339**
Heat: power ratio	2.09	2.16	1.79
CHP efficiency	0.56	0.71	0.60

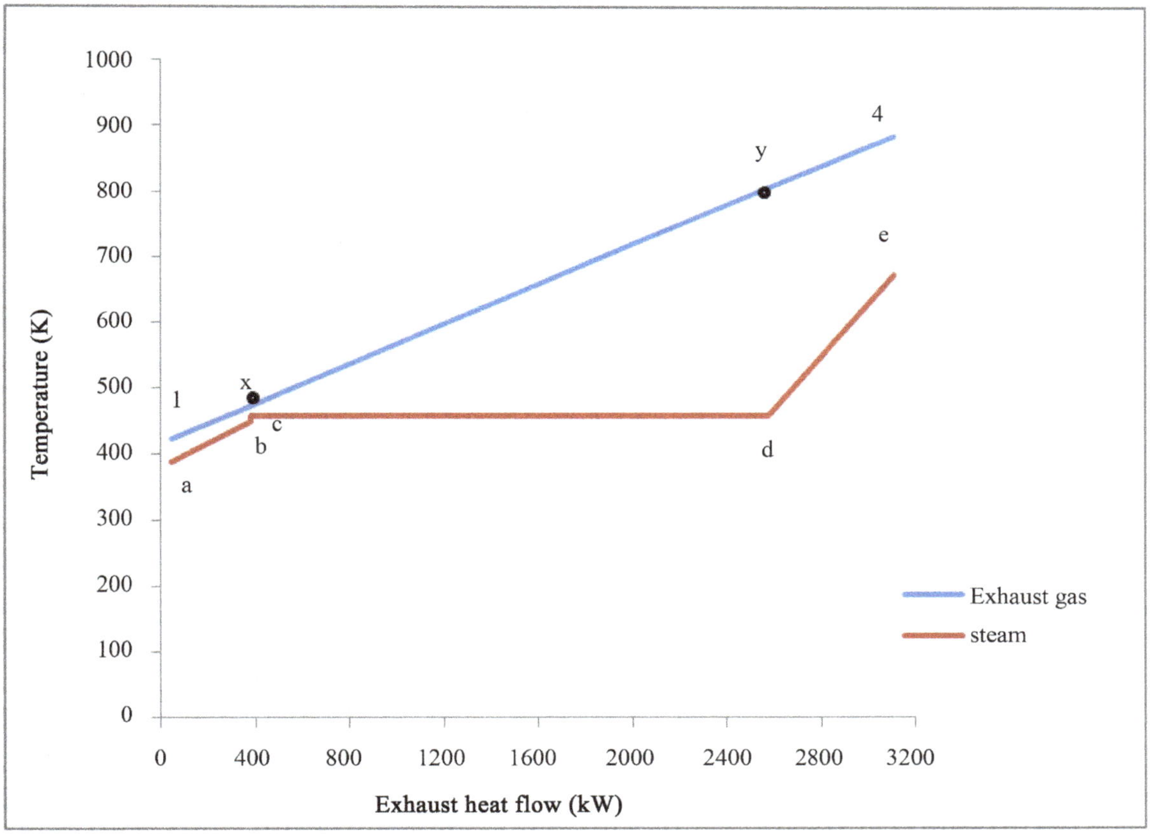

Figure 5. Single steam pressure HRSG temperature/heat profile for the SS-ADIGT-CHP.

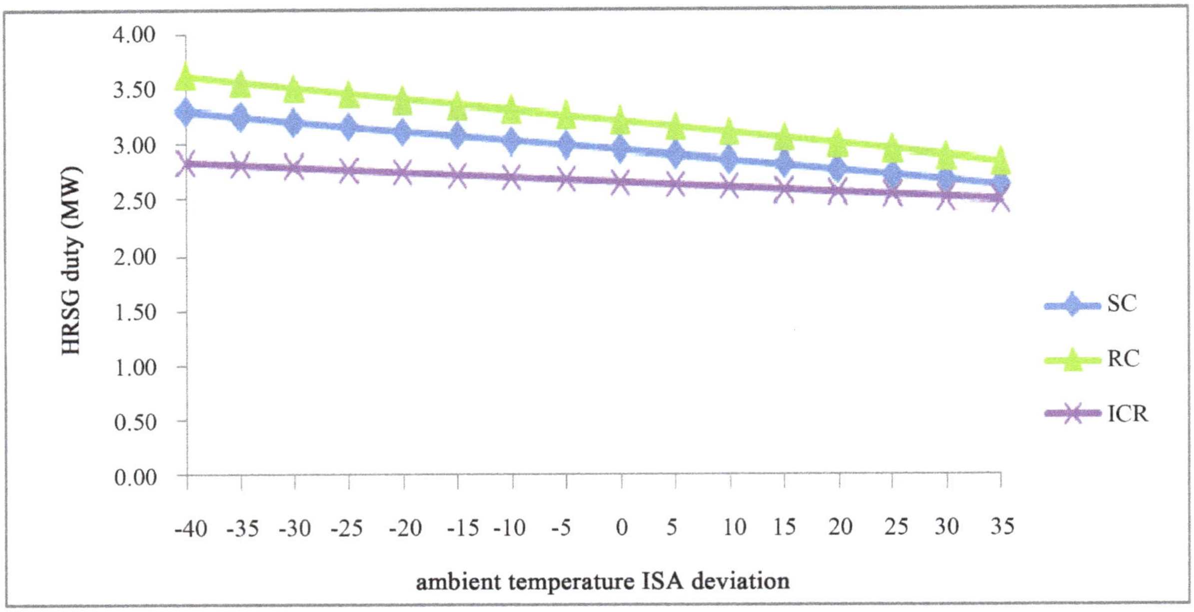

Figure 6. Effect of ambient temperature on HRSG duty.

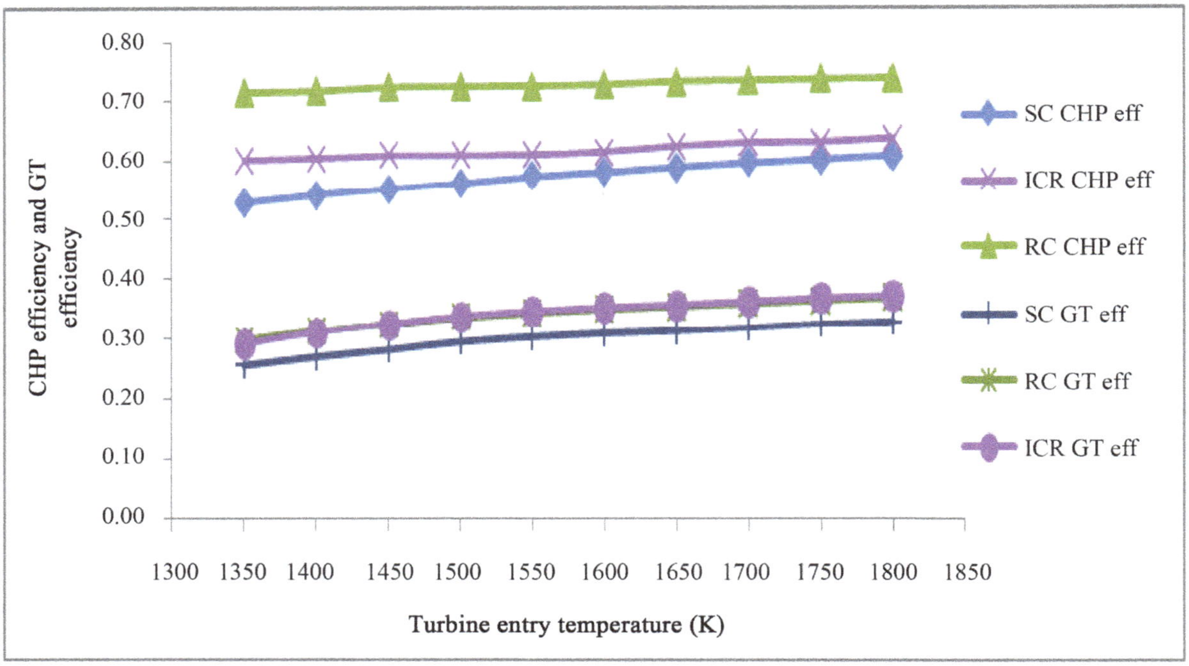

Figure 7. Variation of CHP and GT efficiencies with TET.

SC engine as shown in **Figure 7**, **Figure 9**, and **Figure 10**. As shown in **Table 1** the CHP efficiencies of RC, ICR, and SC SS-ADIGT-CHP cycles were found to be 71%, 60%, and 56% respectively. These results compare favourably with values in the literature. For instance, a 40 kWe CHP plant located within the Queens Building at De Montfort University, was analysed to show an overall (CHP) efficiency of 77% [23]. CHP efficiency can be as high as 80% - 90%. For instance Tervola 0.5 MWe/1.13MWth CHP in Finland was found to exhibit an overall (CHP) efficiency of 81.5% [24]. The CHP efficiencies are observed to increase with increases in turbine entry temperature (TET), gas turbine (GT) power, ambient temperature, and HRSG duty. The percentage increases

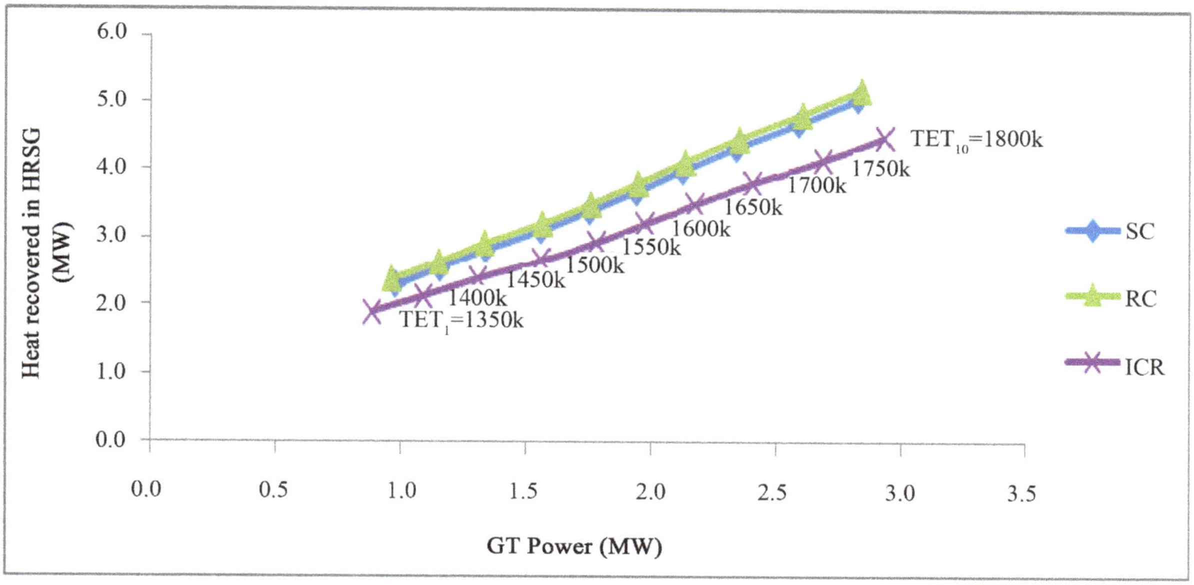

Figure 8. Variation of HRSG duty with GT power at increasing TET.

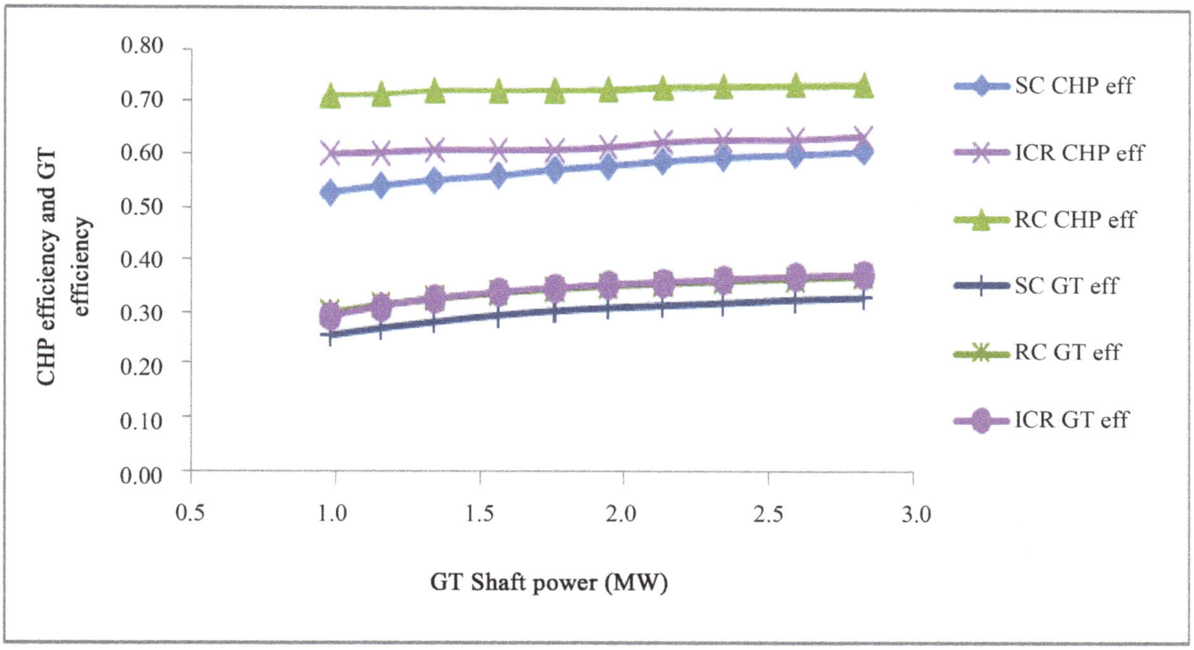

Figure 9. Variation of CHP and GT efficiencies with GT power at increasing TET.

in CHP efficiencies of RC and ICR over SC at design-point (DP) are 16.5% and 3.8% respectively. This superior performance is due to the lower heat input from burning less fuel in the advanced cycle engines. Looking at **Figure 7** and **Figure 9**, the curve "RC GT eff" appears to coincide with that of the "ICR GT eff", whereas actually there is some slight difference between the RC GT efficiency and ICR GT efficiency. This small difference is shown clearly at design point in **Table 1** where ICR GT efficiency is slightly higher than RC GT efficiency by about 0.003.

On the other hand, the SC engine produces more HRSG duty than the ICR cycle as shown in **Figure 6** and **Figure 8**, due to lower exhaust gas temperature and steam rate of the ICR cycle. The highest HRSG duty is produced by the RC engine because of its higher exhaust gas temperature and steam rate.

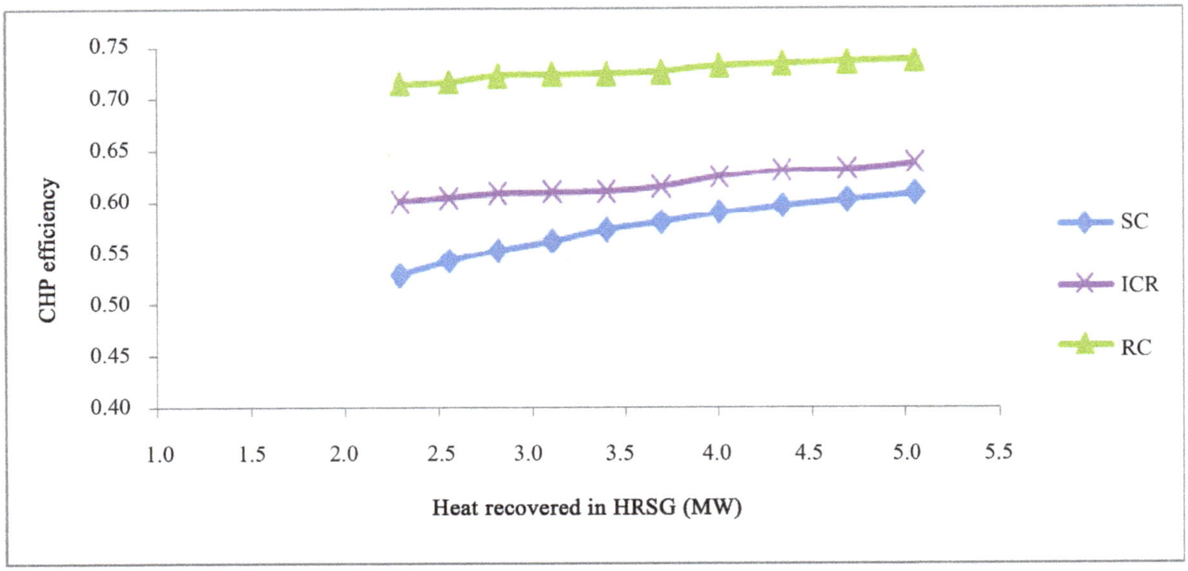

Figure 10. Variation of CHP efficiency with HRSG duty at increasing TET.

Besides, as plotted in **Figure 8**, the heat-to-power ratio is observed to be least in the ICR cycle than SC and RC and highest in the RC cycle at both DP and off-design (OD) conditions. In essence, this means that given a range of power output at any condition of TET and ambience, the RC cycle would generate more steam than the SC and ICR, while ICR would generate the least steam flow. This is so because GT exhaust gas temperature is highest in the RC cycle and exits the HRSG at the least temperature in the RC cycle. This creates huge drop in temperature of exhaust gas from exit of GT to exit of HRSG in the RC cycle. This exhaust gas temperature drop is smaller in SC cycle and least in ICR cycle. Hence, huge amount of heat is extracted from exhaust gas in RC cycle than SC and ICR at same conditions. At design point, the HRSG duties were found to be 3171.3 kW for RC, 2621.6 kW for ICR, and 3063.1 kW for SC as shown in **Table 1**.

4. Conclusion

The foregoing analysis of technical performances of small-scale aero-derivative industrial gas turbine-CHP cycles has led to the conclusions that for small-scale ADIGT-CHP, better CHP efficiency is exhibited by RC and ICR cycles than Simple engine cycle. Also, it was found that the RC engine produces the highest HRSG duty. Therefore, it could be said that performance comparison of simple, recuperated, and intercooled-recuperated small-scale ADIGT cycles in CHP has been achieved. This sort of analysis would actually aid concerned engineers, product developers, and other key decision-makers to logically make good and informed choice of small-scale ADIGT engine cycle option for the purpose of use in CHP application.

Acknowledgements

Very essentially, the authors would want to thank Professor Pericles Pilidis and Dr. Theoklis Nikolaidis of the Department of Power and Propulsion of Cranfield University, United Kingdom, for their invariable contributions to this research.

References

[1] Lebele-Alawa, B.T. and Jo-Appah, V. (2015) Thermodynamic Performance Analysis of a Gas Turbine in an Equatorial Rainforest Environment. *Journal of Power and Energy Engineering*, **3**, 11-23. http://dx.doi.org/10.4236/jpee.2015.31002.

[2] Lebele-Alawa, B.T. (2010) Axial Thrust Responses to a Gas Turbine's Rotor-Blade Distortions. *Journal of Engineering Physics and Thermophysics*, **83**, 991-994. http://dx.doi.org/10.1007/s10891-010-0423-2

[3] International Energy Agency (2007) Tracking Industrial Energy Efficiency and CO_2 Emissions.

http://www.iea.org/textbase

[4] Exxonmobil (2011) Fawley Refinery and Petrochemical Plant. www.exxonmobil.co.uk/UK-English/files/Fawley_2011.pdf

[5] Gielen, D.J., Vos, D. and Van Dril, A.W.N. (1996) The Petrochemical Industry and Its Energy Use: Prospect for the Dutch Energy Intensive Industry. Energy Research Center of the Netherland (ECN), ECN-C-96-029. April 1996, 25-45.

[6] US EPA (2013) Combined Heat and Power Partnership: Efficiency Benefit. http://www.epa.gov/chp/basic/efficiency.html

[7] Nkoi, B., Pilidis, P. and Nikolaidis, T. (2013) Performance of Small-Scale Aero-Derivative Industrial Gas Turbines Derived from Helicopter Engines. *Journal of Propulsion and Power Research*, **2**, 243-253. http://dx.doi.org/10.1016/j.jppr.2013.11.001

[8] Bhargava, R., Blanchi, M., Peretto, A. and Spina, P. R. (2004) A Feasibility Study of Existing Gas Turbines for Recuperated, Intercooled, and Reheat Cycle. *Journal of Engineering for Gas Turbines and Power*, **126**, 531-544. http://dx.doi.org/10.1115/1.1707033

[9] Najjar, Y.S.H. (2000) Gas Turbines Cogeneration Systems: A Review of Some Novel Cycles. *Applied Thermal Engineering*, **20**, 179-197. http://dx.doi.org/10.1016/S1359-4311(99)00019-8

[10] Yang, W. (1997) Reduction of Specific Fuel Consumption in Gas Turbine Power Plants. *Energy Conversion and Management*, **38**, 1219-1224. http://dx.doi.org/10.1016/S0196-8904(96)00151-3

[11] Keller, S.C. and Studniarz, J.J. (1987) Aero-Derivative Gas Turbines Can Meet Stringent NO_x Control Requirements. *Proceedings from the 9th Annual Industrial Energy Technology Conference*, Houston, 16-18 September 1987, 253-260.

[12] Roy, G.K. (2012) Selecting Heavy-Duty or Aero-Derivative Gas Turbines. *Hydrocarbon Processing*, **75**, 57.

[13] Doom, T.R. (2013) Aero-Derivative Gas Turbines. Case Studies on the Government's Role in Energy Technology Innovation. American Energy Innovation Council. http://americanenergyinnovation.org/wp-content/uploads/2013/08/Case-Gas-Turbines.pdf

[14] Center for Sustainable Energy (2014) Combined Heat and Power. http://energycenter.org/self-generation-incentive-program/business/technologies/chp

[15] Bhatt, S.M. (2001) Mapping of General Combined-Heat-and-Power System. *Energy Conversion and Management*, **42**, 115-124. www.elsevier.com/locate/enconman http://dx.doi.org/10.1016/S0196-8904(00)00045-5

[16] Ganapathy, V. (1996) Heat Recovery Steam Generator: Understanding the Basics. *Chemical Engineering Process*, 32-45.

[17] Ganapathy, V. (1990) Heat Transfer—Simplify Heat Recovery Steam Generator Evaluation. *Hydrocarbon Processing*, March 1990, 77-82. http://v_ganapathy.tripod.com/simphrsg.pdf

[18] Koratianitis, T. (2012) DEN107 Thermodynamics Laboratory Exercise [Course Lecture Note]. *Fundamentals of General Thermodynamics*. Queen Mary University of London, London.

[19] Koratianitis, T., Grantstrom, J., Wassingbo, P. and Massardo, A.F. (2005) Parametric Performance of Combined-Cogeneration Power Plant with Various Power and Efficiency Enhancements. *Journal of Engineering for Gas Turbines and Power*. *Transactions of the ASME*, **127**, 65-72.

[20] Pachidis, V.A. (2008) Gas Turbine Performance Simulation [Course Lecture Note]. *Gas Turbine Simulation and Diagnostics*. Cranfield University, Cranfield.

[21] Palmer, J. (1999) The Turbomatch Scheme for Aero/Industrial Gas Turbine Engine Design Point/Off Design Performance Calculation [Course Lecture Note]. *Gas Turbine Simulation and Diagnostics*. Cranfield University, Cranfield.

[22] Nkoi, B., Pilidis, P. and Nikolaidis, T. (2013) Performance Assessment of Simple and Modified Cycle Turboshaft Gas Turbines. *Journal of Propulsion and Power Research*, **2**, 96-106. http://dx.doi.org/10.1016/j.jppr.2013.04.009

[23] Smith, M.A., Few, P.C. and Twidell, J.W. (1995) Technical and Operational Performance of a Small-Scale Combined-Heat-and-Power (CHP) Plant. *Energy*, **20**, 1205-1214. http://dx.doi.org/10.1016/0360-5442(95)00073-P

[24] Kirjavainen, M., Sipila, K., Alakangas, E., Savola, T. and Salomon, M. (2004) Small-Scale Biomass CHP Technologies Situation in Finland, Denmark and Sweden. OPET Report 12, European Commission Directorate-General for Energy and Transport. OPET CHP/DHC Cluster. Espoo, April 2004, 27.

Nomenclature

ADIGT	Aero-derivative industrial gas turbines	
ADIGT-CHP	Aero-derivative industrial gas turbines combined-heat-and-power	
CHP	Combined-heat-and-power	
DP	Design-point	
GT	Gas turbine	
IC	Intercooled cycle	
ICR	Intercooled-recuperated cycle	
ISA	International standard atmosphere	
ISA Dev	International standard atmosphere deviation	
OD	Off-design point	
RC	Recuperated	
SS	Small-scale	
SS-ADIGT	Small-scale aero-derivative industrial gas turbines	
SS-ADIGT-CHP	Small-scale aero-derivative industrial gas turbines combined-heat-and-power	
TET	Turbine entry temperature	
TURBOMATCH	Gas turbine engine performance simulation code	
C	Specific heat	kJ/kg
c_p	Specific heat at constant pressure	kJ/kg
c_{pa}	specific heat at constant pressure of air	kJ/kg
FF	Fuel flow in combustor	Kg/s
H	Specific enthalpy	kJ/kg
h_a	Water specific enthalpy at economiser inlet	kJ/kg
h_c	Saturated water specific enthalpy at evaporator inlet	kJ/kg
h_d	Saturated steam specific enthalpy at evaporator exit	kJ/kg
h_e	Superheated steam specific enthalpy	kJ/kg
LHV	Low heating value of fuel	kJ/kg
$\dot{m}_f$	Fuel mass flow in combustor	Kg/s
P_E	Electrical power	kWe
P_T	Gas turbine power	kW
Q	Heat flow	kW
q_{in}	Heat flow in	kW
q_{out}	Heat flow out	kW
Q_{4x}	Total heat transfer in superheater and evaporator	kW
Q_{comb}	Combustor heat input	kW
Q_{econ}	Economiser duty	kW
Q_{evap}	Evaporator duty	kW
Q_{super}	Superheater duty	kW
Q_{HRSG}	HRSG duty	kW
T	Temperature	K
T_a	Water temperature at HRSG economiser inlet	K

Continued

T_b	Temperature at HRSG economiser exit	K
T_c	HRSG steam saturation temperature	K
T_e	HRSG superheated steam temperature	K
T_x	Exhaust gas temperature at pinch point of HRSG	K
T_y	Exhaust gas temperature at HRSG evaporator exit	K
T_4	Gas turbine exhaust temperature to HRSG	K
ΔT_{4y}	Gas temperature drop in superheater	K
ΔT_{x1}	Gas temperature drop in the economiser	K
η_{th}	Thermal efficiency	%
η_1	First law CHP efficiency	%
η_2	Second law CHP efficiency	%
η_E	Electrical generator efficiency	%
w_g	Exhaust gas mass flow	Kg/s
w_s	Steam mass flow	Kg/s
$\dot{W}_E$	Electrical energy rate	kW
$\dot{W}_{ST}$	Steam energy rate	kW
ΔS_f	Entropy released by fuel combustion	kJ/kg·K
Δh_f	Low heating value of fuel	kJ/kg

Permissions

All chapters in this book were first published in JPEE, by Scientific Research Publishing; hereby published with permission under the Creative Commons Attribution License or equivalent. Every chapter published in this book has been scrutinized by our experts. Their significance has been extensively debated. The topics covered herein carry significant findings which will fuel the growth of the discipline. They may even be implemented as practical applications or may be referred to as a beginning point for another development.

The contributors of this book come from diverse backgrounds, making this book a truly international effort. This book will bring forth new frontiers with its revolutionizing research information and detailed analysis of the nascent developments around the world.

We would like to thank all the contributing authors for lending their expertise to make the book truly unique. They have played a crucial role in the development of this book. Without their invaluable contributions this book wouldn't have been possible. They have made vital efforts to compile up to date information on the varied aspects of this subject to make this book a valuable addition to the collection of many professionals and students.

This book was conceptualized with the vision of imparting up-to-date information and advanced data in this field. To ensure the same, a matchless editorial board was set up. Every individual on the board went through rigorous rounds of assessment to prove their worth. After which they invested a large part of their time researching and compiling the most relevant data for our readers.

The editorial board has been involved in producing this book since its inception. They have spent rigorous hours researching and exploring the diverse topics which have resulted in the successful publishing of this book. They have passed on their knowledge of decades through this book. To expedite this challenging task, the publisher supported the team at every step. A small team of assistant editors was also appointed to further simplify the editing procedure and attain best results for the readers.

Apart from the editorial board, the designing team has also invested a significant amount of their time in understanding the subject and creating the most relevant covers. They scrutinized every image to scout for the most suitable representation of the subject and create an appropriate cover for the book.

The publishing team has been an ardent support to the editorial, designing and production team. Their endless efforts to recruit the best for this project, has resulted in the accomplishment of this book. They are a veteran in the field of academics and their pool of knowledge is as vast as their experience in printing. Their expertise and guidance has proved useful at every step. Their uncompromising quality standards have made this book an exceptional effort. Their encouragement from time to time has been an inspiration for everyone.

The publisher and the editorial board hope that this book will prove to be a valuable piece of knowledge for researchers, students, practitioners and scholars across the globe.

List of Contributors

Bekhada Hamane, Mamadou Lamine Doumbia and Ahmed Chériti
Department of Electrical and Computer Engineering, UQTR, Trois-Rivières, Canada

Hicham Chaoui
Center for Energy Systems Research, Department of Electrical and Computer Engineering, Tennessee Technological University, Cookeville, USA

Mohamed Bouhamida and Mustapha Benghanem
Department of Electrical Engineering, University Mohamed Boudiaf, Oran, Algeria

Lifen Li
School of Control and Computer Engineering, North China Electric Power University, Baoding, China

Huaiyu Zhao
School of Electrical & Electronic Engineering, North China Electric Power University, Baoding, China

Dongmei Zhou and Jennifer A. Eden
Department of Mechanical Engineering, California State University, Sacramento, CA, USA

Oladapo S. Akinyemi and Terrence L. Chambers
Department of Mechanical Engineering, University of Louisiana at Lafayette, Lafayette, USA

Yucheng Liu
Department of Mechanical Engineering, Mississippi State University, Starkville, USA

Jyoti Paudel, Xufeng Xu, Karthikeyan Balasubramaniam and Elham B. Makram
Electrical and Computer Engineering Department, Clemson University, Clemson, USA

Navin Shenoy and R. Ramakumar
School of Electrical and Computer Engineering, Oklahoma State University, Stillwater, USA

Niraj Kumar Choudhary, Soumya Ranjan Mohanty and Ravindra Kumar Singh
Electrical Engineering Department, Motilal Nehru National Institute of Technology Allahabad, Allahabad, India

Luis E. Teixeira, Alexandre Beluco and José Antônio S. Louzada
Instituto de Pesquisas Hidráulicas, Universidade Federal do Rio Grande do Sul, Porto Alegre, Brazil

Johan Caux
Instituto de Pesquisas Hidráulicas, Universidade Federal do Rio Grande do Sul, Porto Alegre, Brazil
Ecole Nationale Superieure de l'Energie, l'Eau et l'Environnement, Grenoble INP, Grenoble, France

Ivo Bertoldo
Companhia Riograndense de Saneamento, Santa Maria, Brazil

Ricardo C. Eifler
Companhia Estadual de Energia Elétrica, Salto do Jacuí, Brazil

Veera Venkata Sudhakar Angatha
EEE Department, SR Engineering College, Warangal, India

Karri Chandram
Department of EEE & Instrumentation, BITS PilaniKK Birla Goa Campus, Goa, India

Askani Jaya Laxmi
EEE Department, JNTUCE, JNTUH, Hyderabad, India

Fouad Amri, Omar Bouattane, Tajeddine Khalili, Abdelhadi Raihani and Abdelkader Bifadene
Lab SSDIA, ENSET Mohammedia, Hassan II University of Casablanca, Casablanca, Morocco

Tajeddine Khalili, Abdelhadi Raihani, Hassan Ouajji, Omar Bouattane and Fouad Amri
Lab. SSDIA, ENSET Mohammédia, Hassan II University of Casablanca, Morocco

Katarzyna Stępczyńska-Drygas, Sławomir Dykas and Krystian Smołka
Institute of Power Engineering and Turbomachinery, Silesian University of Technology, Gliwice, Poland

Mahesh M. Rathore
Research Scholar, Department of Mechanical Engineering, Govt. Engineering College, Aurangabad, India

Ravi M. Warkhedkar
Department of Mechanical Engineering, Govt. College of Engineering, Karad, India

Selvaraj Durai, Srikrishna Subramanian and Sivarajan Ganesan
Department of Electrical Engineering, Annamalai University, Chidambaram, India

Jinxia Duan, Qiu Xiong and Hao Wang
Hubei Collaborative Innovation Center for Advanced Organic Chemical Materials, Hubei Key Laboratory of Ferro & Piezoelectric Materials and Devices, Faculty of Physics and Electronic Science, Hubei University, Wuhan, China

Jinghua Hu
School of Science, Wuhan University of Technology, Wuhan, China

Oyetunde Adeoye Adeaga
Department of Mechanical Engineering, The Polytechnic, Ibadan, Adeseun Ogundoyin Campus, Eruwa, Nigeria

Ademola Adebukola Dare and Kamilu Moradeyo Odunfa
Department of Mechanical Engineering, University of Ibadan, Ibadan, Nigeria

Olayinka Soledayo Ohunakin
Department of Mechanical Engineering, Covenant University, Otta, Nigeria

Nitin Singh and S. R. Mohanty
Department of Electrical Engineering, Motilal Nehru National Institute of Technology Allahabad, Allahabad, India

Mohan P. Thakre, Vijay S. Kale, Koteswara Raju Dhenuvakonda, Bhimrao S. Umre and Anjali S. Junghare
Department of Electrical Engineering, Visvesvaraya National Institute of Technology, Nagpur, India

Guangqing Shang and Chunhua Sun
Department of Mechanic and Electronic Engineering, Suzhou Vocational University, Suzhou, China

Wei Ning
College of Mechanical Engineering, Shanxi University of Technology, Hanzhong, China

Barinyima Nkoi and Barinaadaa Thaddeus Lebele-Alawa
Mechanical Engineering Department, Faculty of Engineering, Rivers State University of Science and Technology, Port Harcourt, Nigeria

www.ingramcontent.com/pod-product-compliance
Lightning Source LLC
Chambersburg PA
CBHW080503200326
41458CB00012B/4074
* 9 7 8 1 6 3 2 3 9 6 4 9 5 *